Chocolate and Health
Chemistry, Nutrition and Therapy

Chocolate and Health
Chemistry, Nutrition and Therapy

Edited by

Philip K. Wilson
Department of History, East Tennessee State University, Johnson City, Tennessee, USA
Email: wilsonpk@etsu.edu

W. Jeffrey Hurst
Hershey Foods Technical Center, Pennsylvania, USA
Email: whurst@hersheys.com

THE QUEEN'S AWARDS
FOR ENTERPRISE:
INTERNATIONAL TRADE
2013

Print ISBN: 978-1-84973-912-2
PDF eISBN: 978-1-78262-280-2

A catalogue record for this book is available from the British Library

© The Royal Society of Chemistry 2015

All rights reserved

Apart from fair dealing for the purposes of research for non-commercial purposes or for private study, criticism or review, as permitted under the Copyright, Designs and Patents Act 1988 and the Copyright and Related Rights Regulations 2003, this publication may not be reproduced, stored or transmitted, in any form or by any means, without the prior permission in writing of The Royal Society of Chemistry or the copyright owner, or in the case of reproduction in accordance with the terms of licences issued by the Copyright Licensing Agency in the UK, or in accordance with the terms of the licences issued by the appropriate Reproduction Rights Organization outside the UK. Enquiries concerning reproduction outside the terms stated here should be sent to The Royal Society of Chemistry at the address printed on this page.

The RSC is not responsible for individual opinions expressed in this work.

The authors have sought to locate owners of all reproduced material not in their own possession and trust that no copyrights have been inadvertently infringed.

Published by The Royal Society of Chemistry,
Thomas Graham House, Science Park, Milton Road,
Cambridge CB4 0WF, UK

Registered Charity Number 207890

Visit our website at www.rsc.org/books

Printed and bound by CPI Group (UK) Ltd, Croydon, CR0 4YY

Theobroma cacao, from Leonhard Ferdinand Meisner's *De Caffe, Chocolatae, Herbae* (1721).
(Courtesy of Hershey Community Archives, Hershey, Pennsylvania, USA).

Preface

Following the positive reception of our jointly authored *Chocolate as Medicine: A Quest Over the Centuries* (Royal Society of Chemistry, Cambridge, UK, 2012)—which we are pleased to share was awarded *Gourmand* magazine's 2012 Best Book on Chocolate from a UK Publisher Award and 2nd Place in 2012 Best Book on Chocolate in the World—we came together again to prepare and edit a companion, multi-authored work, *Chocolate as Medicine: Chemistry, Nutrition and Therapy*. This volume provides a comprehensive overview of our current understanding of the chemistry, nutrition, bioavailability and therapeutic potential of cacao and chocolate.

Chocolate as Medicine: Chemistry, Nutrition and Therapy begins with a brief historical introduction to its thematic coverage, outlining the historical and current medical uses of chocolate and its derivatives. The remainder of the volume is divided into three sections, taking readers through various aspects of the chemical, nutritional and health aspects of cacao. The first section covers the cultivation, sustainability, chemistry and genomic analysis of cacao. The second section discusses the biochemistry and nutritional components of cacao in relation to health, covering its bioavailability and metabolism. The final section provides an overview of the key potential use of chocolate in health and medical care.

We are grateful to each of our contributors, all of whom are recognized experts in their respective areas, in coming together such that we can collectively provide a global perspective of the current and ongoing research in this area. We thank our respective colleagues within East Tennessee State University and the Hershey Company for their support and encouragement. Special thanks are extended to the staff of the Charles C. Sherrod Library at

---

Chocolate and Health: Chemistry, Nutrition and Therapy
Edited by Philip K. Wilson and W. Jeffrey Hurst
© The Royal Society of Chemistry 2015
Published by the Royal Society of Chemistry, www.rsc.org

East Tennessee State University and the Informational Analysis Center at the Hershey Company Technical Center.

Philip K. Wilson
W. Jeffrey Hurst

Contents

Chapter 1	**Chocolate in Science, Nutrition and Therapy: An Historical Perspective**	**1**
	Philip K. Wilson	
	1.1 Introduction	1
	1.2 Chocolate in Science	2
	1.3 Chocolate as Nutrition	6
	1.4 Chocolate as Therapy	12
	References	18
Chapter 2	**Sustainable Cocoa Production: A Healthy Bean Supply**	**28**
	David A. Stuart	
	2.1 Introduction	28
	2.2 Where Did Cocoa Originate?	29
	2.3 Where Does Cocoa Grow Today?	30
	2.4 Early Problems with Cocoa Production	32
	2.4.1 Cocoa Demand in *Nuevo Espana*	32
	2.4.2 Plant Disease	33
	2.5 Recent Concerns with Cocoa Production	33
	2.5.1 Collapse of Malaysian Plantation Cocoa	33
	2.5.2 Witches' Broom Comes Back to American Cocoa	34
	2.5.3 Pest Spread from Malaysia to Indonesia	35
	2.5.4 Panama Conference	36
	2.5.5 1998 View of Cocoa Sustainability	37
	2.5.6 Sustainable Tree Crops Meeting in Washington, DC	38

Chocolate and Health: Chemistry, Nutrition and Therapy
Edited by Philip K. Wilson and W. Jeffrey Hurst
© The Royal Society of Chemistry 2015
Published by the Royal Society of Chemistry, www.rsc.org

		2.5.7	Formation of the World Cocoa Foundation	39
		2.5.8	Child Labor Incident Broadcast in the UK by Channel 4	39
		2.5.9	Child Labor Surveys Planned in West Africa	40
		2.5.10	"Slave Ship" is Reported in the Gulf Guinea	40
		2.5.11	Child Labor Issues Undertaken by the US Congress: The Harkin–Engel Protocol	40
		2.5.12	Child Labor Survey Results	41
		2.5.13	Initiation of Child/Youth-directed Sustainability Projects	42
		2.5.14	Other Child/Youth-directed Projects	42
		2.5.15	WCF/US-AID Winrock Empowering Cocoa Households with Opportunities and Educational Solutions Program	43
		2.5.16	The Cocoa Livelihoods Program	43
		2.5.17	African Cocoa Initiative	44
		2.5.18	The Payson Analysis of Child Trafficking in West Africa	44
		2.5.19	Cocoa Communities Project	45
		2.5.20	Individual Company-funded Efforts	46
		2.5.21	Certification of the Cocoa Supply Chain	47
	2.6	Current View of Cocoa Sustainability		49
		2.6.1	The Progress and Future Directions	50
		2.6.2	Relief from Onerous Cocoa Export Taxes	50
		2.6.3	Develop and Apply Technology to Improve Cocoa Yields	51
	References			52

Chapter 3 Cacao Chemistry   56
W. Jeffrey Hurst

3.1	Flavanols	57
3.2	Methylxanthines	60
3.3	Biogenic Amines	61
3.4	Anandamide	61
3.5	Cocoa Butter	62
3.6	Non-flavanol Polyphenolics	63
3.7	Flavor Chemistry	64
References		65

Contents			xi

Chapter 4 **Applications of Genomics to the Improvement of Cacao** 67
Mark J. Guiltinan and Siela N. Maximova

	4.1	Introduction	67
	4.2	Cacao Genome Sequencing Project	69
		4.2.1 History	69
		4.2.2 Major Findings from the Cacao Genome Sequencing Projects	70
	4.3	Recent Advances in Cacao Genome Analysis	71
		4.3.1 Evolution, Domestication and Germplasm Collections	71
		4.3.2 Breeding	73
		4.3.3 Functional Genomics	74
		4.3.4 Genomics of Cacao Health and Nutrition	75
		4.3.5 Pathogens	76
	4.4	The Future of Cacao Genomics	77
	References		77

Chapter 5 **Nutritional and Physiological Aspects of Chocolate** 82
Michelle A. Briggs, Yujin Lee, Jennifer Fleming,
Christina Sponsky and Penny M. Kris-Etherton

	5.1	Introduction	82
	5.2	Nutrient Composition from Cacao to Chocolate	83
		5.2.1 History of Cacao Consumption	83
		5.2.2 Nutrient Composition of Cacao	84
	5.3	Incorporating Cacao into the Diet	88
	5.4	Physiological Functions of Cacao's Bioactive Compounds	89
		5.4.1 Alkaloids	90
		5.4.2 Opioids	91
		5.4.3 Biogenic Amines	92
		5.4.4 Endocannabinoids	92
		5.4.5 Polyphenolics: Flavanols	92
		5.4.6 Oxalates	93
	5.5	Impact of Processing on the Nutrient/Bioactive Compound Content of Cacao Beans	94
		5.5.1 Cacao Processing	94
		5.5.2 Processing Changes in Nutrient and Bioactive Compounds	94
	5.6	Summary	95
	References		96

| Chapter 6 | Chocolate, the Digestive Tract and Diabetes | 103 |

John W. Finley, Gabriella Crespo and Zuyin Li

6.1	Introduction to Cocoa Phenolics	103
6.2	Cocoa, Cocoa Liquor, Cocoa Powder and Chocolate	104
6.3	Bioavailability	105
6.4	Diabetes, Metabolic Syndrome and Glucose	118
6.5	Cocoa and the Prebiotic Environment	121
6.6	Cocoa, Blood Lipids and Diabetes	122
6.7	Cocoa, Anti-oxidation and Inflammation	124
	References	128

| Chapter 7 | Chocolate and Cardiovascular Health | 132 |

Gabriela Gutiérrez-Salmeán, Eduardo Meaney, Guillermo Ceballos-Reyes and Francisco Villarreal

7.1	Introduction		132
7.2	Historical Remarks		133
7.3	Nutrition and Biochemistry as Related to Health and Medicine		134
7.4	Cardiometabolic Effects of CPs		135
	7.4.1	Systemic Arterial Hypertension	136
	7.4.2	Obesity	137
	7.4.3	Glycemia and Insulin Resistance	138
	7.4.4	Lipids and Atherosclerosis	139
	7.4.5	Heart Failure	140
	7.4.6	Stroke	141
7.5	Conclusions		142
	References		143

| Chapter 8 | Chocolate and Exercise Recovery | 146 |

William R. Lunn and Allyson N. Derosier

8.1	Introduction	146
8.2	Exercise Recovery: Physiology and the Role of Nutrients	147
8.3	Chocolate Milk and Postexercise Recovery	149
8.4	Chocolate Bars and Cocoa	154
8.5	Chocolate and Postexercise Antioxidation	155
8.6	Chocolate and Postexercise Mood State	156
8.7	Conclusion	157
	References	157

Chapter 9	**Chocolate, Cocoa and Women's Health**		**160**
	Eleonora Brillo and Gian Carlo Di Renzo		
	9.1	Introduction	160
	9.2	Health, Cocoa and Chocolate: Scientific Evidence	162
	9.3	Female Mental Well-being	165
	9.4	Cocoa and Reproduction	166
	9.5	Cocoa and Pregnancy	168
	9.6	Cocoa and Menopause	170
	9.7	Cocoa and Beauty	170
	9.8	Conclusions	171
	References		172
Chapter 10	**Chocolate and Skin Health: Effects of Dietary Cocoa Polyphenols**		**179**
	Ulrike Heinrich and Wilhelm Stahl		
	10.1	Introduction	179
	10.2	Skin Structure and Function	180
	10.3	Skin and Nutrition	181
	10.4	Cocoa Constituents with Dermal Activity	182
	10.5	Topical Effects of Cocoa Products	183
	10.6	Methods to Determine Skin Properties and Function	184
		10.6.1 Photoprotection Against UV-induced Erythema	185
		10.6.2 Cutaneous Blood Flow and Oxygen Saturation of Hemoglobin	185
		10.6.3 Skin Structure by Ultrasound Measurements	185
		10.6.4 Evaluation of the Skin Surface	186
		10.6.5 Skin Hydration Measured by Corneometry	186
		10.6.6 Skin Barrier Function Evaluated by the Measurement of TEWL	186
	10.7	Human Studies on Systemic Effects of Cocoa	187
	10.8	Compounds and Biochemical Mechanisms	191
		10.8.1 Antioxidant Activity	191
		10.8.2 UV Absorption: Inflammation	192
		10.8.3 NO: Vasodilation	192
	10.9	Conclusion	193
	References		193

Chapter 11	**Chocolate and Dental Health**		**196**
	Arman Sadeghpour		
	11.1	Introduction	196
	11.2	Theobromine Chemical Structure, Properties and Toxicity	197
	11.3	Background Literature	200
	11.4	An Alternative to Fluoride?	203
	11.5	The Human Mouth	205
	11.6	Room for Innovation	206
	References		208
Epilogue			**211**
Appendix 1	Brief Historical Timeline of the Early Mentions of Chocolate in terms of Science, Nutrition and Medicine		218
Appendix 2	*Theobroma cacao*'s Reputed Medicinal Properties		222
Subject Index			**231**

CHAPTER 1
Chocolate in Science, Nutrition and Therapy: An Historical Perspective

PHILIP K. WILSON

Department of History, East Tennessee State University, Box 70672, Johnson City, Tennessee, USA, 37614-1709
Email: wilsonpk@etsu.edu

> The Chocolate Tree supplied the "raw product for a most delicious, healthy and nourishing drink".
>
> Carl Linnaeus, the 18th-Century Physician who classified this tree with the name *Theoboroma cacao*[1]

1.1 Introduction

As culinary and healing arts schools are increasingly combining efforts to promote an enhanced understanding of and practices around the theme of "food as medicine", chocolate remains at the core of this discourse. This volume, *Chocolate and Health: Chemistry, Nutrition and Therapy*, provides a snapshot in time identifying major areas whereby key bioactive ingredients of chocolate are being increasingly scrutinized to ascertain possibilities and potentials. Of course, snapshots never completely reveal the total scene,[2] though together they can provide something of a synthesis of the total landscape. Over time, current investigations will provide the historical rendering of the nutritional and biomedical pursuits of the early 21st century. As in all science-based research, some leads from previous times meet

---

Chocolate and Health: Chemistry, Nutrition and Therapy
Edited by Philip K. Wilson and W. Jeffrey Hurst
© The Royal Society of Chemistry 2015
Published by the Royal Society of Chemistry, www.rsc.org

roadblocks, thereby diverging efforts onto entirely different paths. Just where chocolate will be featured in nutrition, health and therapy by the middle of the century is unknowable. Still, the pursuit to that eventual placement needs a starting point. This volume serves, among other uses, as that point. Although this chapter's focus is intentionally historical, the references cited throughout this volume provide the respective chapter authors with springboards of earlier work from which to frame their own interpretations and research protocols. As such, they too provide selective historical cornerstones from which the authors' modern accounts are construed.[3]

To better place the following chapters within an historical context, this brief introduction aims to help readers fully appreciate the relatively long-standing quest to identify, validate and promote chocolate's potential in fulfilling nutritional needs, improving health and preventing disease. This historical introduction produces a broader framework for the themes addressed in following chapters. The brief synopsis that follows describes centuries of nutritional and medicinal associations with cacao and chocolate in a manner that corresponds with the three sections of this volume, namely science, nutrition and health. Though considerable history has been noted to be foundational for this volume, much of chocolate's luscious heritage has, alas, been neglected. Readers who wish to delve further into this historical quest are referred to Philip K. Wilson and W. Jeffrey Hurst's *Chocolate as Medicine: A Quest over the Centuries*, as well as a number of other recent writings by authors or editors including Sophie D. and Michael D. Coe; Teresa L. Dillinger *et al.*; Meredith L. Dreiss and Sharon Edgar Greenhill; Martha Makra Graziano; Louis Evan Grivetti and Howard Yana Shapiro; Donatella Lippi; Murdo J. MacLeod; Cameron L. McNeil; Marcia and Frederic Morton; Sarah Moss and Alexander Badenoch; Deanna Pucciarelli and James Barrett; and David Wolfe and Shazzie.[4]

> Of all the products to emerge from the diversity of the tropical rain forest, none approaches the universal appeal and popularity of chocolate.
> Allen M. Young, Tropical Rainforest Zoologist and Museum Coordinator[5]

1.2 Chocolate in Science

The chocolate tree's official name, *Theobroma cacao* (food of the gods), acknowledges both scientific and sacred associations with this plant. The chocolate we consume derives from one of three cacao bean varieties: forastero (foreign born) being the most common variety, which supplies up to 90% of the global use;[6] criollo (native born), which, being more rare, is used in preparing what many deem the finest of chocolates; and trinitario (sent from heaven), being a hybrid of the other two varieties.

Though "chocolate" is commonly referred to in a general sense, several distinctions should be noted. Chocolate itself is the main processed by-product of the cacao bean (or nib or cotyledon). Cacao, the species of the *Theobroma cacao* plant, is typically used in reference to the tree, pod or bean, whereas cocoa refers to the powder made from the processed bean.

Pods of the chocolate tree—historically referred to as *oro negro* (black gold) or *pepe de oro* (seeds of gold)—have long been highly valued and laboriously harvested with machetes or purposefully made cutlasses on long poles. Processing the nibs within the pods has also required intensive skilled labor. The Maya would pound the nibs with stones called *manos* against a hard-surfaced *metate*, facilitating the process by adding a heat source underneath (Figure 1.1).

This treasured plant of the New World found its way to the Old World *via* Hernán Cortés in 1528, with more regular transport after 1585. In these new surrounds, modifications in its use ensued over the centuries. Among the most crucial modifications that revolutionized the industrial processing of chocolate employed James Watts' steam engine in the place of Mayan muscle. An even greater contribution, so David G. Mitchell argued, was the later introduction of commercial refrigeration: "No longer was the [chocolate] industry limited as to geographical location for the longest spell of cool weather; neither was it limited to operation only during the cool season of the year".[7]

In time, further industrialization efforts focused on improving the physical processes of roasting (adding flavor and color), winnowing (separating nibs from their outer husks), milling and conching (kneading in the traditionally shell-shaped machines, together with aerating machines, to increase smoothness, viscosity and flavor), as well as chemically treating components along the production line (*e.g.*, alkalinizing or "Dutching") to enhance the proper solubility, flavor and color.[8]

In areas surrounding the chocolate tree's natural habitat, select members of Olmec, Maya, Zapotec, Mixtec, Toltec and Aztec cultures claimed nutritional and medicinal benefits of their specially prepared *ka-ka-wa* (Olmec)- or *cacao* (Maya)-based drink preparations. Over the centuries, our refined understanding of the chocolate tree has shown that its delectable product requires delicate care throughout the agricultural enterprise.[9] In order to prosper, cacao needs biologically rich and diverse growing areas. The equatorial rainforests provide optimal agronomic conditions, though enhancing its cultivation requires constant attention to planting, pollination, pruning and protection. The significant amount of shade that the chocolate tree requires has been appreciated for centuries.

Accompanying the first known engraved illustration (Figure 1.2) of this tree in Giralamo Benzoni's *La Historia del Mondo* (1565) is an inscription that, in translation, informs us that the chocolate tree

> grows only in hot places, but under shade, for if the sun were to shine on it, it would die. Therefore, they plant it in forests where it is humid, and, afraid that this is not enough, they plant it next to a tree which is higher and which they bend over it, spreading its top so that it covers the cocoa tree, which thus gets shade all over it, so that the sun no longer does any harm.[10]

Figure 1.1 An indigenous American surrounded by a chocolate drinking cup, a molinet, and a chocolate pot, all atop an image of the cacao pod. Frontispiece from Phillippe Sylvestre Dufour's *Traitez Nouveaux et Curieux du Cafe, du The et du Chocolate* (1688). (Courtesy of Hershey Community Archives, Hershey, Pennsylvania, USA.)

Only later was this thick rainforest canopy found to widely support the growth of the *Ceratopogonid* (biting) and *Cecidomyiid* (gall) midges (family Diptera) that consume nectar from and simultaneously pollinate the tiny pink blossoms of the chocolate tree. Each pollinated blossom subsequently produces small cherelles that, upon maturation, form the characteristic rugby ball-shaped pods. Given that not all cherelles from the same cluster of blossoms mature simultaneously, extra care is needed during harvesting.

Figure 1.2 First engraving of the chocolate tree, Giralamo Benzoni's *La Historia del Mondo* (1565), from C.J.J. Van Hall's *Cacao* (1914).

Adding to these agronomic needs are efforts to overcome the diseases to which the chocolate tree is most susceptible. Among this tree's predominant predatory pests are the brown and black capsids (*Sahlbergella singularis* and *Distantiella theobroma*, respectively), both of which damage inner tissue by feeding on the sap. The fungi *Phytophthora megakarya* and *Moniliophthora roreri* induce black pod rot and frosty pod rot, respectively, whereas the broom-like fungal growths of *Moniliophthora (Crinipellis) perniciosa* (commonly known as witches' broom)—first reported in Surinam in 1895—destroy chocolate tree leaf buds, flowers and pods. Pod-boring moths (*Conopomorpha cramerella*) are also known to damage bean development, and mealybugs (*Planococcoides njalensis*) serve as the vector for introducing the cacao swollen shoot virus (family Caulimoviridae, genus *Badnavirus*), which primarily produce stem and root swelling types of destruction. Problems created by cacao or cocoa thrips (*Selenothrips rubrocinctus*), the "enxerto" ant (*Azteca paraensis var. bondari*) and various stem-attacking beetles are considerable, as are losses attributed to rats, squirrels, birds and parasitic plants.[11]

Though many of these plant–pathogen connections were identified in the late 1800s or early 1900s, their wrath became particularly apparent with the "neglect of the [cacao] plantations" during World War II. Following the war, "when demand again rose, it was found that there was a definite shortage of cocoa beans".[12] Consequently, the prices of all chocolate products "rose markedly" for a few years. Once chocolate tree diseases were more stringently addressed, chocolate supply rose to approximate the demand and, in 1949, the rationing allocation that had been instituted by the Internal Emergency Food Committee during the war was finally revoked.

More recently, agroforestry efforts have aimed at establishing more sustainable cacao farming, often in regions beyond cacao's natural habitat. In particular, alterations at the genetic level are being explored in the hopes of increasing plant resistance to disease and producing higher-yield varieties of cacao. These efforts are, in turn, "increasingly threatened" by deforestation measures including farming, grazing, logging and mining—all of which are being touted as responsible measures of "agricultural expansion".[13] Chocolate, which has long held seemingly mystical and magical properties,[14] may have yet another seemingly "magical" role to play in regards to agronomy. It may very well be the increased demand for this precious product that consequently leads to major efforts in saving the natural diversity of rainforest regions.[15]

So, just where are these cacao-growing regions? The first cacao plantations were established in Brazil in 1745, and this region continues to be recognized for its cacao crop. By the early 1800s, cacao was predominantly grown in Central and South America and the West Indies. A century later, African smallholdings and farms had become the predominant growing areas, especially within the rainforest lands of the then-named Gold Coast (now Ghana) and the Ivory Coast.[16] Since then, Malaysia has also become an increasingly important growing region.

As no other product truly mimics the multifaceted cacao, it has become a widely traded commodity since its introduction on the New York Cocoa Exchange in 1925. Only recently, however, have large chocolate companies began devoting significant attention towards acknowledging the rights, welfare and health of laborers whose livelihoods ultimately provide the world with such pleasures of the palate.[17]

> The cacao bean is a phenomenon, for nowhere else has nature concentrated such wealth of valuable nourishment in so small a space.
> Alexander von Humboldt,
> 19th-Century Natural Philosopher and
> Explorer Extraordinaire[18]

1.3 Chocolate as Nutrition

"In its many forms chocolate may be consumed as a beverage, a syrup, a flavoring, a coating or a confection itself", so Norman Potter noted in *Food Science* (1973).[19] But does chocolate have an even greater nutritional value than this comment suggests? Chocolate remains, as tropical rainforest zoologist Allen M. Young claimed, a "gustatory bond between past and present peoples". Indigenous among New World peoples, chocolate was transported to the Old World then back to North America, thereby forming a "bridge between two very distinct spheres of humankind".[20] In terms of nutrition, Old World peoples viewed chocolate as providing "the greatest delicacy for extraordinary entertainments".[21] Speculation about its nutritional value can be based upon an early recipe for drinking chocolate that Antonio Colmenero de Ledesma reported in the first book devoted entirely to chocolate, *Curioso Tratado de la Naturaleza y Calidad del Chocolate* (1631).

> Of cacaos 700 (beans)
> Of white sugar, one pound and a halfe
> Cinnamon 2 ounces
> Of long red peppers 14
> Of cloves, halfe an ounce (the best writers use them not)
> Three Cods of the Logwood or Campeche tree. These Cods are very good, and smell like Fennell.
> [O]r instead of that the Weight of 2 Reals or a shilling of Anniseeds [sic]. As much of Achiote as will give it colour which is about the quantity of a hasell-nut [sic].[22]

Indeed, reputable sources over the centuries have identified the "goodness" of chocolate in terms of food and nutrition. Centuries ago, writers frequently offered opinions on cacao's nutritive value, though their opinions often "differed greatly". Beginning in the mid-19th century, as one early bibliographer of chocolate and nutrition noted, "a greater uniformity of opinion" was found, and writers grew "more and more in accord" regarding its specific nutritional aspects.[23] Consider, for example, the following appellations to chocolate as a food drawn from professional literature worldwide published within the last 150 years.

1860	Chocolate was "among the many articles which have come to be regarded as auxiliarious, if not necessary in our diet".[24]
1864	Chocolate provides a compact and concentrated form of both stimulant and food.[25]
1875	"The value of cocoa as a food is thus apparent, and fully justifies the high eulogiums which have been passed on it".[26]
1879	Cocoa is a "most nourishing article of diet" that contains enough nutrients "to be classified as a food".[27]
1882	Cocoa provides "an indispensable, all-round nursery food".[28]
1897	Chocolate is "a food, which is nourishing for all kinds of people and good for aged persons".[29]
1906	"Ten adjectives sum up the story with regard to good cocoa—it is pure, nutritious, wholesome, strengthening, readily digested, sustaining, invigorating, delicious, refreshing and convenient".[30]

1907	"What has mainly led to the widespread use of cocoa is the understanding that it is not only a food but also a nutritive food".[31]
1908	Chocolate's "food value is highly regarded by all civilized governments".[32]
1910	The theobromine in chocolate "contains 90% nutritive matter, mak[ing] it a very valuable foodstuff ...".[33]
1926	"Breakfast Cocoa ... is nourishing and easily digested. Owing to its concentrated nutriment it becomes, when milk is added, an almost perfect food".[34]
1951	The chocolate industry possesses a product "noted not only as one of the most popular flavors, but also as a valuable food item".[35]

In recent years, chocolate's value in terms of food and nutrition has been regularly highlighted.[36]

1974	Chocolate's "high food value" is widely noted.[37]
1977	Chocolate is characterized as a "thoroughly efficient food, giving you both energy and a representative amount of most of the necessary nutrient" that is servable as a "food for the most philosophical gourmet".[38]
1987	Chocolate is an "emotional food associated with warmth, sweetness and contentment".[39]
1999	Chocolate is defined as a "solid or semiplastic food".[40]
2000	After defining food as that substance which is "required to give us energy", chocolate was identified as being "able to do this relatively rapidly".[41]
2002	No "respectable synthetic [i.e., artificial] substance" exists to mimic cacao's taste and nutritional value.[42]
2005	When "consumed as part of a balanced and varied diet, chocolate can be both a source of nutrients as well as pleasure, and can be considered as being part of a healthful, wholesome diet".[43]

Such passages attest to the long-standing acknowledgement of chocolate's nutritive value. In order to meaningfully appreciate this significance, a working definition of nutrition is helpful. Among the various definitions available, that provided in *The Concise Encyclopedia of Foods & Nutrition* (1995) contains many points common to other characterizations. There, nutrition is defined as "the science of food and its nutrients and their relation to health".[44] Using this broad definition, all of the chapters in this volume address some important aspect of chocolate as nutrition. Questions remain, however, in determining more precisely what type(s) of food chocolate represents and what specifically are chocolate's key nutritive values.

In 1953, Eileen M. Chatt of the British Food Manufacturing Industries Research Association explained that chocolate lacked "accessory factors" whereby it "f[e]ll short of being a perfect food". Still, it was deemed to be a "satisfactory base for the incorporation of supplementary vitamins".[45] Similar language is evident in earlier 20th-Century initiatives into the developing language of food nutrition as a science. As the following select areas of cacao and chocolate research demonstrate, the quest for specialized nutritive knowledge regarding chocolate was underway nearly a century ago. These research areas are drawn from Stroud Jordan's contemporary review of the literature of that period, a work designed "in order that some of the lesser known and understood results will be [made] available" to the greater chocolate manufacturing community.[46]

1922	Chemical analysis of the composition of cocoa butter; determination of the theobromine content of cacao beans.
1928	Quantitative determination of the tannins in cacao.
1929	Identification of the theosterols in cacao and the glycerides in cocoa butter.
1931	Quantification of the fat and phosphate content of cacao; nutritive valuation of commercial cacao; digestibility of cacao's nitrogenous substances.
1932	Caloric determination of cacao; estimation of cacao's solubility; determination of the vitamin A and B content of milk chocolate.
1937	Epicatechins identified in cacao.

An early giant in nutrition science, E.V. McCollum, offered an overview of the increasing emphasis during the early 20th century on approaching nutritional research strictly from a scientific basis in *A History of Nutrition* (1957).[47] Other interwar writers, notably the Kansas City physician–author Logan Clendening, influenced a generation of general readers through his popular representations of the body's biochemical makeup in his *The Care and Feeding of Adults, With Doubts about Children* (1931) and *The Balanced Diet* (1936).[48]

Similar depictions of chocolate's biochemical composition also began to appear at this time. Some of this work transpired at the pioneering Dunn Nutritional Laboratory formed in Cambridge, England, in 1927. There and elsewhere, nutritional authorities investigated claims that chocolate's natural nutritive value was enhanced by adding milk, which was consumable as chocolate milk or cocoa drink or as a milk chocolate bar to eat. Indeed, adding milk elevated chocolate to what many deemed to be a "complete food".[49]

Among the leading chocolate companies of that era, Hershey's made great strides in documenting why chocolate and cacao were "foods which have gained a rightful place in the diet". Such claims were among the concluding remarks of *A Bibliography of the Nutritive Value of Chocolate and Cocoa*, a 1925 Hershey Company publication prepared by the American Food Journal Institute in which noted nutritional expert Edith C. Williams reviewed

Figure 1.3 Cover of Edith C. Williams's *A Bibliography of the Nutritive Value of Chocolate and Cocoa* (1925). (Courtesy of Hershey Community Archives, Hershey, Pennsylvania, USA.)

166 contemporary scientific reports investigating the nutritive value of chocolate (Figure 1.3).[50]

Twenty-five years later, Samuel Hinkle, then Chief Chemist of the Hershey Company, was still touting chocolate's nutritional benefits. All "activities of the human body" were known to "require a constant expenditure of energy" and an "interchange of material". Chocolate products—particularly Hershey's chocolate products—were offered to the public "with the knowledge that they contain the highest grade ingredients prepared under rigid sanitary conditions and ... [prepared with] the finest [chocolate] that can be made". Noting these products to be "sources of highly concentrated food energy", chocolate was deemed to have earned a "rightful place" alongside "all well-known and well-prepared foods".[51]

To further broadcast these nutritive findings, illustrative figures and tables appeared in various popular media rhetorically substantiating the "wise choice" people would make by adding substantial amounts of chocolate to their daily consumption (Figure 1.4). Such charting continued through the century. By 1975, the US Department of Agriculture published tabular comparisons of chocolate (including milk chocolate with and

TABLE OF FOOD VALUES

	CALORIES
1 Pound Hershey's Milk Chocolate,	2335
1 " " Sweet Coating,	2165
1 " " Almond Chocolate,	2600
1 " " Breakfast Cocoa,	1890
1 Cup " Breakfast Cocoa,	185
1 Cup Coffee, with Cream and Sugar,	50
1 Cup Tea,	Little, if any
1 Pound Lean Beef,	1105
1 " White Fish,	475
1 " Oysters,	235
1 " Beets,	160
1 " Potatoes,	295
1 " White Bread,	1200
1 " Apples,	290
1 Dozen Eggs,	1180

A Calorie is the unit measure of food value.

HERSHEY CHOCOLATE COMPANY
HERSHEY, PA., U. S. A.

Figure 1.4 Chocolate's nutritional value, from Bernarr Macfadden's *"Hershey, The Chocolate Town"* (c. 1923). (Courtesy of Hershey Community Archives, Hershey, Pennsylvania, USA.)

without almonds or peanuts, chocolate-coated peanuts and chocolate-coated raisins) with other regular consumables including apples, bananas, cheese crackers, cookies, ice cream, oranges, peanuts, raisins, sunflowers and yogurt. Chocolate came in second (after sunflowers) in terms of total food energy, close to ice cream and yogurt in terms of protein and midrange in terms of carbohydrates. Based upon these findings, the US Chocolate Manufacturers Association (CMA) noted that since "no one food supplies exactly the right balance" of nutrients (specifying water, protein, carbohydrates, fats, vitamins and minerals), "nutritionists recommend [consuming] a variety of foods" as the "best method of assuring good nutrition". Given that chocolate's nutritional value had been scientifically validated by independent investigations across the globe, it is not surprising to find the CMA strongly endorsing the regular consumption of chocolate products as a smart way of "maintaining that daily balance", especially when "combined with other foods such as milk, almonds, peanuts and peanut butter".[52]

> The persons who habitually take chocolate are those who enjoy the most equable and constant health and are least liable to a multitude of illnesses which spoil the enjoyment of life.
>
> Jean Anthelme Brillat-Savarin,
> *Physiologie du Goût, ou Méditations de Gastronomie Transcendante* (1825)[53]

1.4 Chocolate as Therapy

Chocolate, a therapeutic medicine? Yes, and a preventative medicine as well! What an "appealing idea that a food commonly consumed for pure pleasure could also bring tangible benefits for health".[54] Indeed, it is the "epitome of nutritional indulgence" that chocolate "has of late attracted increasing attention because of health effects".[55] Few natural products have been purported to effectively treat such a wide variety of medical disorders as has chocolate, ranging from a "specific" to an aphrodisiac to a panacea. Since the 1990s, investigators have increasingly scrutinized chocolate's potential therapeutic benefits for humans. These claims have been reviewed in a few key multi-authored volumes including edited works by Ian Knight (1999); Rodolfo Paoletti, Andrea Poli, Ario Conti and Francesco Visioli (2012); and Ronald Ross Watson, Victor R. Preedy and Sherma Zibadi (2012).[56] *Chocolate and Health: Chemistry, Nutrition and Therapy* distinguishes itself from these other helpful tomes in covering the vast range of chocolate's science, nutrition and therapeutic potential in a concise yet complete manner.

Promulgating chocolate for its therapeutic claims, however, has a centuries-long heritage, with many claims extending back to Aztec medical practices. There, remedies concocted from cacao beans were used to soothe stomach and intestinal complaints, control childhood diarrhea, reduce fevers, steady the fainthearted, expel phlegm by provoking cough, reduce the passage of blood in stool and promote strength before military conquests as well as before "acts of venery". In later eras, chocolate remedies were used to combat emaciation, decrease "Female Complaints", delay hair growth, promote the expulsion of kidney stones, increase breast milk production, prolong longevity, both encourage and prohibit sleep, clean teeth and diminish one's timidity.[57]

As seen in Appendix 1, a number of early printed monographs appeared in the New World which described the seemingly magical health benefits derived from chocolate. The first book devoted entirely to chocolate, Antonio Colmenero de Ledesma's *Curioso Tratado de la Naturaleza y Calidad del Chocolate* (1631), was widely translated into European languages. Chocolate's perceived medicinal benefits appear on the very title page which states: by the "wise and moderate use whereof health is preserved, sickness diverted, and cured, especially the plague of the guts; vulgarly called the new disease; fluxes, consumptions, and coughs of the lungs, with sundry other desperate diseases. By it also, conception is caused, the birth hastened and facilitated, beauty gain'd and continued".

The physician Henry Stubbe's *The Indian Nectar; or, A Discourse Concerning Chocolata* (1662) included medical cases from cacao growing regions of the New World. For instance, he noted that "English soldiers stationed in ... Jamaica lived [for many months on only] cacao nut paste mixed with sugar ... which they [drank having] dissolved [it] in water". There, women were reported to have eaten chocolate "so much ... that they scarcely consumed any solid meat yet did not exhibit a decline in strength".[58]

In a summary statement, Stubbe noted chocolate to be "one of the most wholesome and pretious [sic] drinks, that [has] been discovered to this day: because in the whole drink there is not one ingredient put in, which is either hurtful in it self, or by commixtion; but all are cordial, and very beneficial to our bodies, whether we are old, or young, great with child, or ... accustomed to a sedentary life."[59] Stubbe referenced Dr Juanes de Barrio's claim that chocolate was "all that was necessary for breakfast, because after eating chocolate, one needed no further meat, bread or drink".[60]

Further information regarding chocolate's use as medicine in Nicaragua, New Spain, Mexico, Cuba and Jamaica is found in William Hughes' *The American Physician, or a Treatise of the Roots, Plants, Trees, Shrubs, Fruit, Herbs &c. Growing in the English Plantations in America; with a Discourse on the Cacao-Nut-Tree ... and All the Ways of Making of Chocolate* (1672). While sailing amongst the West Indies, Hughes "liv'd, at Sea for some Months" on "nothing but Chocolate, yet neither his strength, nor flesh were diminished".[61] Hughes found chocolate to be helpful in balancing "Lean, Weak, and Consumptive Complexions", and he claimed that it "may be proper for some breeding Women, and those persons that are Hypochondriacal, and Melancholly [sic]".[62] Chocolate, when "internally administered", was found to be "good against all coughs, shortness of breath, opening and making the roughness of the Artery smooth", thereby "palliating all sharp Rheums, and contributing very much to the Radical Moisture, being very nourishing, and excellent against Consumptions". The "fat Butter or Oyl" of the cacao bean was reputed to be "very effectual, being externally applied, against all inflammations, *i.e.*, Phlegmons, Erysipelas, St Anthony's Fire, Smallpox, Tumours, Scaldings and Burnings". When applied on the skin, it cooled the "pains proceeding from heat", minimized the "crustiness or scars on Sores, Pimples, chapped Lips and Hands" and "wonderfully refresheth[ed] wearied limbs" and "mitigate[d] the pain of the Gout, and also Aches by reason of old Age".[63] Hughes concluded noting that medicines "whereof the cacao is the principal ingredient" had become "approved of by Learned Physicians, and sufficiently recommended to the world".[64]

Chocolate's usefulness as a medicine prompted its spread throughout Europe. Such claims are noted in the following works:

- Spanish Court Physician Augustin Farfan's *Bref Traité de Médecine* (1579) cited chocolate's usefulness in eliminating kidney stones and purging the gut.
- *Un Discurso del Chocolate* (1624) promoted a wider array of New World views of chocolate's medicinal benefit for Old World audiences.
- Francisco Maria Brancaccio, later Cardinal Brancaccio, described chocolate's usefulness as a medicine that "restores natural heat, generates pure blood, enlivens the heart, [and] conserves the natural faculties".[65]
- René Moreau, in 1643, published his medical dissertation of the healthfulness of chocolate. By 1659, the Paris Faculty of Medicine had bestowed their imprimatur on its use.
- Cornelis Bontekoe, Dutch physician to the Elector Wilhelm of Brandenburg, in 1678 published *Tractaat van het Excellenste Kruyd Thee* in which praise for chocolate's medical qualities increased its consumption throughout Germany.
- Daniel Duncan's *Wholesome Advice Against the Abuse of Hot Liquors, Particularly of Coffee, Tea, Chocolate* (1706) claimed the best benefits were derived from chocolate when drunk in moderation.
- Leonhard Ferdinand Meisner's *Caffe, Chocolatae, Herbae Thee ac Nicotianae: Natura, Usu, et Abusu Anacrisis: Medico-Historico-Diatetica* (1721) described the uses and abuses of chocolate, as did Girolamo Giuntini's *Alto Parere Intorno alla Natura ed all'uso della Cioccolata Disteso* (1728).
- Further remedies containing chocolate which were claimed as useful for fighting disease appeared in François Foucault's *An Chocolatae Usus Salubris?* (1684), in D. Quelus's *Histoire Naturelle du Cacao et du Sucre* (1719), in Pierre Joseph Buc'hoz's *Dissertations sur l'Utilite, et les Bon et Mauvais Effects du Tabac, du Café, du Caco, et du The* (1788), and in Munster College of Medicine Director, Christopher Ludwig Hoffmann's late 18th-Century treatise, *Potus Chocolate*.[80]

The entrepreneurial English physician William Salmon (who some contemporaries dubbed a quack)[66] prepared a special "liquid medicine" he named "chocolate wine", which he prescribed and sold to his patients to be drunk by the glassful. His contemporary, the physician from St Dizier, Champagne, Pierre-Toussaint Navier, also advocated chocolate as medicine in *Observationes sur le Cocao et sur le Chocolate* (1772). Among the benefits Navier noted were chocolate's usefulness against scurvy, consumption, worms, digestive acids and general disorders of the lungs, heart and vessels. In details far beyond the typical descriptions of the day, Navier articulated his view that chocolate's particular usefulness to gut and bowel disorders was due to its being "incorruptible" throughout the digestive tract. Unlike milk and meat, which experimenters of the day had found to easily go rancid, chocolate's cocoa butter content offered it a "high degree of resistance to rancidity".[67] For those who were unable to "stomach" cacao's high fat concentrations, Navier recommended cacao shell infusions that he claimed in medical terms to be "stomachic, balsamic, pectoral and especially aperient" in its properties. In addition to noting the benefits of cacao itself, Navier further described chocolate's medicinal benefit of being used as a vehicle for other types of medicines such as purgatives, attenuants, expectorants, diuretics and incidentia.[68]

A century later, chocolate's potential health benefits were still being touted. For example, Alfred Franklin promoted chocolate's ability to preserve and control health in his 1893 *Le Café, le The et le Chocolate*, as did Edwin Franke in his 1914 *Kakao, Tee und Gewurze*. At times, chocolate prescriptions and recipes appeared in the same work, such as in Thomas Cooper's 1824 *Treatise of Domestic Medicine, to which is added, A Practical System of Domestic Cookery*. Beginning in the 19th century, European and US advertisements revealed an important conceptual change regarding chocolate and health. It was during this time that advertising helped solidify what was to become chocolate's enduring reputation as both a medicine and a food. Soon, chocolate became the medicine handed out by confectioners and the food prescribed by physicians.

By the early 1700s, chocolate had become figuratively and literally linked with milk. London physician and Royal Society President Sir Hans Sloane specifically touted milk chocolate as the new restorative—an additive to his medical armamentarium gleaned from his 1687 voyage to Jamaica. Its benefits were primarily advertised for its "lightness on the stomach" and for its "great use in all Consumptive cases". John Cadbury and his sons George and Richard later purchased Sloane's milk chocolate. The Cadbury's were Quakers who viewed chocolate as a nourishing, healthy alternative to alcohol and promoted it as a healthy "flesh forming substance". In order to enhance its popularity, the Cadbury Brothers promoted their Sloane recipe as a "health food", rhetorically adding that to call it a "medication would not be too strong a term".[69]

Nineteenth-century improvements in cacao drink palatability owed much to Coenraad Johannes Van Houten who, in the 1870s, refined cacao into a more digestible form by extracting the natural fat (cocoa butter) from the bean, leaving only the powder. The powder could then be mixed with potash to darken its color, lighten its flavor and improve its solubility in water or milk. The powder produced by this "Dutching" process is what officially became known as cocoa. Van Houten's mixture soon became advertised as "The Food Prescribed by Doctors". Contemporary French manufacturers promoted their own milk chocolate remedies as being specifically beneficial for individuals with fragile stomachs, as well as more generally for convalescents and children.

Although it was at this time when the ingredients within chocolate became increasingly scrutinized in terms of purity, such claims were found as far back as the 1600s when a "number of quibblers" began to "question the safety of some of the additives commonly found in cocoa and chocolate".[70] Still, the increased use of filler material (*i.e.*, adulterants) in the mid 1800s were depicted as diminishing chocolate's otherwise healthy benefits. The popular *Peterson's Magazine* directed readers in 1891 to two categories of chocolate adulteration:

(1) Those that were "simply fraudulent, but not necessarily injurious to health" by using "some cheap but wholesome ingredient [mixed] with the pure article for the purpose of underselling and increasing profits", and
(2) Those that were injurious to health, by using "drugs or chemicals for the purpose of changing the appearance or character of the pure article, as for instance, the admixture of potash, ammonia, and acids with cocoa to give the apparent smoothness and strength to imperfect and inferior preparations".[71]

The "monstrous fraud perpetuated upon the poor by the adulteration of cocoa cannot be over-estimated".[72] In 1906 the US Pure Food and Drug Act (based on the Heyburn Pure Food Bill) was passed. In it, a food and drug—like chocolate—was considered to be adulterated if "mixed or packed with another substance to lower its quality; if any substance was substituted by another; if any valuable constituent was wholly or in part omitted; or if it was mixed, colored, coated, or covered in any way to conceal inferiority or damage".[73] Following this Act, and the subsequent Pure Food, Drug and Cosmetic Act in 1938, "marketing the healthiness of chocolate products in terms of "purity" rapidly became the norm".[74]

Drawing attention to chocolate's potential therapeutic benefits, the CMA distributed a pamphlet in 1923 devoted to "Chocolate and Health". During the second half of the 20th century, quests for experimental evidence were increasingly undertaken to support claims gathered regarding these potential benefits. Biomedical science began to experimentally assess chocolate's potential for alleviating medical disorders, just as it did with all pharmaceutical products. Randomized control trials of increasingly complex design based at multiple clinical sites were used to identify standards of normalcy and degrees of difference. By the close of the century, claims for chocolate's medical benefits were supported by a growing "science" of chocolate.[75]

By 2000, claims of chocolate's benefits typically focused upon its richness in carbohydrates and fat. Chocolate's natural flavonoid phenolics had been found to prevent the rancification of fat, thereby diminishing the need to add preservatives that might bring their own health risks. The plant-derived, saturated, stearic acid fats were not those guilty of increasing cholesterol levels. Cocoa butter within chocolate products was found to coat the teeth, thereby preventing tooth decay from chocolate's high sugar content. Tannins in cacao were noted to promote healthy teeth as they inhibited dental plaque formation. More recently, investigations have centered on a particular flavonoid, epicatechin. Following chocolate consumption, epicatechin has been found to promote anti-oxidant activity, which decreases low-density lipoprotein cholesterol activity, thereby delaying the onset or progression of atherosclerosis and arteriosclerosis. It may also increase high-density lipoprotein cholesterol levels. Chocolate has also been found to initiate antiplatelet activity, thereby reducing plaque formation and platelet clotting properties. Flavonoids have been demonstrated to stimulate blood flow in the brain, hands and legs due to the regulation of nitric oxide synthesis. Dark (high cacao-concentrated) chocolate also works to reduce blood pressure by promoting blood vessel dilation.

For some time, high-quality dark chocolate's psychoactive attributes have been linked to its high concentration of the stimulant theobromine. Late 20th-century investigators have also explored chocolate's supposed aphrodisiac effects. When people become infatuated or fall in love, the levels of phenylethylamine released from their brain increase. Chocolate was also found to promote this release, though in relatively small quantities. Chocolate appears to promote the neurotransmitter serotonin release as well, thereby producing calming, pleasurable feelings. Finally, an anandamide is also released following chocolate consumption, likely contributing to the euphoria that many claim chocolate induces. All of these psychopharmacological alterations may contribute to chocolate's perceived aphrodisiac effects.

Questions also continue to arise over chocolate's reputed addictive nature. Despite anecdotal evidence that may say "Yes, Yes, Yes", little modern biomedical evidence supports such claims. Since depriving one of chocolate fails to produce scientifically significant signs of withdrawal, it is not technically classed as a physically addictive agent. Further, scientists have not shown a state of dependence regarding chocolate's use. Admittedly, chocolate may pharmacologically stimulate compulsive eating, but this may just as well be the result of a more generalized aesthetic craving for the sweetness and oily richness and complete orosensory experience that chocolate provides. Chocolate's rich natural complexity—a complexity that rivals any other food—makes the actual source of perceived cravings or "chocoholism" exceedingly difficult to ascertain.

Chocolate manufactures continue to expend considerable resources and marketing towards the potential therapeutic benefits of their products. Such efforts were seen in the International Cocoa Organization and the International Cocoa Research and Education Foundation's support of a research symposium and a subsequent publication devoted to "Chocolate and Cocoa: A Review of Health and Nutrition".[76] In recent years, chocolate has also become more apparent in the products of distillers and brewers. This presence is somewhat ironic since, in previous centuries, "chocolate houses" were established as alternatives to the popular Old World public houses (*i.e.*, pubs) that served only alcoholic beverages. The power of chocolate house culture exacerbated chocolate's own stimulant ability. By the 1800s, as Ruth Lopez noted, serving chocolate where working men gathered became "a boon" to the temperance campaigners who relied upon chocolate as central to "keeping workers from alcohol".[77] Such views of chocolate, however, did not entirely dominate the market. By the early 20th-Century, J. Scholz was promoting his own patented "chocolate-health-beer".[78] Chocolate has, in recent years, become widely touted in marketing such products as Young's Double Chocolate Stout, Rogue Ale's Chocolate Stout, Samuel Adams' Chocolate Bock, Sonoran's White Chocolate Ale, Harpoon's Chocolate Stout, Souther Tier's Chokolat and Dogfish Head's Theobroma Ale, among others.

One offshoot of early 20th-century medical reform was the extensive research expended on drug design and development. Indeed, drugs came to dominate the Western medical marketplace. Growing disenchantment with such a reign over medical practice provoked considerable consumer health demands for more holistic, integrative health care.[79] Patients' skepticism over the benefits obtainable from traditionally prescribed remedies alone is evidenced by the increasing demand for more natural medicine.

And what could be more natural than chocolate? As the nomenclature *Theobroma cacao* suggests, we have long viewed chocolate as a "Food of the Gods". Steadily, authorities have been reinforcing chocolate's potential medicinal benefits. According to Harvard Medical School's Norman K. Hollenberg, the "pharmaceutical industry has spent tens, probably hundreds of millions of dollars in search of a chemical that would reverse ... [or ward off vascular diseases]. And God gave us flavanol-rich cocoa which does that".[80]

References

1. M. Morton and F. Morton, *Chocolate: An Illustrated History*, Crown Publishers, New York, 1986, p. 33.
2. S. Sontag, *On Photography*, Farrar, Strauss, and Giroux, New York, 1973.
3. A. Grafton, *The Footnote: A Curious History*, Harvard University Press, Cambridge, MA, 1999.
4. P. K. Wilson and W. J. Hurst, *Chocolate as Medicine: A Quest over the Centuries*, Royal Society of Chemistry, Cambridge, 2012; S. D. Coe and M. D. Coe, *The True History of Chocolate*, Thames and Hudson, London, 1996; T. Dillinger, P. Barriga, S. Escárcega, *et al.*, Food of the Gods: Cure for Humanity? A Cultural History of the Medicinal and Ritual Use of Chocolate, *J. Nutr.*, 2000, **130**(Supplement), 2057S–2072S; M. L. Dreiss and S. E. Greenhill, *The Healing Powers of Chocolate: Folk Medicine, Nutrition, and Pharmacology, in Chocolate: Pathway to the Gods*, University of Arizona Press, Tucson, 2008, pp. 135–151; M. M. Graziano, Food of the

Gods as Mortals' Medicine: The Uses of Chocolate and Cacao Products, *Pharm. Hist.*, 1998, **40**, 132–146; *Chocolate: History, Culture, and Heritage*, ed. L. E. Grivetti and H. Y. Shapiro, John Wiley & Sons, Hoboken, N. J., 2009; D. Lippi, Chocolate in History: Food, Medicine, Medi-Food, *Nutrients*, 2013, **5**, 1573–1584; M. J. MacLeod, Cacao, in *The Cambridge World History of Food*, ed. K. F. Kiple and K. C. Ornelas, Cambridge University Press, Cambridge, UK and New York, 2000, vol. 1, pp. 635–664; M. Morton and F. Morton, *Chocolate: An Illustrated History*, Crown, New York, 1986; D. Pucciarelli, *The Medical Use of Chocolate*, VDM Verlag, Saarbrücken, Germany, 2008; S. Moss and A. Badenoch, *Chocolate: A Global History*, Reaktion Books, London, 2009; C. Orey, *The Healing Powers of Chocolate*, Kensington, New York, 2010; D. Pucciarelli and J. Barrett, Twenty-First Century Attitudes and Behaviors Regarding the Medicinal Use of Chocolate, in *Chocolate: History, Culture, and Heritage*, ed. L. E. Grivetti and H. Y. Shapiro, John Wiley & Sons, Hoboken, N. J., 2009, pp. 653–666; D. Wolfe and Shazzie (Sharon Holdstock), *Naked Chocolate: The Astonishing Truth About the World's Greatest Food*, North Atlantic Books, Berkeley, California, 2005.

5. A. M. Young, *The Chocolate Tree: A Natural History of Cacao*, Smithsonian Institution Press, Washington, DC, 1994, p. x.
6. A. Y. Leung and S. Foster, *Encyclopedia of Common Natural Ingredients Used in Food, Drugs, and Cosmetics*, John Miley & Sons, New York, 2nd edn, 1996, p. 181; For a history of cacao taxonomy, see J. Cuatrecasas, Cacao and Its Allies: A Taxonomic Revision of the Genus *Theobroma*, **35**, Pt 6 of Contributions from the United States National Herbarium, United States National Museum Bulletin, Smithsonian Institution, Washington, D.C., 1964, pp. 379–614, esp. pp. 383–415.
7. D. G. Mitchell, *The Chocolate Industry*, Bellman Publishing Company, Boston, 1951, p. 11.
8. For further historical and modern insight into chocolate processing, see, in particular H. Jumelle, *Le Cacaoyer: Sa Culture et son Exploitation dans tous les Pays de Production*, Augustin Challamel, Paris, 1900; R. Whymper, *Cocoa and Chocolate: Their Chemistry and Manufacture*, J. & A. Churchill, London, 1912; A. W. Knapp, *Cocoa and Chocolate: Their History from Plantation to Consumer*, Chapman and Hall, London, 1920; J. Fritsch, *Fabrication du Chocolat d'Après les Procédés les plus Récents*, Desforges, Paris, 1924; H. W. Bywaters, *Modern Methods of Cocoa and Chocolate Manufacture*, J. & A. Churchill, London, 1930; H. R. Jensen, *Chemistry Flavouring and Manufacture of Chocolate Confectionery and Cocoa*, J. & A. Churchill, London, 1931; S. Jordan, *Chocolate Evaluation*, Applied Sugar Laboratories, New York, 1934; M. Vidal, *Tratado Moderno de Fabricación de Chocolates*, José Montesó, Barcelona, 1935; A. W. Knapp, *Cacao Fermentation*, Bale, Sons & Curnow, London, 1937; H. Damblon, *New Method in the Manufacture of Chocolate, Cocoa Powder and Confectionary*, Joseph Höfer, Cologne, 1939; E. M. Chatt, *Cocoa: Cultivating, Processing, Analysis*, Interscience Publications, New York, 1953;

H. C. L. Mijnoogst, *The Enormous Development in Cocoa and Chocolate Making Since 1955*, Jedermann, Mannheim, 1957; N. W. Kempf, *The Technology of Chocolate*, Manufacturing Confectioner Publishing Company, Oak Park, IL, 1964; R. Desrosiers, *Cacao Handbook*, American Cocoa Research Institute, 1976; G. A. R. Wood and R. A. Lass, *Cocoa*, Longman, London & New York, 1985; N. J. Smith, D. L. Plucknett and J. T. Williams, *Tropical Rainforests and Their Crops*, Comstock Books of Cornell University Press, Ithaca, New York, 1992; B. W. Minifie, *Chocolate, Cocoa, and Confectionary Science and Technology*, Van Nostrand Reinhold, New York, 3rd edn, 1989; S. J. Terrio, *Crafting the Culture and History of French Chocolate*, University of California, Berkeley, 2000; *Industrial Chocolate Manufacture and Use*, ed. S. T. Beckett, Wiley-Blackwell, Oxford, 4th edn, 2009; F. Á. Mohos, *Confectionary and Chocolate Engineering: Principles and Application*, Wiley-Blackwell, Oxford, 2010.
9. J. A. West, A Brief History and Botany of Cacao, in *Chilies to Chocolate: Food the Americas Gave the World*, ed. N. Foster and L. S. Cordell, University of Arizona Press, Tucson and London, 1992, pp. 105–121.
10. E. M. Chatt, *Cocoa: Cultivating, Processing, Analysis*, Interscience Publishing, New York, 1953, p. 4.
11. For further elucidation, both historically and more recently, see B. Head, *The Food of the Gods: A Popular Account of Cocoa*, George Routledge & Sons, E. P. Dutton, London, New York, 1903; H. R. Brinton-Jones, *The Diseases and Curing of Cacao*, Macmillan and Company, New York, 1934; D. H. Urquhart, *Cocoa*, Longman, London, 1955; P. F. Entwistle, *Pests of Cocoa*, Longman, London, 1972; C. A. Thorold, *Diseases of Cocoa*, Clarendon, Oxford, 1975; M. C. Aime and W. Phillips-Mora, The Causal Agents of Witches' Broom and Frosty Rot of Cacao (chocolate, Theobroma cacao) for a New Lineage of *Marasmiacea*, *Mycologia*, 2005, **97**, 1012–1022.
12. D. G. Mitchell, *The Chocolate Industry*, Bellman Publishing, Boston, 1951, p. 12; For further insight into the "Impact of the War on Chocolate", see V. D. Wickizer's chapter of that title in *Coffee, Tea, and Cocoa: An Economic and Political Analysis*, Stanford University Press, Stanford, CA, 1951, pp. 328–346.
13. R. Lopez, *Chocolate: The Nature of Indulgence*, H. N. Abrams in association with the Field Museum, New York, 2002, p. 11.
14. As Julie Pech (The Chocolate Therapist™) quipped, the "story of chocolate is a mirror of itself—rich, complex, mysterious, and enjoyable", *The Chocolate Therapist: Chocolate Remedies for a World of Ailments*, Trafford, Victoria, B.C., 2009, p. 91.
15. C. Bright, Chocolate Could Bring the Forest Back, *World Watch*, 2001, **14**.
16. C. A. Rinzler, *The Book of Chocolate*, St. Martin's Press, New York, 1977.
17. Book-length works specifically addressing chocolate laborers' concerns appeared as early as Henry Woodd Nevinson's *A Modern Slavery*, Harper

and Brothers, London and New York, 1906, with increasing attention being paid over the century through works including Jorge Amado's novels, *Cacaú*, Ariel, Rio de Janeiro, 1933, and *Terras do Sem Fim*, São Paulo, Martins, 1943, and more recently Gillian Wagner, *The Chocolate Conscience*, Chatto and Windus, London, 1987; G. Mikell, *Cocoa and Chaos in Ghana*, Howard University Press, Washington, D.C., 1992; L. J. Satre, *Chocolate on Trial: Slavery, Politics and the Ethics of Business*, Ohio University Press, Athens, OH, 2005; C. Off, *Bitter Chocolate: The Dark Side of the World's Most Seductive Sweet*, The New Press, New York and London, 2008; Ó. Ryan, *Chocolate Nations: Living and Dying for Cocoa in West Africa*, Zed Books, London, New York, 2011; C. Higgs, *Chocolate Islands: Cocoa, Slavery, and Colonial Africa*, Ohio University Press, Athens, Ohio, 2012. All of these works give a more stark meaning to that popular dessert phrase, "Death by Chocolate". See also Kate Blewett and Brian Woods' 2001 documentary, "Slavery: A Global Invasion", which grippingly surveys the life and struggles on cacao farms and plantations. I am grateful to my colleague, History Professor William Douglas Burgess, Jr., for drawing my attention to the powerful video (http://boingboing.net/2014/07/30/watch-a-cocoa-farmer-try-choco.html?fk_bb) of a cacao farmer tasting chocolate for the first time.
18. Cited in M. Morton and F. Morton, *Chocolate: An Illustrated History*, Crown, New York, 1986, p. 28.
19. N. N. Potter, *Food Science*, Avi Publishing, Westport, Connecticut, 1973, pp. 556–557.
20. A. M. Young, *The Chocolate Tree: A Natural History of Cacao*, Smithsonian Institution Press, Washington, DC, 1994, p. x.
21. J. Chamberlayne, *The Natural History of Coffee, Thee, Chocolate, Tobacco, in four several Sections; with a Tract of Elder and Juniper-Berries, Shewing how Useful they may be in our Coffee-Houses: And also the way of making Mum, With some Remarks upon that Liquor. Collected from the Writings of the best Physicians, and Modern Travellers*, Christopher Wilkinson, London, 1682, p. 14.
22. Cited in C. Coady, *Chocolate: The Food of the Gods*, Chronicle Books, San Francisco, 1993, p. 78; Chemical analysis remains ongoing to provide more precise evidence of the ways and recipes by which cacao was commonly used in Mesoamerica. See, in particular, D. Soleri, M. Winter, S. R. Bozarth and W. J. Hurst, Archaeological Residues and Recipes: Exploratory Testing for Evidence of Maize and Cacao Beverages in Postclassic Vessels from the Valley of Oaxaca, Mexico, *Latin American Antiquity*, 2013, **24**, 345–362.
23. E. C. Williams, *A Bibliography of the Nutritive Value of Chocolate and Cocoa with Quotations and Summaries*, prepared for the Hershey Chocolate Company by The American Food Journal Institute, Hershey, PA, 1925, preface, p. iii.
24. J. A. Mann, Cocoa—Its Advantages and Value as an Article of Food, *J. Soc. Arts*, 1860, **8**, 775–800.

25. A. Debay, *Les Influences du Chocolat, du Thé et du Café sur l'Économie Humaine*, E. Dentu, Paris, 1864.
26. Nutritive Value of Chocolate, *Confectioners' J.*, 1875, **1**, 20. This anonymously authored article cites data from John Holm of the Edinburgh Chemical Society.
27. R. F. Fristedt, Om Kakao, *Uppsala Läkaref. Förh.*, 1879, **14**, 105–110.
28. J. M. Fothergill and B. R. Fothergill, *The Food We Eat, Why We Eat It, and Whence It Comes* (1882), as cited by B. Head, *The Food of the Gods: A Popular Account of Cocoa*, George Routledge & Sons, E. P. Dutton, London, New York, 1903, p. 23.
29. H. Schlesinger, Beiträge zur Beurtheilung des Cacaos bei der Ernährung des Menschen, *Dtsch. Med. Wochenschr.*, 1895, **21**, 80–82.
30. J. F. Beale, Jr., Cocoa of To-Day and Yesterday, *Confectioners' J.*, 1906, **32**, 84.
31. L. Pincussohn, Beitrage zur Kakaofrage, *Zentrallblatt für Innere Medizin*, 1907, **28**, 177–186.
32. G. A. Sutherland, *A System of Diets and Dietetics*, Hodder and Stroughton, London, 1908, p. 195.
33. W. B. Snow, The Manufacture of Chocolate—'The Indian Nectar', *Confectioners' J.*, 1910, 82.
34. *The Story of Chocolate and Cocoa: With a Brief Description of Hershey, "The Chocolate and Cocoa Town" and Hershey "The Sugar Town"*, Hershey Chocolate Corp., Hershey, PA, 1926, facing title page.
35. D. G. Mitchell, *The Chocolate Industry*, Bellman Publishing, Boston, 1951, p. 9.
36. M. E. Presilla's *The New Taste of Chocolate: A Cultural and Natural History of Cacao with Recipes*, Ten Speed Press, Berkeley and Toronto, 2001, focuses upon cacao's nutritional aspects, interweaving broader cultural developments over the centuries. Among other key resources that address chocolate's nutritive value, see R. Lecoq, *Cacao, Poudres de Cacao, et Farines Composées Alimentaires avec et sans Cacao*, Vigot Frères, Paris, 1926; H. Labbé, *Le Cacao et le Chocolat au Point de vue Alimentaire et Hygiénique*, Bruxelles, 1930; S. Jordan, Nutritive Value, in *Chocolate Evaluation*, Applied Sugar Laboratories, New York, 1934, pp. 104–112; J. C. Musser, Chocolate—History, Botany, and Preparation for the Market, *Chocolate Handbook: Based on Chocolate Sessions at the Associated Retail Confectioners Annual Short Courses on Candy Making*, Chocolate Manufacturers Association, Washington, D.C., 1968, pp. 1–11; C. A. Rinzler, Chocolate as Food, in *The Book of Chocolate*, St. Martin's Press, New York, 1977, pp. 32–53; C. Coady, *Chocolate: The Food of the Gods*, Chronicle Books, San Francisco, 1993; S. Rössner, Chocolate—Divine Food, Fattening Junk or Nutritious Supplementation?, *Eur. J. Clin. Nutr.*, 1997, **51**, 341–345; *Chocolate: Food of the Gods*, ed. A. Szogyi, Greenwood Press for Hofstra University, Westport, CT, 1997; K. Bruinsma and D. L. Taren, Chocolate: Food or Drug?, *J. Am. Diet. Assoc.*, 1999, **99**, 1249–1256; *Chocolate and Cocoa: Health and Nutrition*, ed. I. Knight, Blackwell Science, Oxford, England, 1999; M. J. MacLeod, Cacao, in *The Cambridge*

World History of Food, ed. K. F. Kiple and K. C. Ornelas, Cambridge University Press, Cambridge, UK and New York, 2000, vol. 1, pp. 635–641; C. L. Keen, Chocolate: Food as Medicine/Medicine as Food, *J. Am. Coll. Nutr.*, 2001, **20**, 436S–439S; M. E. Presilla, *The New Taste of Chocolate: A Cultural and Natural History of Cacao with Recipes*, Ten Speed Press, Berkeley and Toronto, 2001; *Chocolate: History, Culture, and Heritage*, ed. L. E. Grivetti and H. Y. Shapiro, John Wiley & Sons, Hoboken, N. J., 2009; A. M. Beck and K. Damkjoer, Chocolate: A Significant Part of Nutrition Intervention among Elderly Nursing Home Residents, in *Chocolate, Fast Foods and Sweeteners: Consumption and Health*, ed. M. R. Bishop, Nova Science Publishers, New York, 2010, pp. 245–255; M. Rusconi and A. Conti, *Theobroma cacao L.*, the Food of the Gods: A Scientific Approach Beyond Myths and Claims, *Pharmacol. Res.*, 2010, **61**, 5–13; the forty insightful chapters in ed. R. R. Watson, V. R. Preedy and S. Zibadi, *Chocolate in Health and Nutrition*, Humana Press, New York, 2012; D. Lippi, Chocolate in History: Food, Medicine, Medi-Food, *Nutrients*, 2013, **5**, 1573–1584.
37. I. D. Garard, *The Story of Food*, Avi Publishing, Westport, Connecticut, 1974, p. 60.
38. C. A. Rinzler, *The Book of Chocolate*, St. Martin's Press, New York, 1977, pp. 44, and 3.
39. L. K. Fuller, *Chocolate Fads, Folklore & Fantasies: 1,000+ Chunks of Chocolate Information*, Haworth Press, New York, 1994, p. 187.
40. R. S. Igoe and Y. H. Hui, *Dictionary of Food Ingredients*, Aspen Publishers, Gaithersburg, Maryland, 3rd edn, 1999, p. 34.
41. S. T. Beckett, *The Science of Chocolate*, Royal Society of Chemistry, Cambridge, UK, 2000, p. 144.
42. R. Lopez, *Chocolate: The Nature of Indulgence*, H. N. Abrams in association with the Field Museum, New York, 2002, p. 7.
43. J. R. Lupin of the United Nations Food and Agriculture Organization, as cited by Mort Rosenblum, *Chocolate: A Bittersweet Saga of Dark and Light*, North Point Press, New York, 2005, pp. 249–250.
44. M. E. Ensminger, A. H. Ensminger, J. E. Konlande and J. R. K. Robson, *The Concise Encyclopedia of Foods & Nutrition*, CRC Press, Boca Raton, 1995, p. 770.
45. E. M. Chatt, *Cocoa: Cultivating, Processing, Analysis*, Interscience Publishing, New York, 1953, p. 14.
46. S. Jordan, *Chocolate Evaluation*, Applied Sugar Laboratories, New York, 1934; For citations of specific cacao research of that period, see the chapter "Nutritive value", pp. 104–112, and references on pp. 173–188.
47. E. V. McCollum, *A History of Nutrition*, Houghton Mifflin, New York, 1957.
48. L. Clendening, *The Care and Feeding of Adults, With Doubts About Children*, Alfred A Knopf, New York, 1931; *The Balanced Diet*, D. Appleton-Century Co., NY & London, 1936.
49. Such an account appeared in *The Story of Chocolate and Cocoa: With a Brief Description of Hershey, "The Chocolate and Cocoa Town" and Hershey "The Sugar Town"*, Hershey Chocolate Corp., Hershey, PA, 1926, p. 22.

50. E. C. Williams, *A Bibliography of the Nutritive Value of Chocolate and Cocoa with Quotations and Summaries*, prepared for the Hershey Chocolate Company by The American Food Journal Institute, Hershey, PA [1925]. Hershey was the "first chocolate and confectionary company to voluntarily provide nutritional labeling on food labels". *"Good Nutrition Makes Good Sense"*, Hershey Foods Corporation Collection, Hershey Community Archives, Accession 87006, Box B-11, Folder 40, 1982.
51. S. F. Hinkle, *Fuel Values of Foods*, Hershey Foods Corporation Collection, Hershey Community Archives, Accession 87006, Box B-11, Folder 36, *ca*. 1936–1949. Hinkle, a Penn State University alum, became President of Hershey Chocolate Corporation in 1956. In 1963, as a result of a "$50 million phone call" to then Penn State University President Eric Walker and speaking on behalf of the Hershey Trust Company Board of Directors, Hinkle created the foundation of what is now the Penn State Hershey College of Medicine and M.S. Hershey Medical Center. Founding Penn State Hershey Professor C. Max Lang published an historical overview of this medical school's founding as *The Impossible Dream: The Founding of the Milton S. Hershey Medical Center of the Pennsylvania State University*, AuthorHouse, Bloomington, IN, 2010.
52. *The Story of Chocolate*, Chocolate Manufacturers Association of the U.S.A., McLean, Virginia, 1960, p. 29.
53. Cited in R. Whymper, *Cocoa and Chocolate: Their Chemistry and Manufacture*, J. & A. Churchill, London, 1912, p. ix.
54. K. A. Cooper, J. L. Donovan, A. L. Waterhouse, *et al.*, Cocoa and Health: a Decade of Research, *Br. J. Nutr.*, 2008, **99**, 1.
55. D. L. Katz, Health Effects of Chocolate, in *Nutrition in Clinical Practice: A Comprehensive Evidence-Based Manual for the Practitioner*, Lippincott Williams & Wilkins, Philadelphia, 2nd edn, 2008, p. 391.
56. *Chocolate and Cocoa: Health and Nutrition*, ed. I. Knight, Blackwell Science, Oxford, England, 1999; *Chocolate and Health*, ed. R. Paoletti, A. Poli, A. Conti and F. Visioli, Springer Verlag Italia, Milan, 2012; *Chocolate in Health and Nutrition*, ed. R. R. Watson, V. R. Preedy and S. Zibadi, Humana Press, New York, 2012.
57. T. Dillinger, P. Barriga, P. S. Escárcega, *et al.*, Food of the Gods: Cure for humanity? A Cultural History of the Medicinal and Ritual Uses of Chocolate, *J. Nutr.*, 2000, **130**, 2057S–2072S.
58. H. Stubbe, *The Indian Nectar; or, A Discourse Concerning Chocolate Wherein the Nature of the Cacao-nut … is Examined … the Ways of Compounding and Preparing Chocolate are enquired into; Its Effects, as to its Alimental and Venereal Quality, as well as Medicinal (Specially in Hypochondriacal Melancholy) are Fully Debated*, A. Crook, London, 1662, p. 31.
59. H. Stubbe, *The Indian Nectar; or, A Discourse Concerning Chocolate Wherein the Nature of the Cacao-nut … is Examined … the Ways of Compounding and Preparing Chocolate are enquired into; Its Effects, as to its Alimental and Venereal Quality, as well as Medicinal (Specially in*

Hypochondriacal Melancholy) are Fully Debated, A. Crook, London, 1662, pp. 83–84.

60. H. Stubbe, *The Indian Nectar; or, A Discourse Concerning Chocolate Wherein the Nature of the Cacao-nut ... is Examined ... the Ways of Compounding and Preparing Chocolate are enquired into; Its Effects, as to its Alimental and Venereal Quality, as well as Medicinal (Specially in Hypochondriacal Melancholy) are Fully Debated*, A. Crook, London, 1662, p. 84.

61. J. Chamberlayne, *The Natural History of Coffee, Thee, Chocolate, Tobacco, in four several Sections; with a Tract of Elder and Juniper-Berries, Shewing how Useful they may be in our Coffee-Houses: And also the way of making Mum, With some Remarks upon that Liquor. Collected from the Writings of the best Physicians, and Modern Travellers*, Christopher Wilkinson, London, 1682, p. 17.

62. J. Chamberlayne, *The Natural History of Coffee, Thee, Chocolate, Tobacco, in four several Sections; with a Tract of Elder and Juniper-Berries, Shewing how Useful they may be in our Coffee-Houses: And also the way of making Mum, With some Remarks upon that Liquor. Collected from the Writings of the best Physicians, and Modern Travellers*, Christopher Wilkinson, London, 1682, p. 17.

63. W. Hughes, *The American Physitian or A Treatise of the Roots, Plants, Trees, Shrubs, Fruit, Herbs &c. Growing in the English Plantations in America: Describing the Place, Time, Names, Kindes, Temperature, Vertues and Uses of them, either for Diet, Physick, &c. Whereunto is added A Discourse of the Cacao-nut Tree, and the use of its Fruit; with all the ways of making of Chocolate. The like never extant before*, J. C. for William Crook, London, 1672, section on "Use" following "Of the Simple Cacao-Kernels".

64. W. Hughes, *The American Physitian or A Treatise of the Roots, Plants, Trees, Shrubs, Fruit, Herbs &c. Growing in the English Plantations in America: Describing the Place, Time, Names, Kindes, Temperature, Vertues and Uses of them, either for Diet, Physick, &c. Whereunto is added A Discourse of the Cacao-nut Tree, and the use of its Fruit; with all the ways of making of Chocolate. The like never extant before*, J. C. for William Crook, London, 1672, section on "Name" in the chapter "Of the Cacao-Tree and Fruit".

65. S. D. Coe and M. D. Coe, *The True History of Chocolate*, Thames and Hudson, London, 1996, p. 154.

66. P. K. Wilson, W. Salmon, in *New Dictionary of National Biography*, ed. H. C. G. Matthew, Oxford University Press, Oxford, 2004, vol. 48, pp. 734–735.

67. P. T. Navier, *Bemerkungen über den Cacao und die Chocolate, worinnen der Nutzen und Schaden untersuchet wird, der aus dem Genusse dieser nahrhaften Dinge entsehen kann: Alles auf Erfahrung und zergliedernde Versuche mit der Cacao-Mandel gebauet; Nebst einigen Erinnerungen über das System des Hrn. De-La-Müre, betreffend das Schlagen der Puls-adern*, Saalbach, Leipzieg, 1775, p. 95.

68. M. M. Graziano, Food of the Gods as Mortals' Medicine: The Uses of Chocolate and Cacao Products, *Pharm. Hist.*, 1998, **40**, 136.

69. M. M. Graziano, Food of the Gods as Mortals' Medicine: The Uses of Chocolate and Cacao Products, *Pharm. Hist.*, 1998, **40**, 136.
70. C. A. Rinzler, *The Book of Chocolate*, St. Martin's Press, New York, 1977, p. 8.
71. *Peterson's Magazine*, 1891, p. 269, as cited by L. P. Brindle and B. F. Olsen, Digging for Chocolate in Charleston and Savannah, in *Chocolate: History, Culture, and Heritage*, ed. L. E. Grivetti and H. Y. Shapiro, John Wiley & Sons, Hoboken, N. J., 2009, p. 626; For insight into food fraud history in general, see B. Wilson, *Swindled: The Dark History of Food Fraud, from Poisoned Candy to Counterfeit Coffee*, Princeton University Press, Princeton, NJ, 2008.
72. W. C. Saunders, Adulteration of Cocoa and Chocolate, *Confectioners' Journal*, 1895, **21**, 64.
73. P. P. Gott and L. F. Van Houten, *All About Candy and Chocolate: A Comprehensive Study of the Candy and Chocolate Industries*, National Confectioners' Association of the United States, Chicago, 1958, p. 23; For more on food purity legislation in the USA, see C. A. Coppin and J. High, *The Politics of Purity: Harvey Washington Wiley and the Origins of Federal Food Policy*, University of Michigan, Ann Arbor, 1999; By the 1910s, biochemical analysis and ultraviolet rays were used to identify the adulterated substances within chocolate, as reported in R. Wasicky and C. Wimmer, Eine neue Methode des Nachweises der Schalen im Kakao, *Zeitschrift für Untersuchung der Nahrungs-und Genussmittel*, 1915, **30**, 25-27. Relatedly, the purity of milk in milk chocolate was improved once compulsory pasteurization of milk became legislated beginning in 1908.
74. P. K. Wilson and W. J. Hurst, *Chocolate as Medicine: A Quest over the Centuries*, Royal Society of Chemistry, Cambridge, 2012, p. 105.
75. Key reviews of this literature are found in ed. J. N. Parker, P. Parker, *Chocolate: A Medical Dictionary, Bibliography and Annotated Research Guide to Internet References*, ICON Health Publications, San Diego, CA, 2003, http://www.netLibrary.com/urlapi.asp?action=summary&v=&bookid=99889; K. A. Cooper, J. L. Donovan, A. L. Waterhouse, *et al.*, Cocoa and Health: A Decade of Research, *Br. J. Nutr.*, 2008, **99**, 1–11; F. Visioli, H. Bernaert, R. Corti, *et al.*, Chocolate, Lifestyle, and Health, *Crit. Rev. Food Sci. Nutr.*, 2009, **49**, 299–312; F. Visioli, E. Bernardini, A. Poli and R. Paoletti, Chocolate and Health: A Brief Review of the Evidence, in *Chocolate and Health*, ed. R. Paoletti, A. Poli, A. Conti and F. Visioli, Springer Verlag Italia, Milan, 2012, pp. 63–75; J. L. Donovan, K. A. Holes-Lewis, K. D. Chavin and B. M. Egan, "Cocoa and Health", in *Teas, Cocoa and Coffee: Plant Secondary Metabolites and Health*, ed. A. Crozier, H. Ashihara and F. Tomás-Barberan, Wiley-Blackwell, Oxford, 2012, pp. 219–246; M. Castell, F. J. Pérez-Cano and J.-F. Bisson, Clinical Benefits of Cocoa: An Overview, in *Chocolate in Health and Nutrition*, ed. R. R. Watson, V. R. Preedy and S. Zibadi, Humana Press, New York, 2012, pp. 265–275; S. Ellam and G. Williamson, Cocoa and Human Health, *Ann. Rev. Nutr.*, 2013, **33**, 105–128.

76. *Chocolate and Cocoa: Health and Nutrition*, ed. I. Knight, Blackwell Science, Oxford, England, 1999.
77. R. Lopez, *Chocolate: The Nature of Indulgence*, H. N. Abrams in association with the Field Museum, New York, 2002, p. 62.
78. P. Zipperer, *The Manufacture of Chocolate*, Spon and Chamberlain, New York, 3rd edn, 1915, p. 306; William Salmon had recommended his "chocolate wine" by the glassful in the 18th century, as Norah Smaridge noted in *The World of Chocolate*, J. Messner, New York, 1969, p. 81.
79. B. M. Berman, *et al.*, The Public Debate over Alternative Medicine: The Importance of Finding a Middle Ground, *Altern. Ther. Health Med.*, 2000, **6**, 98–101.
80. L. Quaid, Chocolate Gets Healthy Support [Harrisburg, Pennsylvania, USA], *Patriot News*, 19 August, 2005, p. C-6.

CHAPTER 2

Sustainable Cocoa Production: A Healthy Bean Supply

DAVID A. STUART

Food & Nutrient Impact, LLC, 391 Vesper Road, Hershey, Pennsylvania 17033, USA
Email: davidstuart17033@gmail.com

2.1 Introduction

Whereas the terms "sustainable" or "sustainably produced" are widely used to promote many consumer products and services, no common set of principles exists by which consumers define sustainability. This is in part because each good or service is unique and faces different challenges. Over the last 20 years, cocoa sustainability programs have increasingly been discussed worldwide. The chocolate industry began its efforts to help farmers raise cocoa trees more efficiently in the face of the pests and disease that are common in the rainforest environment where cocoa thrives. These efforts were designed to improve the economic viability of the most important raw material for chocolate manufacture—the cocoa bean. As is well known, carob is not a substitute for the aroma and flavor of chocolate. However, virtually all chocolate manufacturers, large and small, neither grow their own cocoa beans nor buy them directly from farmers, except perhaps in small quantities. Instead, cocoa beans are collected by local merchants and truckers and taken to a portside community where they are consolidated into lots that are then graded, cleaned and prepared for shipment. Several decades ago, most cocoa beans were exported to Europe and the developed world, after which they were processed into the key ingredients of chocolate, namely chocolate liquor, cocoa butter and cocoa powder. Increasingly, cocoa

Chocolate and Health: Chemistry, Nutrition and Therapy
Edited by Philip K. Wilson and W. Jeffrey Hurst
© The Royal Society of Chemistry 2015
Published by the Royal Society of Chemistry, www.rsc.org

beans are being processed into these ingredients in the major port-of-origin countries. Regardless, the beans or the cocoa ingredients are not processed on the farms but rather in factories distant from the farms. This process creates a long supply chain whereby the farmer, chocolate manufacturers and the final consumers of chocolate have little idea what each other does. The final stages of chocolate production occur in Europe, North America and other areas of the developed world that do not grow cocoa. And yet, if problems or controversies arise in the production of cocoa beans—whether due to increased commodity prices that increase candy bar prices, lack of farmer education, protection of the tropical environment or other issues—it is the retailer of the chocolate or the branded manufactures who are typically the first to hear consumer complaints. More importantly, chocolate manufacturers know that the long-term success of their business depends on a vibrant and healthy supply of cocoa beans. Thus, the US and global chocolate industry has taken increasing responsibility for and engaged in efforts to ensure that there are not only good, reliable sources of cocoa beans, but also that the supply chain supports a good living for the cocoa farmer and her/his family.

This chapter focuses on cocoa as an agricultural output from millions of small farms growing and harvesting cocoa trees throughout the tropics. Key events prompting the chocolate industry to address sustainable cocoa production in the 1990s, the evolution of cocoa sustainability organizations, the current programs of the global cocoa initiative and a vision of what needs to be done in the future are all discussed below.

2.2 Where Did Cocoa Originate?

The origins of the cocoa tree (*Theobroma cacao*) are thought to be in the upper reaches of the Amazon watershed on the eastern slopes of the Andes bordering what is now Brazil, Ecuador and Peru.[1] Cacao is thought to have spread from this region by "macro-fauna" such as birds, squirrels and monkeys. Later, around 10 000 BC, humans participated in the spread of cacao north and eastward. At the time of the expeditions of Columbus and Cortés—the first Europeans to see cacao beans and to taste chocolate drinks—cacao had been spread to the Pacific forests of Peru, Ecuador and Columbia, northward in South America to Guyana and Venezuela and into Mesoamerica, northward to Southern Mexico.[2]

Cacao is considered to be a New World crop, just like corn (maize), tomato and potato. Before the exploration of the New World, Europeans had neither seen nor experienced cocoa drinks. When early Spanish explorers first saw cacao, they called the beans *almendras*—Spanish for almonds—because the cacao seed, if unfermented, bears a striking resemblance to the shape and color of almonds with which the Spaniards were familiar.[3]

Cacao trees are about 40 feet tall and can grow in the tropical rainforests around the world. The tree evolved in the tropics where the environment lacks the extremes of extended dry periods and freezing, typical of more temperate climates. Cacao seeds do not undergo the dormancy typical of temperate crops. Instead, the seeds germinate over time without drying out. They remain trapped within the robust cocoa husk—the surrounding pod cover—unless the pod is broken open by birds, squirrels, monkeys or humans. If the pod is left intact, the seedlings will germinate and die within the confines of the husk.

The Spanish were the first to colonize Mexico—*Nuevo Espana*.[4] As the population grew in Mexico City, they soon ran out of the cacao supplies that were being transported from southern farming regions in the tropics.[5] The Spanish then began transporting the plants to their Caribbean Island colonies of Hispaniola (now Haiti and the Dominican Republic) in 1560, and Trinidad in 1575.[6] Over long distances, cocoa is transported either as a pod-derived seedling or as a plantlet. Historical records indicate that the Spanish, who were the first of the European colonialists to move cocoa, did so by transporting cacao seedlings in mini-greenhouses onboard ships.[7] In Southeast Asia, the Spanish were also colonizing the Philippine Islands and decided to transport cacao from Mexico westward across the Pacific in 1660–1670.[8]

Africa is now the largest world producer of cacao. In 1822, the Portuguese were the first to move cacao from Brazil, then its colony, to Principe and Sao Tome off the coast in West Africa. Important movement from Fernando Po to Ghana by Tetteh Quarshie occurred in 1879.[9] Quarshie became a national hero and established cacao first in regions east of Accra and then in the central region of Ghana, native lands of the Ashanti tribe who have adopted cacao and cocoa farming into its culture. Another major movement of cacao occurred from Fernando Po to Ivory Coast in 1912.[10]

2.3 Where Does Cocoa Grow Today?

About 100 years ago, the majority of cocoa was produced mainly in the Americas with 73 700 tons originating in Central and South America (85%), 15 300 tons originating in Africa (15%) and 5100 tons from Southeast Asia (5%).[11] Recent data from the International Coffee and Cocoa Organization note that the world distribution has changed dramatically.[12] In the 2011/12 production year, total world production was 4.05 million tons, with Africa accounting for 71.4% of the total production, within which Ivory Coast was the largest producer at 36.4%, and Ghana at 21.6%. The native home of cacao—the Americas—now accounts for only 15.5% of world production, with Brazil producing 5.4% and Ecuador producing 4.7% of world cocoa (See Table 2.1). Southeast Asia accounts for 13.1% of world production, led by Indonesia at 11.1%.

Table 2.2 documents the use or import of beans and cocoa-containing ingredients into the top 20 countries that use these raw materials for manufacturing finished products.[14] Europe dominates the world use with about 40% of the total consumption of beans and cocoa-containing ingredients, followed by the USA with about 25%.

Table 2.1 World cocoa production data from 2010/11 harvest season (data in 1000 tons).[13]

Region	Regional Production (% of World Production)	Country	Country Production (#Rank)
Africa	3224 (74.8%)	Ivory Coast	1511 (#1)
		Ghana	1025 (#2)
		Nigeria	240 (#4)
		Cameroon	229 (#5)
		Other African	220
Americas	561 (13.0%)	Brazil	220 (#6)
		Ecuador	161 (#7)
		Other Americas	201
Southeast Asia/Oceania	526 (12.2%)	Indonesia	440 (#3)
		Papua New Guinea	48
		Other Southeast Asia and Oceania	39

Table 2.2 Importation of cocoa beans and cocoa-containing ingredients by country in 2009.

Nation	Rank
United States	1
Netherlands	2
Germany	3
France	4
Belgium	5
United Kingdom	6
Malaysia	7
Russian Federation	8
Italy	9
Canada	10

Yield data indicate that the Malaysian plantation production systems discussed below show the highest productivity of all commercial systems at about 2000 kg ha^{-1}. However, using experimental systems that optimize watering, fertilization, the best genetics and crop management systems, it is possible to double these rates of production to as much as 4000 kg ha^{-1}. Indonesia is one of the most highly productive areas in the world with about 900 kg ha^{-1} production.[15] In West Africa, average production estimates in the Ivory Coast are about 350 to 400 kg ha^{-1}. Ghana has slightly higher production rates due in part to government-sponsored training, fertilization and crop management support. In the Americas, yields average about 300–350 kg ha^{-1}.

In looking at these averages, one might surmise that a single factor such as soil fertility could cause these differences. That is not the case, although soil fertility can improve yields significantly. Rather, these yields reflect local, traditional growing practices in the respective regions as well as plant genetics, tree spacing and crop management, agricultural extension services, access to market information and government taxation on exported beans. For example, Ivory Coast and Ghana place significant export taxes on raw cocoa beans. These taxes were originally levied to support prices in the origin countries. While the provisions of taxation still exist without effective price supports, the export taxes are relaxed if the cocoa beans are converted in-country prior to export into products such as chocolate liquor, cocoa butter and cocoa powder. Most of the benefit from the tax relief accrues to the manufacture of the cocoa products, with little to none directly going back to the farmer.

2.4 Early Problems with Cocoa Production

Before discussing cocoa sustainability, it is worthwhile to chronicle some pertinent events in cocoa production that have driven key changes over time.

2.4.1 Cocoa Demand in *Nuevo Espana*

During the landing of Columbus on his third voyage to Mesoamerica in 1497, the Spanish had their first glimpse of cocoa beans but did not taste the drink or take cocoa beans back to Spain. In 1519, Hernán Cortés met Moctezuma and was said to have tasted the drink and learned of the exploits of the Aztec royalty as they celebrated cocoa.[17] But again, Cortés only chronicled the cocoa drink, and his party did not take actual cocoa beans back to Spain after their three-year conquest of the Aztec empire.

However, Spain did set up a major colony in what they called *Nuevo Espana* in which Spanish émigrés became familiar with the native cocoa drink. Demand for the drink sky-rocketed in the colony as its population grew to about one million inhabitants.[18] Supplies of cocoa beans were grown many miles south of the old Aztec capital of Tenochtitlán, which the Spanish renamed Mexico City. The cocoa beans had to be transported up to the main population center in the capital. The demand of cocoa beans from this region quickly out-stripped the supply. First, local farmers were susceptible to European diseases, and massive deaths of peasant farmers notably occurred between 1540 and 1600.[19] Second, *Nuevo Espana* colonists loved the chocolate drink, thereby driving demand for more beans. Finally, the introduction of cocoa to Spain itself in 1585 drove demand for beans even higher.[20] Overall, a serious shortage of beans existed by 1600. The Spanish responded by exporting cocoa trees to set up plantations in their Caribbean colonial islands and setting out to explore further lands within Mesoamerica and South America for more cocoa.[21] In seeking new sources of cocoa, they discovered wild cocoa groves in the Guayas River Valley of Ecuador in 1620.[22] As history shows, without vibrant and healthy farm communities, it is impossible to keep up with a growing demand for cocoa beans.

2.4.2 Plant Disease

A second problem with cocoa early on was brought about by the movement of cocoa from the Amazon basin, which spread fungal disease to cocoa growing on the shores of the Caribbean. The local crop that consisted of high-flavor criollo cocoa was decimated by this disease in the 1720s. Capuchin monks returning from Amazonian missions passed through Trinidad with Amazonian cocoa plants, which they called forastero, meaning foreigner.[23] Soon plant crosses occurred, allowing hybridization to create a more disease-resistant variety. The hybrid between criollo and the Amazon cocoa (forastero) acquired disease resistance, allowing them to grow well in the presence of the local witches' broom infestation. These hybrid lines of cocoa are today known as trinitario cocoa, and they appear to be intermediate in shape between the criollo and the forastero cocoa varieties.[24] Breeding good cocoa has long been aimed at incorporating disease and pest resistance as well in improving yield. In other food crops, plant breeding has significantly improved the productivity of grains, oil seeds and numerous fruits and vegetables.

2.5 Recent Concerns with Cocoa Production

2.5.1 Collapse of Malaysian Plantation Cocoa

Moving forward into the second half of the 20th century, cocoa production has continually increased to meet world demand. Malaysia grew cocoa under some unique, intensive farm management systems on very large plantations being worked with paid labor that yielded, in some instances, 2000 lbs of cocoa per acre. Malaysia was on a track to become a dominant supplier of cocoa to the world. Concurrently, farmers on Sabah, a northerly island in Malaysia, were carving out small cocoa farms from the rainforest to raise cocoa at about 800 kg ha^{-1}. Compare this to the 350 lbs per acre in the Americas and in West Africa (Table 2.3). But then, several things happened

Table 2.3 Yield of cocoa in various regions throughout the world.[15,16]

Region/farm	Production System	Average Yields in Region (kg ha^{-1})
Experimental farms	Optimized	3500–4000
Malaysia	Intensified plantations	*ca.* 2000
Indonesia	Small growers	700–900
West Africa	Small growers	350–400
South America	Small growers	300–350

to the large Malaysia plantations. First, local economic development in the Malaysian manufacturing sector created better-paying jobs in the cities, and workers who had tended the cocoa farms migrated to the cities for better pay. As the shortage of labor on the large cocoa farms grew, it drove plantation owners to begin pulling out their cocoa and planting other high-yielding, profitable crops such as oil palm, which required considerably less labor. The large plantation system that had been so successful in 1980 had virtually disappeared by 1995, thereby diminishing Malaysian production. As with any human endeavor, farmers and farm workers are forced to make economic choices, whether it is farming on a large scale, on a small farm or a choice between work on the farm *versus* work in the city. Here, the attraction of manufacturing jobs in the cities drew the labor force away from cocoa plantations.

2.5.2 Witches' Broom Comes Back to American Cocoa

In 1989, two specific outbreaks of witches' broom infection occurred when infected plants were taken from outside the region and planted in the Amazon region of Brazil—Rondonia.[25] Following this, the disease spread rapidly. Although this area was remote from the Bahia cocoa-growing region, in several years it had spread throughout both the Amazon and the Bahia regions. Apparently either a new strain of the disease had mutated and was transported into Rondonia or the planting stock that was developed by the Brazilian government's food and agriculture organization—CEPLAC—had not maintained adequate genetic resistance to new varieties of fungus. At the start of the outbreak, Brazil produced the second-highest amount of cocoa in the world: 400 000 tons of annual bean production. Within seven years, the crop was devastated, and yields fell by over 70%. In 1996, the Brazil cocoa industry, which had been self-sufficient in cocoa, was faced with importing cocoa from Ghana in order to meet local chocolate manufacturing demand. Farmers in the port of Salvador rioted at the sight of ships carrying Ghanaian cocoa, and they prevented the off-loading of cocoa beans for 40 days. These findings demonstrate that the cocoa industry must work with coordinated efforts not only to develop new varieties of cocoa that are resistant to pests and disease, but also to establish safeguards that prevent the spread of noxious diseases and pests from other parts of the world. Sadly, there is not a disciplined and coordinated international effort to develop disease resistance and increase yield. There are, however, quarantine programs that hold plants in isolation before shipping them to other regions of the world, but these programs are used mainly by scientists for plant transport. In response to the witches' broom devastation of the Brazil crop, the US Chocolate Manufacturers Association (CMA) through its scientific arm, the American Cocoa Research Institute (ACRI), began initial programs to support cocoa crop protection and the breeding programs already underway in Brazil and in Centro Agronomico Tropical de Investigacion y Ensenanza (CATIE)

(Costa Rica), a site of one of the major cocoa germplasm collections in the world. The outreach for expertise that the chocolate manufactures initiated in these and subsequent efforts was an important step in developing solutions to complex problems of sustainable cocoa agriculture.

2.5.3 Pest Spread from Malaysia to Indonesia

In the late 1980s, Indonesia was a rapidly growing new source of cocoa that expanded after the collapse of the large Malaysian plantations. In 1993, pods from Indonesia were taken from Sabah and transplanted to Sulawesi, Indonesia, an area that was ideal for cocoa production. The pods harbored a minute insect that quickly spread to the surrounding plantings of cocoa. This insect, the cocoa pod borer, is only 6 mm long as an adult. It lays microscopic eggs on immature pods. The eggs hatch into larvae that bore into the developing pods. Within the pod, the larva eats the developing placenta that nourishes the developing cocoa bean, at times even eating the seed itself. This insect spread rapidly and caused a severe decline in otherwise very good yields (900 kg ha^{-1}) of cocoa beans.

In October 1995, ACRI, the British Chocolate Cookie and Cake Association and ASKINDO, the agricultural service of Indonesia, organized a scientific gathering specifically devoted to cocoa pod borer control.[26] Meetings were held for four days to develop strategies on how to effectively deal with the problem. Fieldtrips to infested sites were held after the meetings. Present at the meeting were United Nation's Food and Agriculture Organization (UN-FAO) representatives who had been training farmers on how to grow rice, a representative expert in the life-cycle as well as the control of this pest, experts from local and international universities, representatives from the local agricultural extension, representatives from agricultural non-governmental organizations (NGOs) working on the ground to assist in rural development and experts in biological control who used attractants, the spread of sterilized male pod borers and fungal pathogens of pod borers, as well as general plant breeders. The meeting resulted in a multi-pronged research program that included farmer training, thorough harvesting of pods to break the insect life-cycle, several bio-control programs and a program to cover young pods to prevent egg laying and larval growth, as well as plant selection of cocoa varieties resistant to the disease. Funding of these programs was multilateral and included support from American Cocoa Research Institute (ACRI), which was the scientific arm of the Chocolate Manufacturers of America, the FAO, the local and national Indonesian government, US-AID and other groups. This was one of the first multiple stakeholder efforts to plan, set up and attack a critical problem in a cocoa-growing area. The program has all the hallmarks of a sustainability program, although at the time that description was not attached to the Pod Borer Program. This effort was one of the first to use a broad-brush approach to an agricultural problem engaging chocolate manufacturers, cocoa processors and local governments, and to draw upon multilateral funding—in this case from the US Government—in order to engage multi-lateral groups such as the United Nations, which supports the FAO. Key to the success of this program were: (1) recognizing problems early enough to act before they devastated the local industry; and (2) engaging a broad array of expertise beyond the chocolate-manufacturing or cocoa-processing companies.

2.5.4 Panama Conference

In 1988, a unique event was held at the Smithsonian Institution of Washington DC's administrative facilities in both Panama City and at the Barro Colorado Island research station in the Panama Canal Zone. The conference was attended by the world's leading confectionery branded companies, chocolate processors, representatives of some of the cocoa-growing countries, agricultural development experts, ecologists including biodiversity, bird and large mammal experts, bio-control experts, cocoa breeders and developmental NGOs specializing in rural development. After the four-day meeting, field excursions to cocoa-growing regions in the Bocas del Toro region of Panama were conducted to examine cocoa farming in action as well as to view firsthand the devastating effects of frosty pod on a farm that essentially had no yield. The meetings resulted in what is believed to be the first statement of sustainability for cocoa as follows:

Panama Cocoa Sustainability Principles:[27] A sustainable, biologically diverse system of growing cocoa will:

- Be based on cocoa grown under a diverse shade canopy in a manner that sustains as much biological diversity as is consistent with economically viable yields of cocoa and other products for farmers.
- Use constructive partnerships that are developed to involve all stakeholders with special emphasis on small farmers.
- Build effective policy frameworks to support these partnerships and address the particular needs of small farmers for generations to come.
- Encourage future cocoa production that rehabilitates agricultural lands and forms part of a strategy to preserve remnant forests and develop habitat corridors.
- Maximize the judicious use of biological control, techniques of integrated management of pests, disease and other low-input management systems.

Action plans arising from this conference were as follows:

- Address socio-economic issues relating to small-scale farmers.
- Research ecological priorities.
- Improve the long-term profitability of cocoa production, especially for small farmers.

- Develop biologically sensible ways to control pests and diseases.
- Improve cocoa genetics for yield and pest resistance.

Important to this statement was the recognition of three basic interests in a robust sustainability agenda; namely economic, social and environmental needs. First and foremost was that ecological interests must support agriculturally diverse farm management, rehabilitation of agricultural lands and the use of shade systems to increase biological diversity. The statement emphasizes that a sustainable cocoa production system must be economically viable. There was little mention of specific social needs other than that used to support small farms as part of partnerships in order to meet small holder needs into the future.

2.5.5 1998 View of Cocoa Sustainability

Figure 2.1 diagrammatically shows the relative importance of the three main drivers of cocoa sustainability at that time. This agenda placed a heavy emphasis on the environment, perhaps due to the Smithsonian Institution's interest and that of many of the invited conference participants. Obviously, the economics of production was of great interest to the chocolate industry and cocoa farmers. Sustainable cocoa supplies in the future would come to see devastating losses due to pests and disease, accompanied by the loss of major growing areas. Notably, there was little mention of the social aspects of the small holders and of the communities that depend on cocoa for their livelihood. This may be due to industry knowledge that small farmers can make a good living if they produce high yields and can afford better inputs for their crop such as in Indonesia (Table 2.3) where they have better homes,

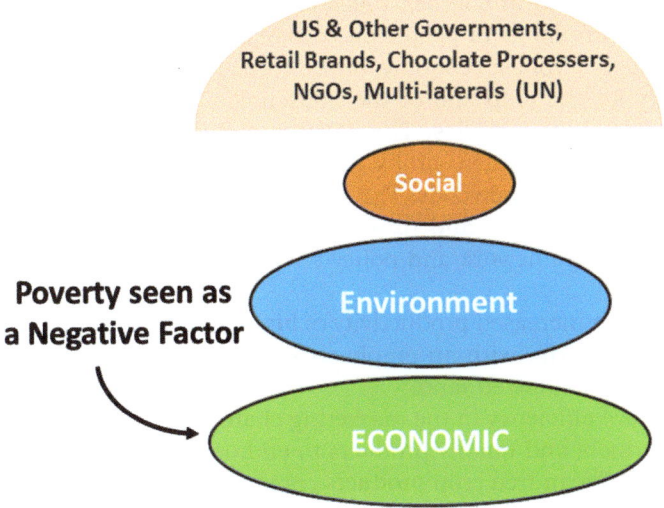

Figure 2.1 Concept of cacao sustainability after the Panama Conference.

personal transportation and television and can afford schooling and medical care for their children. This diagram also makes clear the importance of partnerships between local and international organizations who hold expertise to drive change in the programs.

2.5.6 Sustainable Tree Crops Meeting in Washington, DC

In October 1999, a large meeting was held in Reston, Virginia in the USA, at which members of the chocolate industry met with agricultural development experts from the World Bank, the Consultative Group for International Agriculture Research, US-AID, government agricultural representatives from West Africa and a number of the NGOs that had been present at the Smithsonian Panama meeting.[28] This gathering was designed to discuss and plan for a new and very novel African development program named the Sustainable Tree Crops Programme (STCP). The STCP concept is borne out of the outstanding success that "food security" programs had in making developing countries self-sufficient in their own food production. Because of the declining need for food security programs, the concept was to design a program focused on "economic security" in countries that had enough food but were still agriculturally based and needed better farm productivity in order to generate cash for their rural families. For cocoa, it was clear from studies of household economics that as much as 50% of the cash income was derived from the sale of cocoa. This cash income would benefit the three to four million small farms in West Africa with households of eight or nine persons each. Indeed, over 18 million people could have been impacted by this program! The extra cash generated would enable farmers to buy food, improve housing and pay for educational and medical expenses. But it also became clear that cocoa "crop security" issues, such as the rapid spread of disease or pests like that seen in Southeast Asia and South America, could destroy the livelihoods of millions of people on cocoa farms. The idea behind the STCP was to develop a robust economy for the cocoa-farming sector leading to self-sufficiency for the crop and of the households that depend upon the cash income that cocoa supplies. The STCP was designed to improve not only cocoa, but also other tree crops in the region, including coffee, cashews, tropical fruits and hardwoods. The stated program goal was:

To promote public and private sector partnerships to provide an organizational framework and policy environment that is necessary to:

1. Maintain increased productivity of high-quality tree crop products, over the long term, with an emphasis on farm rehabilitation and reclamation of deforested land;
2. Improve efficiency in the marketing chain, so that it delivers fair prices to farmers and quality products to end users;
3. Make African tree crop products competitive in international markets;
4. Improve the socio-economic situation of farmers; and
5. Conserve the natural resource base and biodiversity.

The STCP program launched in late 2000 in Ghana, Ivory Coast, Cameroon and Nigeria and was designed as a demonstration or "pilot program" to impact about 2% of the cocoa farms in the West African growing region.

2.5.7 Formation of the World Cocoa Foundation

Prior to the Indonesian Pod Borer Project and the STCP, no formal organization existed to provide research and implementation for sustainable cocoa production. The industry funding and coordination was done on an *ad hoc* basis through the US, Canadian and European chocolate trade organizations. There was a clear need to create a global organization with the expertise and mission to manage agricultural development projects. In 1997, the CMA had attempted to form a stand-alone organization for the purpose of helping cocoa growers improve yield and provide pest and disease resistance, but the fundraising was directed mainly towards US-based companies and wealthy donors, and the effort failed.

In 2000, the timing for the formation of such an organization proved much better, and the World Cocoa Foundation (WCF) was established.[29] It is headquartered in Washington DC and operates under the following mission:

The WCF promotes a sustainable cocoa economy through economic and social development and environmental stewardship in cocoa-growing communities.

In 2000, with the formation of the WCF, projects that had been initiated by CMA–ACRI for cocoa breeding and to combat destructive diseases and pests were transferred to the WCF. Additional programs such as the farmer economic development programs, including the STCP, were also moved to the WCF. WCF donations are tax exempt under its unique funding status and mission, which are separate from the mission of CMA or the National Confectioners Association, thereby enabling the WCF to work with a broad array of national and international research, development and implementing agencies. In 2013, membership of the WCF had grown to 100 companies that represented over 70% of the global cocoa/chocolate industry.

2.5.8 Child Labor Incident Broadcast in the UK by Channel 4

In a 28 September 2000 documentary aired on Channel 4, reporters and a camera crew in West Africa were searching for a story on farm labor. There they investigated a cocoa farm on which a boy aged 12–14 years old was found marked with significant burns.[30] They interviewed the boy, and he said that he was not free to leave and that he was working very hard every day under the direction of an older man on the farm—the Big Man. The boy who was interviewed was never remanded to local authorities or to child protection services. The report was televised in the UK later that summer, and the show garnered considerable attention regarding cocoa and its growth. The report was immediately challenged the next day by an Ivorian Minister.[31]

2.5.9 Child Labor Surveys Planned in West Africa

In the fall of 2000, a small number of individuals from the top US branded chocolate companies met with representatives of US-AID and other representatives of the US Government. Here, they learned that neither the West African governments nor agencies that support international development in the region had collected information on farm labor, specifically child labor, on West African cocoa farms. At that point, the US industry, in cooperation with US-AID, the UN International Labor Organization and US Department of Labor, agreed to field a representative sampling of the cocoa-growing region of Ghana, Ivory Coast, Cameroon and Nigeria, where about 70% of the world's cocoa grows, in order to determine the general status of children on cocoa farms in that region. Surveys launched early in 2001 took as random a sampling as possible, much like the political polling process. The surveys were administered by people familiar with the local customs and tribal languages. All people administering the questionnaires were trained in how to recognize children and youth in indentured servitude, in abusive conditions or in slavery, and they were provided with local child protection services that could provide safety to any children found to be in danger. The survey results were due in late summer 2001.

2.5.10 "Slave Ship" is Reported in the Gulf Guinea

On 30 March 2001, reports emerged of a freight ship of Nigerian registry, the *MV Etireno*, sailing in the Gulf of Guinea that was rumored to have between 28 and 250 child slaves onboard. UNICEF urged local authorities to close ports so that this ship could not leave port if it had landed or prevent its landing to offload its cargo without a proper investigation. The ship sailed for 18 days amidst a media frenzy in the US and UK. The ship finally disembarked in Cotonou, Togo, and offloaded children who were either under the supervision of their parents or guardians who were close relatives.[32] No slave children were found.

2.5.11 Child Labor Issues Undertaken by the US Congress: The Harkin–Engel Protocol

In the spring of 2001, the African child labor reports reached the USA as the September 2000 Channel 4 program was being re-aired by CNN. The issue was picked up by activists and radio/TV talk shows and was eventually discussed in the US Congress. A rider, sponsored by Representative Eliot Engel of New York and Senator Tom Harkin of Iowa, was attached to a US appropriations budget that called for the labeling of chocolate products that contained beans processed from West Africa, especially Ivory Coast, specifying that the product contained beans produced by slave labor, unless there was proof otherwise. Although the US chocolate industry found this type of labeling to be ill advised, there were no data available to contradict the claims of massive child slavery in West Africa, especially in Ivory Coast.

Seeking a compromise to avoid such labeling, the congressmen and the US chocolate industry entered into a multiyear agreement, called the Harkin-Engel Protocol, to address child labor issues in West Africa.[33] The Protocol called for specific steps to be taken in a timely manner and was outlined in a document on the United Nations International Labor Organization (ILO) website.[34] The protocol was scheduled to be signed on 11 September 2001, but more pressing needs created by airplane attacks on New York City, Washington DC and Pennsylvania that day delayed the signing for eight days.

2.5.12 Child Labor Survey Results

The results of the cocoa farm child labor surveys were reported in a 24 July 2001 press release co-authored by United States Agency for International Development (US-AID), United States Department of Labor (US DOL) and the ILO.[35] It is important to note that the definition of children for most of the international community is girls/young women and boys/young men below the age of 18. It is also important to point out that education in West Africa is compulsory through the sixth grade, after which secondary education is optional. Thus, it is normal to graduate from a rural grade school and then work for the family or in the community between the ages of 12 and 18 years.

Results of the survey projected to the populations found in the region are shown in Tables 2.4 and 2.5. The survey found that no children had been enslaved. Rather, children working on the farms were typically the offspring or relatives of the farm owners. An estimated 284 000 children were found working on cocoa farms in all activities throughout the region (Table 2.4). The highest number involving children working on cocoa farms was in Ivory Coast, the country with more cocoa farms and the highest world cocoa production, and fewer were found in Ghana, Cameroon and Nigeria.

Table 2.4 Estimates of child labor by selected characteristics in study areas of West Africa. Survey results of cocoa child labor surveys (24 July 2001).

Characteristic	Ivory Coast	Nigeria	Cameroon	Ghana
Children who carry out all tasks	129 410	–	–	–
– Apply pesticides	13 200	4600	5500	–
– Use dangerous tools	71 100	9300	35 200	38 700
Paid child workers	5121	1220	0	0
Children with no family ties	11 994	–	–	–

Table 2.5 Estimates of working children at high risk according to selected activities and characteristics in study areas of West Africa. Survey results of cocoa child labor surveys (24 July 2001).

Characteristic	Ivory Coast	Nigeria	Cameroon	Ghana
Recruited through intermediaries	2100	354	0	0
Use of machetes (children under 15)	109 300	2325	16 192	18 189

The largest category of work was using dangerous tools, which included machetes, harvesting knives and axes. Fewer, but still significant numbers of children were found applying pesticides throughout the region. Some working children in Ivory Coast and Nigeria were paid, and none were paid in Cameroon and Ghana. An estimated 12 000 children in the Ivory Coast worked on farms where there was no clear relative in residence.

Table 2.5 identifies the numbers of children at higher risk. An estimated 2100 (1.6% of all children in work activities) in Ivory Coast and 354 (2.3% of children in work activities) of children recruited through intermediaries might be considered as being trafficked into the area. The use of machetes by children under 15 years of age was common to all of the cocoa areas studied. However, machetes are commonly used by youth to trim the grass at home and to tend their school garden, as well as to trim the local football field (pitch), most often under the supervision of a teacher or adult.

2.5.13 Initiation of Child/Youth-directed Sustainability Projects

Programs specifically designed to assist children were established under the cooperative support of the ILO, and their specialized child labor unit, the International Programme to Eliminate Child-labor (IPEC), which would serve as the project managing partner, together with the US DOL and the global chocolate industry as funding partners. ILO–IPEC was charged with promulgating child labor-monitoring systems at the community and district levels with the aims of eventually:

(1) Rolling out a general picture of strenuous or hazardous child labor in the cocoa sector;
(2) Providing local education regarding the issues of child labor and how it impacts youth development; and
(3) Identifying hot spots where a focus of effort could be placed.

These are some of the first efforts of their type in cocoa-growing regions around the world. The program known as the West African Cocoa Agriculture Program (WACAP), launched in 2003, was designed to rescue youth at high risk on cocoa farms in Ghana and Ivory Coast and to rehabilitate them by helping children/youth re-enter school.[36] At the end of 2004, 3000 children had been counselled and removed from the work environment and re-entered into either local elementary or vocational schools. Over the three years of this program's work, WACAP intervened in the lives of 9700 children under 14 years of age.

2.5.14 Other Child/Youth-directed Projects

An industry-funded radio broadcasting program was initiated in Ghana's western region in 2003 that included public service announcements regarding the need for improving education and to route more kids into schools. This program received its primary funding from the global cocoa industry.

An additional program was sponsored by the International Foundation for Education and Self Help (IFESH) that provided computer equipment, printers, internet access and books for four teacher's colleges—two within each of the cocoa-growing regions of Ghana and Ivory Coast. The program was designed to support in-service teachers and student teachers who were about to graduate from the colleges. This program provided them with contemporary and varied curricula, improved their library holdings and allowed teachers to print study lessons for students who could not afford books. This program was launched in 2004.

2.5.15 WCF/US-AID Winrock Empowering Cocoa Households with Opportunities and Educational Solutions Program

Building on the IFESH pilot education programs mentioned above, the WCF Empowering Cocoa Households with Opportunities and Educational Solutions (ECHOES) program launched in October 2007 strengthened cocoa-growing communities by expanding opportunities for youth and young adults through education.[37] Its vision was to provide youth who had completed compulsory primary education with opportunities to continue their education along pathways designed to serve the rural sector. In other words, it is a post-elementary school training program in occupational and business skills that are relevant to employment in the rural sector. ECHOES is a scalable educational model that supports 38 cocoa communities in Ivory Coast and 41 cocoa communities in Ghana. The overarching objectives are to: (1) empower communities through basic and adult education—education that is drastically lacking in the rural sector; (2) improve household income in the rural cocoa sector; and (3) strengthen the capacity of community-based organizations by supplying skills relevant for building rural businesses.

2.5.16 The Cocoa Livelihoods Program

The Cocoa Livelihoods Program launched in 2009 and is a cooperative project between the WCF and the Bill and Melinda Gates Foundation.[38] Its goal is to double the income of about 200 000 small-holder households that grow cocoa in the targeted West African countries of Ivory Coast, Ghana, Nigeria, Cameroon and Liberia.

The overall program objectives are to:

1. Improve marketing efficiency;
2. Improve cocoa production efficiency and quality at the small farm level; and
3. Improve cocoa production efficiency and quality at the farm level.

In addition to increasing farm income, this program aims to strengthen the local service economies that support cocoa growers, including agricultural extension, market information, access to newly improved cocoa tree planting stock and farm inputs such as fertilizer, as well as improving plant husbandry. The concept is that in an economically healthy cocoa economy, youth will have expanded opportunities for education and jobs, including being in a better position to take over the family farm when the time comes. To date, 106 000 farmers have been trained and 13 cocoa business centers have been formed.

2.5.17 African Cocoa Initiative

The African Cocoa Initiative is a partnership between the WCF and the US-AID-sponsored "Feed the Future" program. The program, launched in March 2011, benefits 100 000 cocoa farmers in Ghana, Ivory Coast, Nigeria and Cameroon.[39]

This program's goals are to:

1. Improve cocoa farmer incomes;
2. Alleviate poverty;
3. Strengthen government and regional institutions; and
4. Advance food security in these communities.

The overall program aims to:

1. Strengthen national public–private partnerships for investing in agriculture and in cocoa;
2. Improve cocoa productivity through better planting material;
3. Enhance public–private sector extension and farmer education; and
4. Foster market-driven farming input supply.

This program is conceptually designed to primarily enhance household income by improving the efficiency and profitability of each cocoa household. In so doing, the family will be better able to improve their cocoa farms and to have sufficient income to support a healthy cocoa community through such expenditures as better housing, improved education for children and improved access to medical treatment, as well as more effectively meeting the general needs of cocoa households.

2.5.18 The Payson Analysis of Child Trafficking in West Africa

An independent survey of child labor activities on cocoa farms specifically in Ivory Coast and Ghana was performed by the Payson Center for International Development and Technology Transfer (The Payson Center) of Tulane

University, New Orleans, Louisiana.[40] This work was funded by the US DOL. Their survey found that 819 000 children in the Ivory Coast cocoa sector and 997 000 children in the Ghana cocoa sector did some sort of work, the vast majority of which was household work. Of the working children, about 5% (41 000) in Ivory Coast and about 10% (98 000) in Ghana were paid for their work. There was evidence that some children were exposed to dangerous work conditions and that some of the children might have been trafficked or obtained through a forced labor situation. The report went on to say that industry activities were not adequate and that more should be done to prevent such atrocities. The report indicated that the efforts to certify cocoa supplies and to commit to having full cocoa certification of company supplies by 2022 using independent certification agents like the Rainforest Alliance, UTZ[†] and Fair Trade, for example, "is a large step in the right direction …". However, the report goes on to state that outreach to an estimated 3463 communities in Ghana and an estimated 3608 communities in Ivory Coast was still necessary, but due to the political instability of Ivory Coast, this objective would be "impossible to implement" at this time.

One view is that the participating members of the cocoa industry, in their attempt to certify 100% of their bean supplies by 2022, set an example for the entire cocoa sector in Ghana and Ivory Coast (see "Certification of the Cocoa Supply Chain" section below). The supply system is so complicated and remote that it is unreasonable to expect bean users in North American and Europe to fix the whole cocoa bean system. Cocoa beans from Ivory Coast and Ghana supply almost 70% of the world's supply, of which even the large chocolate companies use but a fraction of those beans. Much of the responsibility for fixing this sector rests with the sovereign governments of Ghana and Ivory Coast who make and enforce the laws of their countries and support the infrastructure in the rural sector, including schools, teachers and local law enforcement personnel. If chocolate processors and manufacturers of cocoa beans independently set up their own certification systems to ensure that their respective cocoa bean supplies comply with child labor conventions and consumer expectations, then these companies have met their responsibility.

The Payson Center has a report in preparation regarding the incidence of child trafficking from Burkina Faso and Mali into the cocoa-growing regions of Ghana and Ivory Coast. This report is due in 2014.

2.5.19 Cocoa Communities Project

The Cocoa Communities Project is a cooperative effort between the global cocoa industry members, the International Cocoa Initiative of Geneva (ICI) and the US DOL, which will use ILO–IPEC as the implementing partner.[41] This project, initiated in September 2011, is a child protection and education

---

[†]The UTZ Certification Program originated in the Netherlands and was named after the Mayan words for good coffee "Utz Kapah" and currently certifies coffee and cocoa.

program. It is designed to withdraw and prevent children from engaging in hazardous and exploitative labor, which constitute the worst forms of child labor, in the cocoa-producing areas of Ivory Coast and Ghana. The program will provide education and/or occupational skills training to the children and livelihood services to their families. The project also works with cocoa-producing communities to develop community action plans to sustainably reduce child labor and to reinforce government efforts to develop and implement a vast child labor-monitoring system. This project aims to eliminate the worst forms of child labor in broader geographical zones, not just within individual communities. In Ghana, for example, there are forty communities in four zones that are targeted for these interventions.

In addition, the ICI is working in cooperation with corporations that have on-the-ground activities in Ivory Coast and Ghana. These programs participate in broad-ranging activities aimed at strengthening farmer cooperative development, building new schools, establishing child labor-monitoring systems in communities, training field child labor workers and distributing literature on the prevention of child labor in the cocoa sector.

2.5.20 Individual Company-funded Efforts

Since 2005, an increasing number of projects have been funded by individual companies in West Africa. The projects are so numerous that here they can only be mentioned in a broad sense.

General categories of projects include:

Community development: building sanitation facilities for villages and drilling wells for clean water.

Supplemental education: building or expanding schools, supplying books and uniforms, providing internet access and computers, adopting a village and performing information exchanges and supporting vocational education for teenage youth.

Technology: supplying cell phones and programing and providing internet access and computers for teachers and villagers.

Medical assistance: building or remodeling hospitals and supplying contraceptives and anti-malarial drugs to community inhabitants.

Economic assistance: providing loans for school supplies and microfinancing for small businesses.

Agronomic development: this is probably the largest category of assistance that is provided with the understanding that if farmers can be more profitable, they will be able to send their children to school, *etc*. Efforts include cloning large numbers of high-yielding cocoa trees, supporting local genetic improvement crop programs, establishing seed and seedling gardens, supporting regional or national agricultural research efforts, training farmers in ways to improve yield and supplying fertilizers.

For further information, see the websites noted in the references.[42]

2.5.21 Certification of the Cocoa Supply Chain

Considerable interest exists on the part of government, industry and consumers to certify the sustainability of the cocoa supply chain. For governments, there is an immediate interest in knowing the extent of unreasonable child labor, of pesticide use and of water availability and quality, as well as performing agricultural farmer education and agricultural extension services to assist the growers. Corporations are interested in assuring themselves and their customers that they are advancing the cause of sustainability, not just for cocoa, but for all of their products. For example, the sustainability of manufacturing operations and their package recyclability, together with the health and wellness of their employees, are all key areas of concern. These programs are usually lumped under the category of corporate social responsibility (CSR), an area that most corporations have established over the last 15 years. Annual self-ratings and an independent rating of a corporation's practices are performed, and in many programs, their ratings are independently verified. Consumers increasingly express interest in purchasing products that they believe are sustainably and responsibily produced. They will often look for logos or trademarks from independent certifying organizations to guide their shopping choices. Others are not as interested in shopping this way, or may only selectively purchase certified products, especially if the product has a higher price, as is typically that case with organically grown foods.

Cocoa certification can be classified into three general categories: (1) government certification; (2) self-certification; and (3) third-party certification. We have already noted government-funded surveys such as the US-AID/DOL/ILO surveys of child labor in 2002,[43] the Payson example[44] and the child labor monitoring surveys being conducted by ILO.[45] The Harkin–Engel Protocol in fact calls for industry and host governments to design and implement child labor-monitoring programs in each country being investigated. Ivory Coast and Ghana have, at the urging of the chocolate industry, formed their own national child labor-monitoring committees, and though not formal certification schemes, the data they generate serve to classify countries, regions or communities on the occurrence—from high to low—of certain types of child labor. Here, the survey provides a snapshot of labor practices that can then be used to classify not only types of labor practices, but also precise locations where these practices occur. The WACAP[46] and the Cocoa Communities Program[47] child labor-monitoring systems also represent certification systems that have been used to identify school attendance, data that are useful for classifying how student attendance, drop-out rates and apparent gender bias can be tracked over time to focus upon improvement or erosion in specific areas.

Industry self-certifies many of its activities regarding sustainability *via* an annual or biannual social responsibility report (SRP). One can simply look at any corporation's SRP for more detail, and all of the branded cocoa companies and suppliers of cocoa ingredients have these. These reports cover critical raw materials, supplier quality programs, product safety, product quality, product packaging, manufacturing efficiency, hiring employee education and organized labor programs, employee medical programs and community giving. Companies prefer these types of certification programs because they can focus on items that are deemed unique and key to the success of their business, they focus on any shortcomings they may have regarding sustainability and they highlight their specific areas of progress. Many SRPs are audited and all are reviewed and rated by independent academic or third parties specializing in CSR, including investment funds. Two excellent raw material self-certification programs have emerged in the last decade, including the C.A.F.E. Practices[48] program for coffee, which undergoes third-party certification,[49] and the Roundtable for Sustainable Palm Oil initiative.[50]

The other main way to certify raw materials is through established certification organizations.

For cocoa, the three main parties are:

The Rainforest Alliance,[51] whose stated objectives are to support farmers and communities, protect land and waterways and improve growers' incomes;
UTZ,[52] a Netherlands-based organization serving primarily Europe, whose stated objectives are to enable farmers to learn better farming methods, thus improving farm income to allow for improving working conditions and taking better care of their children and the environment; and
Fair Trade,[53] whose stated objectives are to produce quality products, improve lives and protect the environment.

Each organization has unique components within their respective programming, with the Rainforest Alliance and UTZ specializing in improving farm productivity and crop diversity in order to boost household income. Fair Trade favors returning part of the certification premium that they charge back to the farmers' organizations, which are mainly cooperatives.

In 2009, Cadbury was the first major chocolate manufacturer to adopt broad certification in the UK market. They focused on the Cadbury Dairy Milk bar, the most popular chocolate bar in the UK.[54] Within several weeks, two competitors followed suit. Today, many of the large branded chocolate manufacturers have stated goals to work with these third-party certifiers in order to achieve 100% coverage of their cocoa bean needs by 2022.

Several cautions are required when relying on any certification scheme. First, the practical implementation of certification for West African cocoa is now only done through farmer cooperatives. Yet when one looks at cooperative organizations, only 15–20% of the farmers in the region belong to cooperatives. If the level of cooperative participation does not increase, this will set a practical limit on the amount of certified cocoa from the West African region to about 560 000 tons annually, or only 14% of world's cocoa at today's production levels from West Africa. Systems are under development to map the 80% of farms that are not part of the cooperatives using global positioning satellites technology to pinpoint their location.[55] Surveying these remote farms, assessing their needs, improving access to child schooling and improving agricultural production practices remains a formidable challenge for certification programs. Moreover, it would be a shame if this gap is not adequately bridged, since it would put remote family farms out of reach of important world markets for their cocoa beans.

A second cautionary reminder is that despite claims of some certification agencies strictly prohibiting slave or child labor, no certification system can provide an absolute guarantee. This was what prompted the Harkin–Engel Protocol in the first place.[56] Trafficked farm workers and other alleged forms of slavery have been reported in the developed world, including the USA and the UK. Ensuring that such forms of illicit labor do not occur in remote areas within Ivory Coast or Nigeria is simply impossible to guarantee. Instead, it is more accurate and reasonable to state that the certification efforts are designed to "improve the labor situation" in these developing countries.

2.6 Current View of Cocoa Sustainability

Figure 2.2 depicts a current view of cocoa sustainability. Here, a present-day view of the role of family income and farm profitability is shown. Increasing income makes it possible for a family to provide for its own social needs. In the last 12 years, interested parties have come to understand that fixing child

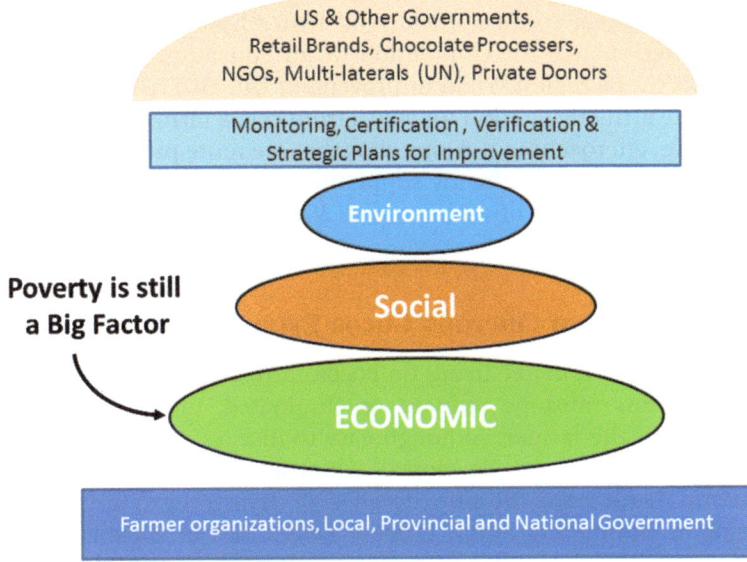

Figure 2.2 Cocoa sustainability concept today.

labor issues will be difficult without first improving family income, which can be accomplished by improving the yield and profitability of the farm. With that income, local schools can be built, teachers hired and children clothed in traditional uniforms and equipped with books, pencils and computers supplies. Support of the local and national governments is essential to assist in building schools and training teachers. Local and national governments can also help to either lower their cocoa bean export taxes or reinvest some of those taxes back into the rural cocoa communities. In all likelihood, as these efforts expand and change with time, cocoa sustainability will continue to evolve to ensure a healthy supply of cocoa.

2.6.1 The Progress and Future Directions

The improvement in the sustainability agenda of the cocoa industry since 2001 has been stunning, particularly in regard to the large-scale multi-sectorial investments in West Africa. The report cards on child labor for Ivory Coast and Ghana alone, as judged by the US DOL, have been most encouraging indications that both countries have made "moderate advancements" in this area.[57] But underpinning these advancements is the fact that West African farmers are producing low yields of cocoa per hectare compared to what farmers in Indonesia can yield. Doubling West African annual yields to 700 kg per hectare would greatly expand the supply of cocoa for the future and enhance the profit for those farmers, assuming that there is no worldwide oversupply. Increasing yield, and hence farm profitability, would improve the social circumstances for millions of youth on cocoa farms throughout West Africa region. The profit realized could help communities build better homes and give their children a better chance for completion of an education, especially for the young girls in the region. Yield improvement will also make staying on the farm a more attractive option for youth from cocoa farms. If cocoa yields do not improve, encroachment from crops that are more profitable than cocoa will occur. Oil palm has already supplanted cocoa in Malaysia, and rubber is currently a highly profitable crop in West Africa. Two things must be changed.

2.6.2 Relief from Onerous Cocoa Export Taxes

In order to make a better living on cocoa farms, the onerous export taxes on cocoa beans must be reduced or eliminated, thereby enhancing the profitability of the farmer. Although easy to discuss, these export taxes are so entrenched that the developed countries, the United Nations and the World Bank have, despite decades of trying, been unsuccessful in reducing or eliminating export taxes in Ivory Coast or Ghana. These taxes, which were originally developed to help support prices, now only serve to tax the farmer.

2.6.3 Develop and Apply Technology to Improve Cocoa Yields

Training cocoa farmers in improved methods to manage their farms, like those mentioned in the STCP and the Cocoa Livelihoods Program, provide an excellent start. In the long term, a concerted effort is required to improve the planting stock available to farmers and to incorporate better pest and disease resistance, thereby improving overall yield. In a true "classic", G.A.R. Wood and R.A. Lass's *Cacao*, first published in 1955, cites that cocoa suffers from 40% losses due to pest and disease.[58] Today, this is still the estimated loss due to pests and disease that most experts on cocoa production cite. This speaks to the fact that cocoa has undergone little genetic improvement against major pests and disease for nearly 60 years. Concerted regional and global research and development efforts for cocoa genetic improvement must be undertaken. This effort should identify and incorporate, using conventional plant breeding technology, the genes for yield and pest- and disease-resistant cocoa varieties that are so needed by cocoa farmers. Funding for this effort should be shared by the chocolate manufactures and processors, the trading community and the origin countries. Notably, this process must be viewed as a long-term effort, because conventional breeding of tree crops takes well over a decade. The 16 year effort of Wilbert Philips-Mora at CATIE in Costa Rica resulted in the development of frosty pod (*Moniliophthora roreri*) resistance.[59] The process of changing low-yielding and susceptible cocoa variety material can be enhanced by identifying resistant materials and cloning them, thereby providing single-loci resistance. Cloning of cocoa from stem cuttings, scions grafted onto root stock and somatic embryogenesis have all been proven to work on the large scale, although these have yet to be implemented broadly in West Africa. Furthermore, breeding for multiple resistances to pests and diseases will require more than a decade if the methods of Phillips-Mora are followed. The breeding process might be accelerated somewhat by using resistance markers available from the cocoa genome and modern selection methods, though it will still take about 15 years to perform crosses and to conduct field tests to ensure robust pest and disease resistance. If these efforts are not started now, the new resistant varieties will be delayed for decades. Organizational models for international plant variety development already exist for essential food crops like rice and corn. A similar research and breeding program must be established for cocoa. Research sites should be situated near major germplasm centers and satellite research stations should be placed in important origin regions like in West Africa. The cocoa genome has recently been described and is being used to develop a better understanding of cocoa yield improvement.[60] It would be a shame if that tool and its understanding were not put to use to globally improve yield and pest/disease resistance in the coming decade.

If concerted efforts are not made, the future for cocoa farmers may not look so bright. In seeking to better their lives, today's farmers and the next generation will have to decide whether to migrate to where they can make a better living or to stay put and decide whether to raise cocoa or replace it with a more profitable crop. Cocoa prices are on the rise, but the age of farmers in West Africa is also rising and is estimated to be about 50 years on average. The outlook projects ever-higher prices for cocoa in the near future as the demand for cocoa increases in emerging economies. If cocoa supplies are to keep pace with demand, then the productivity of cocoa farms must also increase. One option is to clear rainforest areas for more cocoa farms, but this option is increasingly untenable as preservation of the rainforest is a top priority for multi-lateral groups, NGOs and local governments. Without cocoa farmers to farm the new land, the whole effort would be futile. So, the only clear option is to improve cocoa profitability on established farms in order to satisfy the needs of the farmer, the cocoa communities and the markets for cocoa beans.

References

1. J. C. Motamayor, *Heredity*, 2002, **89**, 380–386.
2. S. D. Coe and M. D. Coe, *The True History of Chocolate*, Thames and Hudson, New York, 1996.
3. S. D. Coe and M. D. Coe, *The True History of Chocolate*, Thames and Hudson, New York, 1996, p. 107.
4. S. D. Coe and M. D. Coe, *The True History of Chocolate*, Thames and Hudson, New York, 1996.
5. S. D Coe and M. D. Coe, *The True History of Chocolate*, Thames and Hudson, New York, 1996, p. 83.
6. B. G. D. Bartley, *The Genetic Diversity of Cacao and Its Utilization*. CABI Publications, CAB International, Wallingford, Oxfordshire, 2005.
7. *Chocolate: History, Culture, and Heritage*, ed. L. E. Grivetti and H. Y. Shapiro, John Wiley & Sons, Hoboken, N. J., 2009, p. 504.
8. B. G. D. Bartley, *The Genetic Diversity of Cacao and Its Utilization*. CABI Publications, CAB International, Wallingford, Oxfordshire.
9. Tetteh Quarshie. http://www.ghanadistricts.com/home/?_=49&sa=4642.
10. *Chocolate: History, Culture, and Heritage*, ed. L. E. Grivetti and H. Y. Shapiro, John Wiley & Sons, Hoboken, N. J., 2009, p. 504.
11. F. Hardy, Cacao Manual, *Marketing of Cocoa*, Inter-American Institute of Agricultural Sciences, Turrialba, Cost Rica, 1960, ch. 28.
12. *ICCO Cocoa Production Data*. ICCO Quarterly Bulletin of Cocoa Statistics, 2011–12, 39, **No. 4**.
13. *ICCO Cocoa Production Data*. ICCO Quarterly Bulletin of Cocoa Statistics, 2011–12, 39, **No. 4**.
14. comtrade.un.org/pb/FileFetch.aspx?docID=3620&type.
15. ADM Advantage Cocoa College. http://www.adm.com/en-US/products/Cocoa/adm_cocoa_advantage/Pages/default.aspx.
16. Personal Communication with Gordon Patterson.

17. S. D. Coe and M. D. Coe, *The True History of Chocolate*, Thames and Hudson, New York, 1996, p. 108.
18. S. D. Coe and M. D. Coe, *The True History of Chocolate*, Thames and Hudson, New York, 1996, pp. 183–186.
19. S. D. Coe and M. D. Coe, *The True History of Chocolate*, Thames and Hudson, New York, 1996, pp. 183–186.
20. S. D. Coe and M. D. Coe, *The True History of Chocolate*, Thames and Hudson, New York, 1996, p. 133.
21. S. D. Coe and M. D. Coe, *The True History of Chocolate*, Thames and Hudson, New York, 1996, p. 185.
22. S. D. Coe and M. D. Coe, *The True History of Chocolate*, Thames and Hudson, New York, 1996, p. 188.
23. S. D. Coe and M. D. Coe, *The True History of Chocolate*, Thames and Hudson, New York, 1996, p. 200.
24. F. L. Bekele, "The History of Cocoa Production in Trinidad and Tobago", *APASTT Seminar Series on the Re-vitalization of the Trinidad & Tobago CocoaIindustry*, Univ. West Indies, St. Augustine, Trinidad, 2004, 20.
25. http://www.worldcrunch.com/business-finance/facing-cocoa-shortage-swiss-chocolate-makers-aim-to-boost-african-production/, and Chocolate Could Bring Back Forest. World Watch 2001 http://www.worldwatch.org/node/.
26. Indonesia Cocoa Pod Borer conference and demonstration plots. http://www.new-ag.info/00-1/develop/dev01.html.
27. H. Y. Shapiro and E. M. Rosenquist, *Agrofores. Syst.*, 2004, **61**, 453–462; See also the 1998 Smithsonian Panama Conference on Sustainable Cacao Production. http://nationalzoo.si.edu/scbi/migratorybirds/research/cacao/principles.cfm.
28. Sustainable Tree Crops Programme—STCP. http://www.cocoafederation.com/issues/stcp/.
29. World Cocoa Foundation formed. http://worldcocoafoundation.org/about-wcf/history-mission/.
30. L. Blunt, The Bitter Taste of Slavery. http://news.bbc.co.uk/2/hi/africa/946952.stm.
31. Response from Kouadio Adjoumani, Ivprian ambassador to the UK. http://news.bbc.co.uk/2/hi/africa/948876.stm.
32. Suspected slave ship docks with no sign of children. USA TODAY 16 April 2001. http://usatoday30.usatoday.com/news/world/2001-04-16-boat.htm.
33. Africa: Child Labor in Cocoa Fields/ Harkin–Engel Protocol 8 July 2001. http://www.ilo.org/washington/areas/elimination-of-the-worst-forms-of-child-labor/WCMS_159486/lang–en/index.htm.
34. http://www.cocoainitiative.org/images/stories/pdf/harkin%20engel%20protocol.pdf.
35. http://www.dol.gov/ilab/media/reports/iclp/cocoafindings.pdf.
36. Reference to the DOL ILO–IPEC projects in Ivory Coast and Ghana. http://www.dol.gov/ilab/programs/ocft/cocoa/2012-CLCCG-Report.pdf,

and ILO-IPEC Combating child labour in cocoa growing. http://www.ilo.org/public//english/standards/ipec/themes/cocoa/download/2005_02_cl_cocoa.pdf.
37. ECHOES Program details. http://worldcocoafoundation.org/echoes/.
38. Cocoa Livelihood Program details. http://worldcocoafoundation.org/wcf-cocoa-livelihoods-program/.
39. African Cocoa Initiative (ACI). http://www.ifdc.org/Media_Center/IFDC_in_the_News/March_2012/World_Cocoa_Foundation_African_Cocoa_Initiative_Wi/.
40. W. E. Bertrand, 2011 Oversight of Public and Private Initiatives to Eliminate the Worst Forms of Child Labor in the Cocoa Sector in Cote d'Ivoire and Ghana. Available on the internet at http://www.childlabor-payson.org/Tulane%20Final%20Report.pdf.
41. Cocoa Communities Project. http://www.dol.gov/ilab/programs/ocft/cocoa/2012-CLCCG-Report.pdf.
42. ADM SERAP Farmer Training. http://www.adm.com/en-US/products/Cocoa/news_trends/_layouts/StoryDetail.aspx?ID=39&l=/en-US/products/Cocoa/; Amajaro Source Trust Foundation. http://www.sourcetrust.org/media/assets/file/6987_Source_trust_West_Africa_art.pdf; Barry Callebaut Cocoa Horizons. www.barry-callebaut.com/cocoa-horizons; Blommer OLAM partnership. http://www.blommer.com/blommer-new/_documents/16.4.14-OLAM-AND-PARTNERS-CONSEIL-SAN-PEDRO-BLOMMER-CHOCOLATE-AND-COSTCO-WHOLESALE.pdf; Cargill Cocoa Promise. http://www.cargillcocoachocolate.com/sustainable-cocoa/the-cargill-cocoa-promise/; Lindt/Ghiradelli Sustainable Cocoa Sourcing. www.lindt.com/swf/eng/company/social-responsibility/sustainably-sourced/; Hershey CocoaLink. http://worldcocoafoundation.org/cocoalink/; Mars West African Cocoa Research Center. http://www.mars.com/global/about-mars/mars-pia/our-supply-chain/cocoa.aspx; Mondelez Cocoa Life. http://www.mondelezinternational.com/Newsroom/Multimedia-Releases/Mondelez-International-Launches-Cocoa-Life-Sustainability-Program-in-Cote-dIvoire; Nestle Multiple programs including farm yield enhancement and distribution of high yielding plantlets. http://www.nestle.com/csv/case-studies/allcasestudies/pages/the-cocoa-plan.aspx.
43. http://www.dol.gov/ilab/media/reports/iclp/cocoafindings.pdf.
44. W. E. Bertrand, 2011 Oversight of Public and Private Initiatives to Eliminate the Worst Forms of Child Labor in the Cocoa Sector in Cote d'Ivoire and Ghana. Available on the internet at http://www.childlabor-payson.org/Tulane%20Final%20Report.pdf.
45. Reference to the DOL ILO–IPEC projects in Ivory Coast and Ghana. http://www.dol.gov/ilab/programs/ocft/cocoa/2012-CLCCG-Report.pdf; ILO–IPEC Combating child labour in cocoa growing. http://www.ilo.org/public//english/standards/ipec/themes/cocoa/download/2005_02_cl_cocoa.pdf.

46. http://www.ilo.org/public//english/standards/ipec/themes/cocoa/download/2005_02_cl_cocoa.pdf.
47. Cocoa Communities Project. http://www.dol.gov/ilab/programs/ocft/cocoa/2012-CLCCG-Report.pdf.
48. Starbuck's Café Practices website. http://www.scsglobalservices.com/starbucks-cafe-practices.
49. Starbuck's C.A.F.E. Practices third party verification. http://www.conservation.org/global/celb/Documents/2011.04.08_SBUX_CAFE_Results_Assessment_ExecSummary.pdf.
50. RSPO website. http://www.rspo.org/.
51. Rainforest Alliance website. http://www.rainforest-alliance.org/work/agriculture/cocoa.
52. UTZ website. https://www.utzcertified.org/en/aboututzcertified/how-utz-works.
53. Fair Trade USA. http://fairtradeusa.org/products-partners/cocoa.
54. Fair trade and Cadbury reach an agreement to support Cadbury Dairy Milk in the UK. BBC World News Report. http://news.bbc.co.uk/2/hi/business/7923385.stm; Fair Trade Press Release http://www.fairtrade.org.uk/press_office/press_releases_and_statements/archive_2009/march_2009/cadbury_dairy_milk_commits_to_going_fairtrade.aspx.
55. Cote d'Ivoire 2012 Findings on the worst forms of Child Labor. http://www.dol.gov/ilab/reports/child-labor/cote_divoire.htm.
56. Africa: Child Labor in Cocoa Fields/Harkin–Engel Protocol 8 July 2001. http://www.ilo.org/washington/areas/elimination-of-the-worst-forms-of-child-labor/WCMS_159486/lang–en/index.htm, and http://www.cocoainitiative.org/images/stories/pdf/harkin%20engel%20protocol.pdf.
57. Ghana 2012 findings by DOL. http://www.dol.gov/ilab/reports/child-labor/ghana.htm; Interim report on the progress in Ghana towards the goals of the Harkin-Engel Protocol. http://www.dol.gov/ilab/programs/ocft/20111005Interim.pdf.
58. G. A. R. Wood and R. A. Lass, *Cocoa*, Blackwell-Wiley, Oxford, UK, 1955.
59. W. A. Phillips-Mora and J. Morera, *Selection and generation of moniliasis resistant cocoa genotypes. 1995 to 2001*. Research supported by American Cocoa Research Institute.
60. X. Argout, J. Salse, J. M. Aury, M. J. Guiltinan, G. Droc, J. Gouzy, M. Allegre, C. Chaparro, T. Legavre, S. N. Maximova, M. Abrouk, F. Murat, O. Fouet, J. Poulain, M. Ruiz, Y. Roguet, M. Rodier-Goud, J. F. Barbosa-Neto, F. Sabot, D. Kudrna, J. S. S. Ammiraju, S. C. Schuster, J. E. Carlson, E. Sallet, T. Schiex, A. Dievart, M. Kramer, L. Gelley, Z. Shi, A. Berard, C. Viot, M. Boccara, A. M. Risterucci, V. Guignon, X. Sabau, M. J. Axtell, Z. R. Ma, Y. F. Zhang, S. Brown, M. Bourge, W. Golser, X. A. Song, D. Clement, R. Rivallan, M. Tahi, J. M. Akaza, B. Pitollat, K. Gramacho, A. D'Hont, D. Brunel, D. Infante, I. Kebe, P. Costet, R. Wing, W. R. McCombie, E. Guiderdoni, F. Quetier, O. Panaud, P. Wincker, S. Bocs and C. Lanaud, *Nat. Genet.*, 2011, **43**, 101–108.

CHAPTER 3
Cacao Chemistry

W. JEFFREY HURST

Hershey Foods Technical Center 3, 1025 Reese Avenue, Hershey, Pennsylvania 17033, USA
Email: whurst@hersheys.com

The first reported citation identifying compounds in cocoa was a 1912 paper published in the *Journal of Chemical Society* entitled "The Essential Oil of Cocoa".[1] This paper listed 12 compounds. In his 1920 *Cocoa and Chocolate: Their History from Plantation to Consumer*, A.W. Knapp described the composition of cacao as 9.3% water, 8.2% minerals, 18.81% albuminoids, 13.85% fiber and 46.1% digestible carbohydrates.[2] We do not know exactly what these values refer to, but the assumption has always been the bean. The beans are contained in a pod called the shell or husk and while they are in the pod, they are suspended in a viscous material—the pulp. Other studies have focused on the chemical composition of the cacao plant's leaves, though these will not be discussed herein. After harvest, the cocoa bean undergoes a fermentation process that affects the chemistry of bean. Following this, it is dried and prepared for eventual use. These processes vary by region and other parameters. Table 3.1 provides compositional data regarding fermented and dried cocoa beans.

With an excess of 500 compounds in cacao (see Table 3.2), there is sufficient information to prepare an entire volume on this topic rather than a single chapter. But rather than attempt to be all encompassing, this contribution selectively addresses several key facets. Advances in analytical technologies over the past few years have led to an explosive growth in knowledge about the chemical composition of cacao. This chapter will focus on several areas of this advanced knowledge: (1) the flavanols; (2) compounds that may have some pharmacological or physiological significance;

Chocolate and Health: Chemistry, Nutrition and Therapy
Edited by Philip K. Wilson and W. Jeffrey Hurst
© The Royal Society of Chemistry 2015
Published by the Royal Society of Chemistry, www.rsc.org

Cacao Chemistry

Table 3.1 General composition of cocoa beans after fermentation and drying.

Parameter	% Max
Water	3.2
Fat	57.0
Ash	4.2
Total nitrogen	2.5
Theobromine	1.3
Caffeine	0.7
Starch	9.0
Crude fiber	3.2

Table 3.2 Partial list of chemical compounds found in cacao.

Acetic acid, aesculetin, alanine, alkaloids, alpha-sitosterol, alpha-theosterol, amyl-acetate, amyl-alcohol, amyl-butyrate, amylase, apigenin-7-o-glucoside, arabinose, arachidic acid, arginine, ascorbic acid, ascorbic acid oxidase, asparigninase, beta-carotene, beta-sitosterol, beta-theosterol, biotin, caffeic acid, caffeine, calcium, campesterol, catalase, catechins, catechol, cellulase, cellulose, chlorogenic acid, chrysoeriol-7-o-glucoside, citric acid, coumarin, cyanidin, cyanidin-3-beta-l-arabinoside, cyanidin-3-galactoside, cyanidin-glycoside, cycloartanol, D-galactose, decarboxylase, dextrinase, diacetyl, dopamine, epigallocatechin, ergosterol, ferulic acid, formic acid, fructose, furfurol, galacturonic acid, gallocatechin, gentisic acid, glucose, glutamic acid, glycerin, glycerophosphatase, glycine, glycolic acid, glycosidase, haematin, histidine, *i*-butyric-acid, idaein, invertase, isobutylacetate, isoleucine, isopropyl-acetate, isovitexin, kaempferol, L-epicatechin, leucine, leucocyanidins, linalool, linoleic acid, lipase, luteolin, luteolin-7-o-glucoside, lysine, lysophosphatidyl-choline, maleic acid, mannan, manninotriose, mannose, melibiose, mesoinositol, methylheptenone, *n*-butylacetate, *n*-nonacosane, niacin, nicotinamide, nicotinic acid, nitrogen, nonanoic acid, o-hydroxyphenylacetic acid, octoic acid, oleic acid, oleo-dipalmatin, oleopalmitostearin, oxalic acid, *p*-anisic acid, *p*-coumaric acid, *p*-coumarylquinic acid, *p*-hydroxybenzoic acid, *p*-hydroxyphenylacetic acid, palmitic acid, palmitodiolen, pantothenic acid, pectin, pentose, peroxidase, phenylacetic acid, phenylalanine, phlobaphene, phosphatidyl-choline, phosphatidyl-ethanolamine, phosphatidyl-inositol, phospholipids, phosphorus, phytase, planteose, polygalacturonate, polyphenol-oxidase, polyphenols, proline, propionic acid, propyl-acetate, protocatechuic acid, purine, pyridoxine, quercetin, quercetin-3-o-galactoside, quercetin-3-o-glucoside, quercitrin, raffinase, raffinose, reductase, rhamnose, riboflavin, rutin, rutoside, saccharose, salsolinol, serine, sinapic acid, stachyose, stearic acid, stearodiolein, stigmasterol, sucrose, syringic acid, tannins, tartaric acid, theobromine, theophylline, thiamin, threonine, trigonelline, tyramine, tyrosine, valerianic acid, valine, vanillic acid, verbascose, verbascotetrose, vitexin

(3) members of the polyphenol class identified in cacao that are not flavan-3-ols; and (4) compounds related to flavor.

3.1 Flavanols

The flavanols—catechin and epicatechin—are likely the most studied aspect of cocoa's chemistry over the last decade, as evidenced by the sheer volume of research papers published. Topics have ranged from basic chemistry and stereochemistry to those examining putative physiological effects and, for this reason, a substantial portion of this chapter will focus on these compounds. The discovery of flavan-3-ols and their procyanidin polymeric forms in cocoa can be traced back as early as 1909.[3] These flavan-3-ol compounds were later identified as catechins. In this case, catechin was purified by crystallization to obtain what turned out to be (+) (2R3S)-catechin. According to A.G. Perkin and Y. Yoshitake (1902), the flavanol catechin had been previously described by Nees van Esenbeck in 1832.[4] Subsequent studies were performed independently by the renowned Swedish chemist J.J. Berzelius in 1837.[5] The catechins were likely first isolated from the thorny, deciduous tree *Acacia catechu*, native to Burma and India. The catechin is extracted from the heartwood by distillation and has long been used in tanning to harden leather and in medicinal preparations.[6] The plant contains between 2 and 12% catechins in a mixture of (+) and (−) catechin and (+) and (−) epicatechin, though percentages in the compounded extract are boosted to between 20 and 60% catechins.

Much of the 19th-century work was confused by the occurrence of stereoisomers from plant extracts, the nature of which were not understood since Perkin and Yoshitake noted that five different catechins were described in earlier literature. The empirical formula was established by combustion analysis, whereby its melting point was measured by the comparative depression of the melting point of naphthalene. Other chemical measures used to determine its structure included methylation with alkali to give a tetramethyl ether and methylation with acetic anhydride to give a pentacetate. These data indicated the presence of four phenolic and one alcoholic hydroxys. Oxidation of the methyl ether resulted in the formation of veratric acid and phloroglucinol dimethyl ether, both of which were known compounds. The correct structure, 3-hydroxybenzopyran, was favored by Perkin and Yoshitake largely on the basis of a likely biogenetic relationship with quercetin. This assumption was confirmed by the synthetic work of Freudenberg.[7] In 1939, leucoanthocyanin phenolic compounds were identified,[8] and in 1955, fractionation and characterization of these compounds was reported.[9]

The procyanidins in cocoa have more recently been fractionated into monomers through decamers, with even higher forms existing.[10] The most abundant polyphenols present in cocoa are the flavan-3-ol monomers epicatechin and catechin, which also serve as building blocks for the polymeric procyanidin forms.[11] The makeup of the polymeric forms is determined by structure of the flavan-3-ol starter unit and its companion compound. Two primary forms of procyanidins occur—A-type and B-type—which differ by the linkage between the individual compounds. The A-type procyanidins form 2–7 cross links and can be found in cranberries. The B-type procyanidins form 4–8 cross links. The B-1 through B-4 types differ only in the arrangement of their catechin and epicatechin units, with procyanidin B-1 found in grapes, sorghum and cranberries, B-2 in apples, cocoa and cherries, B-3 in strawberries and hops and B-4 in raspberries and blackberries.

The catechins are members of the class of flavanoids called flavan-3-ols, which are derivatives of 1,3-diphenylpropane, with the entire class numbering into the thousands. The term flavonoid (*i.e.*, with flavan meaning yellow) was first used to describe the family of yellow-colored compounds with a flavone moiety. It was later expanded to include less intensely colored polyphenols, colorless compounds and even red and blue anthocyanidins.

The flavanoids are secondary plant metabolites responsible for, among other things, the color of flowers, fruits and sometimes leaves. They are based on a very few core structures and are formed only in plants from the amino acids phenylalanine and tyrosine. These compounds are benzo-gamma pyrone derivatives and consist of a benzene ring (commonly named the A ring) attached to a heterocycle (named the C ring) that at C-2 carries a phenyl group (named the B ring) as a substituent. In the case of the flavanols, the C ring is a 2,3-dihydro derivative. A sample structure can be seen in Figure 3.1.

These compounds can also appear in glycosylated forms, where they are bound to a sugar moiety. This normally occurs at position 3, but occasionally at position 7. The predominant sugar residue is glucose, with galactose, rhamnose and xylose having been detected. They differ in their substituents—mostly hydroxyl and methoxy groups—and in the nature and position of the sugar residues. Over 8000 flavonoids have been reported, forming the largest group of any naturally occurring polyphenols. The most common flavonoid subclasses are flavones, flavonols, anthrocyanidins, flavanones, catechins, isoflavonoids and procyanidins. Together, these classes represent more than 80% of the known flavonoids. The nomenclature of the flavan-3-ols is somewhat confusing since it is based on the actual structure, the chemical identification and derivations. For the procyanidins, a highly systematic nomenclature exists based on the structures of the monomers and attachment sites. An example procyanidin is seen in Figure 3.2.

With respect to cocoa, Miller *et al.* reported on the polyphenol content of a variety of commercial chocolates.[12] With this information as a background, a number of compounds in cocoa have been shown to be of either pharmacological or physiological significance. These compounds can be divided into hydrophilic and lipophilic compounds. Since the entire scope of

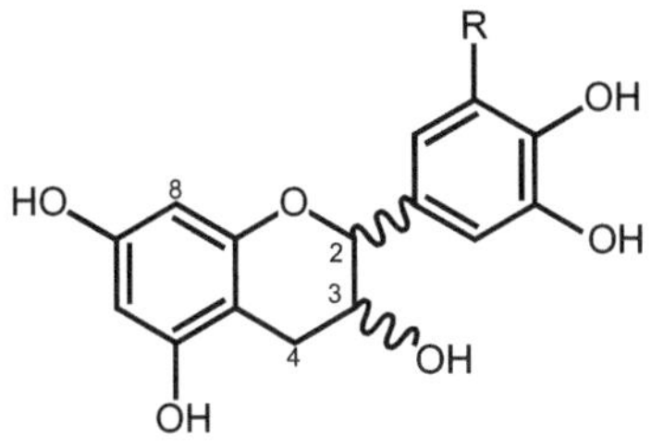

Figure 3.1 Flavan-3-ol nucleus.

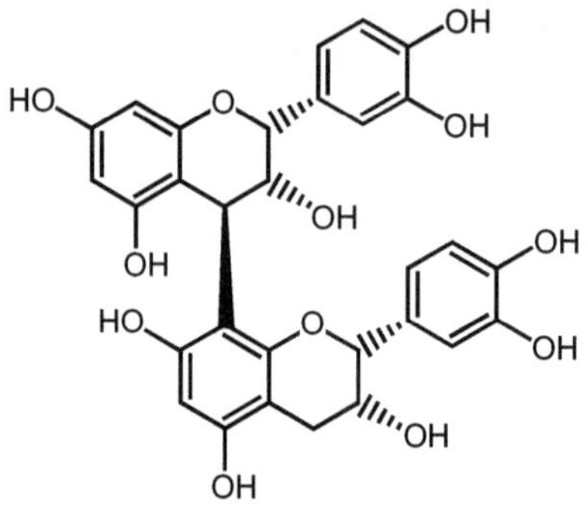

Figure 3.2 Example procyanidin structure.

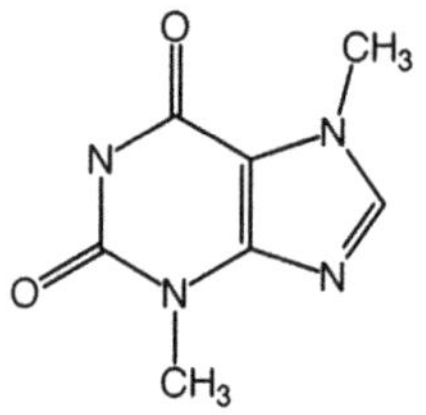

Figure 3.3 Theobromine (3,7-dimethylxanthine).

hydrophilic compounds is substantial, only the three pertinent subgroups will be discussed below.

3.2 Methylxanthines

The first of these groups is the methylxanthines. The primary methylxanthine in cocoa is theobromine (3,7-dimethylxanthine), which can be seen in Figure 3.3. Theobromine occurs in cocoa (powder or cocoa beans) at approximately 2% by weight. In addition to theobromine, caffeine (1,3,7-trimethylxanthine) also occurs at approximately 0.2% by weight, with other members of this class occurring in relatively minor amounts.[13]

The methylxanthines, primarily caffeine, exhibit a variety of effects on various bodily physiological systems, including the cardiovascular system, gastrointestinal (GI) tract, respiratory system, endocrine system and renal system. In the case of the cardiovascular system, these compounds affect functionality by modifying the contractibility of the heart and blood vessels and indirectly affecting neurotransmission in the central nervous system (CNS) and peripheral nervous system The effects in the GI tract are varied and include heartburn, stimulation of gall bladder contraction, pancreatic hormone secretion and, in high doses, increased levels of cyclic AMP and decreased levels of branched chain and aromatic amino acids in the plasma. In the respiratory system, one effect of the methylxanthines is the stimulation of respiration. Other noted effects of this class of compounds are CNS stimulation and diuresis.[14]

3.3 Biogenic Amines

The second hydrophilic subset is the "biogenic amines". These compounds are deamination products of amino acids and have been associated—anecdotally at least—with a host of physiological effects, including the strong desire for chocolate, migraine headaches and self-medication for "lovelorn" ladies.[15] One of the compounds that has been increasingly noted in the popular press is 2-phenylethylamine and can be seen in Figure 3.4.

During the 1980s, The Hershey Company in Hershey, Pennsylvania, USA, conducted research on 2-phenylethylamine. When compared to other 2-phenylethylamine-containing foods, chocolate was found to have only minimal levels. Given the low levels found in chocolate, these compounds were determined not to be the agents that are most likely for the self-reported desire for chocolate.[16]

3.4 Anandamide

The discovery of a compound called anandamide in cocoa represents another member of this class of compounds. Anandamide (*N*-arachidonyl-ethanolamine) is a brain lipid that binds to cannabinoid receptors with high affinity and mimics the psychoactive effects of plant-derived cannabinoid drugs.[17] The "ananda" portion of the compound name is derived from Sanskrit language and means blissful. The levels found in cocoa are minor but its discovery has prompted intense research interest. The structure of this compound can be seen in Figure 3.5.

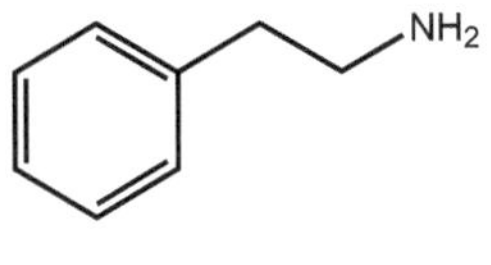

Figure 3.4 2-Phenylethylamine.

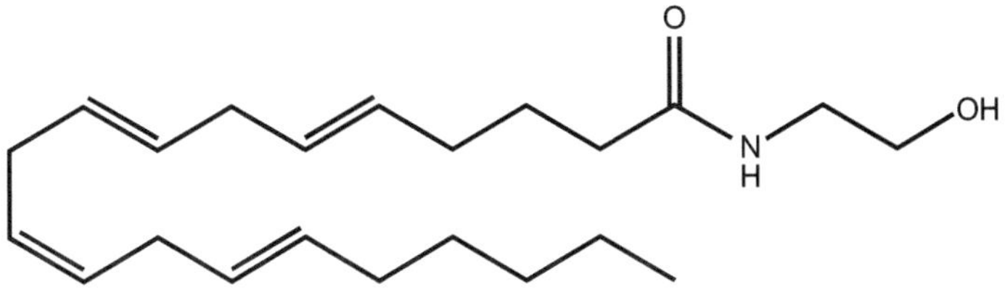

Figure 3.5 Anandamide.

3.5 Cocoa Butter

Given that most cocoa powder contains 11–12% fat, fewer lipophilic compounds exist in cocoa. About a third of the fat in cocoa butter is comprised of stearic acid. While stearic acid (Figure 3.6) is a saturated fatty acid, it has been shown to exert a neutral cholestrolemic effect in humans.[18]

Table 3.3 provides some general values on the glyceride composition of cocoa butter.

According to analytical research conducted by The Hershey Company, conjugated linoleic acid (CLA) has been detected in samples of milk chocolate; however, it is likely that the CLA is a component of the added milk.[19] CLA is a collective name given to the mixture of geometric and positional isomers of conjugated dienoic derivatives of linoleic acid (Figure 3.7). CLA has received substantial attention as a result of animal studies that report it having anti-carcinogenic, anti-atherogenic and anti-diabetic properties, as well as its modulation of body composition and immune function.[20] The human data have mainly been drawn from investigations into the effects of CLA on body composition. In rodents, CLA produces robust decreases in body fat mass and increases in lean body mass.[21] A recent human clinical trial provided 180 overweight participants

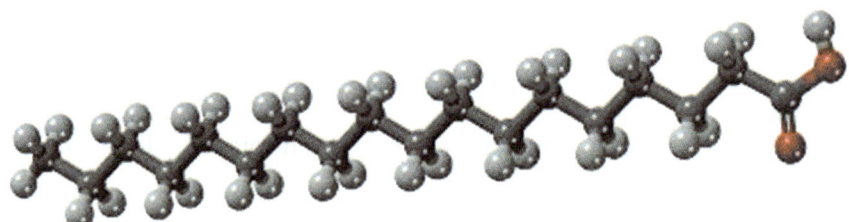

Figure 3.6 Stearic acid.

Table 3.3 Typical composition of cocoa butter.

Glyceride	Percentage
Trisaturated	2.5–3.0
Triunsaturated triolein	1.0
Diunsaturated	
Stearodiolein	6.0–12.0
Palmitodiolein	7.0–8.0
Monounsaturated	
Oleodistearin	19.0–22.0
Oleopalmitostearin	52.0–54.0
Oleodipalmitin	4.0

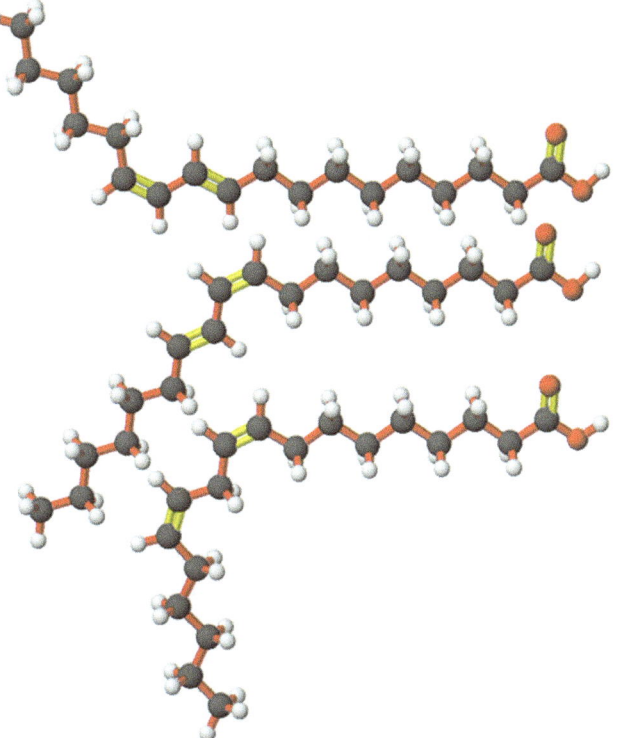

Figure 3.7 Structures of t-10, c-12-conjugated linoleic acid (Top), c-9, t-11-conjugated linoleic acid (Center) and ordinary linoleic acid, c-9, c-12-octadecadienoic acid (Bottom). The molecules are aligned at their carboxyl end to show the influence of the double bonds (lighter-colored bonds) on the molecule.

4.5 g per day of CLA in two forms compared to 4.5 g per day olive oil capsules for 12 months. Participants receiving both forms of CLA showed significant reductions in body fat mass (average of 7.8% lower than placebo) and increases in lean body mass (4.3% compared to placebo). The CLA groups also had a slight increase in low-density lipoprotein and high-density lipoprotein cholesterol. No other adverse effects were reported.[22]

3.6 Non-flavanol Polyphenolics

A comprehensive paper by Snachez-Rabaenda *et al.* discussed the development of a LC/MS/MS method to analyze a target list of 31 compounds found in cocoa.[23] In addition to catechin, epicatechin and procyanidins, a substantial number of other compounds were found. Additional research

Table 3.4 Other Polyphenolic Compounds Found in Cocoa.

Luteolin-8-*C*-glucoside (orientin)
Luteolin-6-*C*-glucoside (isooreintin)
Apigenin-8-*C*-glucoside (vitexin)
Apigenin-6-*C*-glucoside (isovitexin)
Quercetin-3-*O*-galactoside (hyperoside)
Quercetin-3-*O*-glucoside (isoquercitin)
Luteolin-7-*O*-glucoside
Quercetin-3-*O*-arabinoside
Prunin
Luteolin
Naringenin
Apigenin

---

has been performed on the stilbene fraction of cocoa, with Counet and Hurst focusing on resveratrol and piceid, while Jerkovic *et al.* reported on the discovery of a new resveratrol hexoside in cocoa.[24] Resveratrol and its derivatives have been singled out as sources of putative health effects related to longevity and endurance. These compounds are listed in Table 3.4.

It is worth noting that an examination of the parent compounds indicates that cocoa contains members of the other flavonoid classes (*e.g.*, quercetin [flavonol], apigenin and luteolin [flavones], as well as narigenin and prunin [flavonones], with the four latter compounds strongly linked to citrus fruits). These compounds have different physiological actions that, though not discussed here, offer exciting opportunities for further exploration.[25]

3.7 Flavor Chemistry

Flavor chemistry in cocoa is complex and diverse, involving a series of volatile and non-volatile compounds. Compounds in the former group include fatty acids. Regarding the latter category, a series of compounds have been deemed to be responsible for the astringent taste in unfermented cocoa beans.[26] These compounds are a series of nine *n*-phenylpropenoyl amino acids. Among these, clovamide and deoxyclovamide were previously reported by Wollgast *et al.*[27] Elsewhere, Park reported that caffedyme, a phenylpropenoic acid amide found in cocoa, exhibited inhibition of COX-1 activity by 43% (Table 3.5).[28]

Frauendorfer and Schieberle concentrated on changes to key aromatic compounds of criollo beans during roasting.[29] For further insights, see Appendix 2, which contains a list of chemicals from cocoa and their associated physiological actions.

Table 3.5 Thirty one odor-active compounds in unroasted and roasted criollo cocoa beans.

Acetic acid	3-methylbutanal
Methylpropanoic acid 9	3-methylbutanoic acid
2-phenylethanol	2-phenylacetic acid
Phenylacetaldehyde	2-methylbutanal
2-methylbutanoic acid	4-hydroxy-2,5-dimethyl-3(2H)-furanone
2-heptanol	2-phenylethyl acetate
2,3,5-trimethylpyrazine	Butanoic acid
2-methoxyphenol	Linalool
δ-decenolactone	2-ethyl-3,6-dimethylpyrazine
Dimethyl trisulfide	Ethyl-2-methylbutanoate
Ethyl methylpropanoate	δ-octenolactone$b$
2-ethyl-3,5-dimethylpyrazine	3-hydroxy-4,5-dimethyl-2(5H)-furanone
4-methylphenol	2-acetyl-1-pyrroline
2,3-diethyl-5-methylpyrazine	(E,E)-2,4-nonadienal
1-octen-3-one	2-methoxy-3-isobutylpyrazine
2-methyl-3-(methyldithio)-furan	

References

1. J. C. Bainbridge and S. H. Scott, *J. Chem. Soc., Trans.*, 1912, **101**, 2209–2221.
2. A. W. Knapp, *Cocoa and Chocolate: Their History from Plantation to Consumer*. Chapman and Hall, London, 1920.
3. G. Spiller, "Basic Metabolism and Physiological Effects of the Methylxanthines", in *Caffeine*, ed. G. Spiller, Boca Raton, CRC Press, 1998, pp. 225–231.
4. A. G. Perkin and Y. Yoshitake, *J. Chem. Soc., Trans.*, 1902, **81**, 1160–1173.
5. N. van Esenbeck, *Ann. Pharm.*, 1832, **1**, 243.
6. J. J. Berzelius, *Jahresber*, 1837, **14**, 235.
7. K. Freudenberg, *Ber. Dtsch. Chem. Ges.*, 1920, **53**, 416–421.
8. E. C. Bate-Smith, *Phytochemistry*, 1975, **14**, 1107–1113.
9. L. Y. Foo and L. Porter, *Phytochemistry*, 1980, **19**, 1747–1754.
10. C. Priuer, J. Riguad, V. Cheynier and M. Moutounet, 1994, **3.6**, 781–784.
11. W. J. Hurst, B. Stanley, J. A. Glinski, M. Davey, M. J. Payne and D. A. Stuart, *Molecules*, 2009, **14**(10), 4136–4146.
12. K. B. Miller, D. A. Stuart, N. L. Smith and C. Y. Lee, *J. Agric. Food Chem.*, 2006, **45**, 1025–1062.
13. J. Apgar and S. Tarka, "Methylxanthine Composition and Consumption Patterns of Cocoa and Chocolate Products", in *Caffeine*, ed. G. Spiller, Boca Raton, CRC Press, 1998, pp. 163–193.
14. G. Spiller, "Basic Metabolism and Physiological Effects of the Methylxanthines", in *Caffeine*, ed. G. Spiller, Boca Raton, CRC Press, 1998, pp. 225–231.
15. P. Rozin, *et al.*, *Appetite*, 1991, **17**, 199–212.

16. W. J. Hurst and P. B. Toomey, *Analyst*, 1981, **106**, 394–404; W. J. Hurst, R. A. Martin, Jr., B. L. Zoumas and S. M. Tarka, Jr., *Nutr. Rep. Int.*, 1982, **26**, 1081–1086; W. J. Hurst, *J. Liq. Chromatogr.*, 1990, **13**, 1–23.
17. D. Piomelli, *Nature*, 1996, **382**, 677–678.
18. F. M. Steinberg, M. M. Bearden and C. L. Keen, *J. Am. Diet. Assoc.*, 2003, **103**, 215–223.
19. W. J. Hurst, S. M. Tarka, G. Dobson and C. M. Reid, *J. Agric. Food Chem.*, 2001, **49**, 1264–1265.
20. S. Tricon, G. C. Burdge, C. M. William, P. C. Calder and P. Yaqoob, *Proc. Nutr. Soc.*, 2005, **64**, 171–182.
21. M. E. Evans, J. M. Brown and M. K. McIntosh, *J. Nutr. Biochem.*, 2002, **13**, 508–516.
22. J. M. Gaullier, J. Halse, K. Hoye, K. Kristiansen, H. Fagertun, H. Vik and O. Gudmundsen, *Am. J. Clin. Nutr.*, 2004, **79**, 1118–1125.
23. F. Sanchez-Rabenda, O. Jauregui, I. Casals, C. Andres-Lacueva, M. Izquierdo-Pulido and R. Lamela-Raventos, *J. Mass Spectrom.*, 2003, **38**, 35–42.
24. C. Counet, D. Callemien and S. Collin, *Food Chem.*, 2006, **98**, 649–657.
25. W. J. Hurst, J. A. Glinski, K. B. Miller, J. Apgar and M. Davy, *J. Agric. Food Chem.*, 2008, **64**, 234–256.
26. V. Jerkovic, M. Bröhan, E. Monnart, F. Nguyen, S. Nizet and S. Collin, *J. Agric. Food Chem.*, 2010, **58**, 7067–7074.
27. J. Wollgast, L. Pallaroni, M. E. Agazzi and E. Anklam, *J. Chromatogr.*, 2001, **926**, 211–220.
28. J. B. Park, *J. Agric. Food Chem.*, 2007, **55**, 2171–2175.
29. F. Frauendorfer and P. Schieberle, *J. Agric. Food Chem.*, 2008, **56**, 10244–10251.

CHAPTER 4

Applications of Genomics to the Improvement of Cacao

MARK J. GUILTINAN* AND SIELA N. MAXIMOVA

Department of Plant Science, The Pennsylvania State University, 422 Life Sciences Building, University Park, Pennsylvania 16802, USA
*Email: mjg9@psu.edu

> "[Decoding the human genome sequence] is the most significant undertaking that we have mounted so far in an organized way in all of science. I believe that reading our blueprints, cataloging our own instruction book, will be judged by history as more significant than even splitting the atom or going to the moon".
> Francis S. Collins, Former Director, U.S. National Institutes of Health[1]

4.1 Introduction

The genomes of living organisms encompass the totality of DNA necessary to carry out the complex genetic program of all life forms. The size of genomes varies from 1759 base pairs, which encode a viral genome, to over 670 billion base pairs, found in a tiny amoeba. Comparatively, the human genome is relatively modest, at 3.2 billion bases of DNA. Each cacao cell contains a mere 430 million base pairs. Decoding a genome reveals the locations and sequences of all of the genes that encode the instructions for the production of all of the proteins that carry out the biochemistry of life. It turns out that for humans, the number of genes is only about 20 000 (8000 less than in cacao), but the complexity of the interactions between the proteins they encode, and the biochemistry they carry out, is far more complex than anyone ever imagined before this window into the molecular basis of life was opened.

Several transformative technologies have been developed over the past three decades that have revolutionized the study of life at the molecular level *via* molecular biology. These, along with corresponding advances in computer science, have dramatically increased our ability to collect and analyze data, and have reduced the costs of doing so by many orders of magnitude. One of the most radically advanced of these technologies is DNA sequencing, which today allows the sequence of an individual's personal genome to be ordered online at a cost of $10 000 which results in a 99.99% accurate coverage of 90% of the entire genome. It costs 50 000-times less to sequence DNA than it did ten years ago (www.genome.gov/sequencingcosts/). The costs continue to decline and the accuracy is increasing as new and even more advanced technologies are being developed. This capability has accelerated the pace of biomedical research and the discovery of the basis of human genetic variation, including the identification of the actual base pair changes underlying many devastating diseases. These discoveries are leading to promising new therapies and diagnostic tests, as reflected in the quote from Francis Collins above. For cacao, the impact of genomic technologies will be no less profound to the future of the cocoa value chain.

To a large degree, humankind's motivation to eliminate human disease has been the driving force for these developments, but they can be equally applied to agricultural research. The field of plant science has taken advantage of all genomics era tools and developed a few new ones of its own, too. This new era began with the full genomic sequencing of the model plant *Arabidopsis* in 2000.[2] This plant—a tiny weed—serves as the "lab rat" of the plant science world, because of its tiny size, fast life cycle and its relatively small genome. Since then, a growing number of plant genomes have been sequenced, including (but not limited to) rice, grape, corn, poplar, soybean, papaya, strawberry and *Theobroma cacao*, the chocolate tree. The genome of a criollo variety of cacao was published by the International Cocoa Genome Sequencing Consortium,[3] and the sequence of a forastero variety was published by a United States Department of Agriculture (USDA)-led collaboration,[4] providing a detailed window into two cacao genomes at high resolution. In addition to these landmark papers, a growing number of manuscripts on cacao genomics and genetics have made use of the incredible tools and data resources now available to scientists working in the field and by other scientists interested in comparative genomics.

In light of these advances, very important questions became even more relevant: how will research on cacao genomics, genetics and molecular biology lead to improvements of farmers' lives, protect the environment and contribute to a balanced cocoa market chain? The development of a sustainable cocoa chain requires advances in every link, starting with the low productivity of cacao farms, aging tree populations, poor cocoa quality, a general lack of interest in cacao farming by children in producing regions, unsustainable farming systems and in aspects of the downstream marketing and value chain. While still in its infancy, the study of the cacao genome is poised to greatly advance our understanding of the basic biology of this species and to accelerate progress on many of the most critical facets of cocoa production and its value chain.

This review is aimed at summarizing information on the general advances in the field since the genome sequence was obtained and to provide a vision of how the field will progress in the near future. Rather than focus on details, it will attempt to describe the basic themes and concepts of the field. Research on cacao genomics prior to 2011 has been reviewed elsewhere,[5-8] and thus this review will be mainly limited to more recent publications that were obtained by searching PubMed (National Library of Medicine) and The Web of Science covering 2011–2013 in September of 2013.

4.2 Cacao Genome Sequencing Project

4.2.1 History

The human genome sequencing project (1987–2003) opened up a new potential of whole-genome sequencing. In 1998, a workshop held at Penn State University brought together scientists from academia, industry and government to discuss the idea of sequencing the cacao genome, which was estimated at that time to require over $80 000 000. With insufficient funding, the cacao genetics community set goals to lay the basic groundwork that would later support sequencing the genome and became organized through the International Group for the Genetic Improvement of Cacao by forming a Molecular Biology working group. In 2006, at the International Cocoa Research Conference held in Costa Rica, the International Cacao Genome Sequencing (ICGS) consortium was formally organized, under the leadership of Clair Lanaud of Centre de Coopération Internationale en Recherche Agronomique pour le Développement (CIRAD), France. In 2008, the Cocoa Genome Consortium (CGS)—an industry-funded partnership—was formed, led by Raymond Schnell of the USDA. These two projects set out to sequence the genomes of two quite distinct types of cacao: a member of the criollo genetic group and a member of the amelonado genetic group, respectively. It was reasoned that the comparison of these two quite distinct types of cacao would provide a far deeper understanding of the structure and function of the cacao genome compared to a single genome alone. Both projects benefitted from the rapid advances and plummeting costs of genome technology. Interdisciplinary teams of scientists were organized with specialists in each of the complex steps of the projects. The major steps in the process included: the creation of detailed genetic maps; DNA and RNA sequencing at a large scale; the assembly of a large amount of short sequence information into entire chromosome segments (pseudochromosomes); annotation of the genome information (identifying all of the genes, comparing them to other species and describing other features of the genome); and the creation of web-based relational databases making it easy to visualize and analyze the massive amount of resulting information. Two key manuscripts were published on this work involving a very large number of scientists from many different locations.[3,4]

4.2.2 Major Findings from the Cacao Genome Sequencing Projects

The ICGS sequenced the genome of a criollo type (B97-61/B2) collected by Vish Mooleedhar in 1995 deep in the Maya mountains of Belize. The CGS sequenced Matina 1-6, a Costa Rican variety, grown for many years in the Matina river valley. Both varieties were chosen for their highly homozygous genome, which greatly facilitated the final genome assemblies. The genomes of the two sequenced types of cacao were found to differ in their sizes by about 3.4% (430 million bp (Mbp) and 445 Mbp for B97-61/B2 and Matina 1-6, respectively). Differences in genome size within a single species are commonly found, and subsequent analysis showed that most of this difference was accounted for by the increased amount of repetitive DNA and transposons found in the Matina genome. Transposons, the so-called jumping genes, are DNA sequences that have the ability to excise and move to new locations. They were first identified in corn by Barbara McClintock in her 1983 Nobel Prize-winning discovery.[9] Using automated gene-finding software and by comparison of the cacao genome to other species, approximately 29 000 cacao genes have been predicted, which is similar in number to the genes found in the model plant *Arabidopsis*, and surprisingly, larger than that found in humans (about 20 000). Over 700 genes were found in cacao that have not been found in other plant genomes sequenced to date, and these may be involved in specializations within the cacao lineage (family Malvaceae). Overall, the global organization of the two genomes is very similar, although interestingly, 12 relatively small regions were found to be located on different chromosomes in the two varieties, possibly as a result of transposon activity.[4]

By comparing the data to genomes of other plant species, hundreds of genes predicted to be involved in major traits of commercial importance for cacao were identified. These include the genes for disease resistance and seed traits such as cocoa butter biosynthesis, flavonoid biosynthesis, polyphenol biosynthesis and terpenoid biosynthesis (molecules contributing to the aromatic qualities of cocoa).[3] Using an innovative approach involving association genetics, a candidate gene that may control the color of cocoa pods was identified.[4] The gene was predicted to encode a transcription factor that may regulate anthocyanin production. This discovery may have utility in cocoa production by marking pods of specific genotypes with different colors, making it easier to segregate cocoa by variety. This result demonstrates the synergistic power of traditional genetics and genomics, leading to the rapid identification of a gene of interest. In the future, many other genes will be identified with similar approaches, greatly accelerating the pool of knowledge of the mechanisms of cacao growth, development and disease resistance, which, in turn, will help accelerate breeding programs.

The cacao genome sequence projects have resulted in a wealth of resources and tools that can be used to further understand the genetics of cacao and to help breed for better varieties.[10–16] These include a greatly expanded number of molecular markers for mapping and genotyping and genome sequence databases that house all of the information found in the cacao genome to date (http://cocoagendb.cirad.fr/ and http://www.cacaogenomedb.org/). These achievements constitute a quantum leap for the world of cacao science; however, there is still much to be done. The next step to advance breeding programs will be to further map the important genes and to understand their functions. This will provide breeders with enhanced knowledge about the actual mechanisms and genes they are working with and allow the selection of new promising genotypes by molecular screens, as well as provide very precise tools for the identification and accumulation of desirable forms of genes into the cacao varieties of the future.

4.3 Recent Advances in Cacao Genome Analysis

4.3.1 Evolution, Domestication and Germplasm Collections

Using DNA marker technologies and building on earlier work reviewed elsewhere,[5,7,8] a number of key findings published in the past few years have begun to unravel an unprecedented level of detail of the evolutionary and domestication history of cacao.[17–29] Two studies examined the genetic diversity of collections found at the largest cacao germplasm collection in ICGB-Trinidad.[25,29] These two studies have provided a large DNA fingerprinting dataset that can be used to assess the diversity in the collection and to perform detailed population genetics studies. Both studies reported a significant high rate of potential mislabeling of trees, in the range of 30–40%. This finding highlights the importance of DNA fingerprinting in correcting errors propagated in germplasm collections and the crucial need to perform this upon the entire collection. Eliminating mislabeling and reducing redundancy are important for reducing the large maintenance costs of this important collection and guiding more efficient use of the genetic diversity that can be found within.

Using DNA sequence changes found in chloroplasts of the cacao accessions from Trinidad, Yang *et al.* explored the origin of the famed trinitario cultigen thought to be derived from crosses between criollo and forastero types of cacao long ago.[17] Their work supports this hypothesis and further suggests that the trinitario cultivar group is perhaps more complex than previously thought, being derived from at least three different introduction events. Using genomic DNA polymorphisms represented by single-nucleotide changes, Ji *et al.* explored the origins of 84 fine-flavored farmer varieties collected in Honduras and Nicaragua.[19] They showed that the farmer accessions represented five genetic groups: ancient criollo, amelonado, trinitario (including Nicaraguan trinitario and Honduran trinitario) and Upper Amazon forastero (only one accession). Interestingly, the Nicaraguan and Honduran trinitario varieties are clearly distinguishable from one another. Surprisingly, the local traditional variety indio was found to be identical to the well-known amelonado type of cacao. Two other publications described additional work in further understanding the distribution of and genetic variability in cacao germplasm grown by Nicaraguan farmers.[26,27] These findings note a relatively low degree of genetic diversity and the discovery of linkages to useful sources of disease resistance genes. Several trees found within uncultivated forest regions in Nicaragua seem to have been derived directly from the criollo cacao cultivated by the ancient Maya.[27]

Similar studies of genetic diversity were conducted with plants growing in the Dominican Republic, Bolivia and Ecuador.[20,21,24] Solorzano *et al.* explored the origin of the famed nacional variety grown in Ecuador.[24] The highest genetic similarity was observed between the nacional pool and some wild genotypes from the southern Amazonian region of Ecuador, sampled along the Yacuambi, Nangaritza and Zamora rivers in the Zamora Chinchipe province. The LCT-EEN85, LCT-EEN86 and LCT-EEN91 genotypes were shown to be the most likely parents of the nacional variety. Researchers suggest that the nacional variety first derived from the southern Amazon region of Ecuador and later migrated to the coastal provinces. This finding provides clues as to where new genetic diversity may be collected in order to expand the genetic diversity of this important source of highly valued cocoa.

Boza *et al.* reported an extensive genetic diversity study of the germplasm found in Dominican Republic.[20] In this study, 955 trees from germplasm collections and in farmers' fields were compared using simple sequence repeats (SSRs). Overall, a high degree of useful genetic variation was observed with varieties from amelonado, trinitario and criollo found to be widely distributed in both the collections and farmers' fields. A similar report published by Zhang *et al.* described the genetic diversity found in germplasm grown in Bolivia, the so-called Cacao Nacional Boliviano (CNB) group.[21] Using SSR markers with 164 accessions, the CNB group was shown to be uniquely distinguishable from all other known cacao germplasm groups described to date in South America. Furthermore, the cultivated and wild cacao trees in Bolivia are similar in genetic makeup, suggesting that CNB is of indigenous origin in Bolivia.

The work of Thomas *et al.* looked more broadly at the ancient evolution and distribution of cacao germplasm throughout the neotropics and investigated the role of glaciation and human dispersal in determining the current distribution.[22] Using a dataset developed and made publically available by the USDA, the authors observed the highest levels of genetic diversity in the Upper Amazon areas from southern Peru to the Ecuadorian Amazon and the border areas between Colombia, Peru and Brazil, consistent with earlier studies.[30] These results suggest that cacao was widely distributed in the Western Amazon prior to the last Ice Age (22 000–13 000 years ago) and that glaciations restricted cacao to several regions, resulting in a number of genetic clusters related to the original wild cacao populations.

The analyses also suggest that the later genetic diversity of cacao has been significantly altered by human domestication and the resulting genetic bottlenecks.

As a whole, progress in the use of genomics technologies to study the genetic diversity and origins of cacao is impressive. This work has shed light on many facets of cacao genetics, especially in guiding the efforts of cacao germplasm collectors, germplasm curators and the work of cacao breeders. Armed with this powerful information, the future of cacao germplasm conservation and utilization will be much more efficient, resulting in further acceleration of cacao breeding progress and in reducing the associated costs. Furthermore, this work illuminates the role of geological and human forces in cacao genome evolution and has provided in-depth insights into the general principles of plant evolution overall. Genomics will play a continuing role in this area as new germplasm is collected and additional collections are characterized in the future.

4.3.2 Breeding

The period reviewed in this chapter focusses upon only a few major peer-reviewed publications in the literature of cacao breeding.[31–36] A larger number of reports presented at the International Cocoa Research Conference in 2012 and elsewhere are not included here. Based on our review, we predict that the future of cacao breeding is likely to change dramatically in the coming years as new findings emerge realizing the full impact of the genomic data. One example is the catalog describing the release of a series of new cacao varieties by Centro Agronómico Tropical de Investigación y Enseñanza (CATIE) in Costa Rica, which includes details of the DNA fingerprints that can be used to identify the new high-yielding and disease-tolerant varieties.[37] To help facilitate the application of genomics to cacao breeding, Livingstone et al. reported the development of a DNA testing technique that can be used under field conditions in cacao-growing regions.[34] Such a simple and robust technique is necessary before breeders can harness the full potential of cacao genomics. A review of breeding efforts in Brazil was published by Lopes et al., highlighting how cacao genomics has been integrated into the long-standing breeding programs in Bahia.[35] An excellent meta-study reviewing all earlier work on cacao gene mapping for disease resistance and other traits was published by Lanaud et al.[38] A recent paper by Royaert et al.[36] focused on self-incompatibility (SI), an important trait for tree productivity. These authors identified that a single genetic locus, located on chromosome 4, is strongly associated with SI, and this information can help select for self-compatible trees.

Significantly, three of the recent manuscripts on cacao breeding focused on the search for resistance to important pathogens of cacao.[31–33] Moderate to good levels of disease resistance were identified for both *Ceratocystis* wilt[32] and *Phytophthora*.[33] Combined with prior research uncovering sources of genetic resistance against diseases, major progress is being made. However, in the meta-analysis performed by Padi *et al.*, which reviewed the progress in breeding for resistance to cocoa swollen shoot virus (CSSV) over seven decades,[31] the conclusions are not as promising. The authors state that, "Overall, except for the somewhat better resistance of introduced Upper Amazon germplasm in the 1940s compared with local amelonado, little genetic gains for resistance have been achieved from breeding for resistance to CSSV over the past seven decades". Since cacao and CSSV did not co-evolve, it is not surprising that specific resistance genes have not been discovered; however, it is likely that with further intensive breeding efforts, multi-locus horizontal resistance may be obtainable.

4.3.3 Functional Genomics

Although it is not absolutely necessary for breeders to understand the specific function of the genes they are selecting for, it is an advantage. Functional genomics is the field of study that deciphers the precise mechanisms each gene participates in and the interactions between the gene products they encode. Armed with this knowledge, breeders can design better strategies, accelerate their progress and improve their final results. For example, if a breeder knows that two different genetic loci encode for proteins that work to fight a pathogen by two independent mechanisms, they will have confidence that a combination of both genes together will most likely give a more robust and durable form of resistance then either gene alone. Conversely, if two genes are linked together in a single resistance mechanism, combining these two genes may not give the desired synergistic effect. With a deep understanding of gene function, breeders can breed for precise molecular functions and the predictive power of their breeding strategies will be greatly enhanced. The ultimate goal of functional genomics is to provide breeders with the knowledge of gene function, the identity of the most important genes for desired traits, discovery of accessions with the most desirable alleles (*i.e.*, variants) of these key genes and the provision of molecular markers that are useful for their selection in breeding programs. With the cacao genome sequence resources available, progress in this field has accelerated. Research prior to 2010 has been reviewed elsewhere.[5] Since then, a growing body of literature has begun to identify the functions of many important cacao genes,[11,16,39–52] some of which are reviewed here.

One level of understanding gene function is to identify genes that respond to specific stimuli, for example, pathogen infection. Gene expression levels (transcription and mRNA concentrations) can rapidly change (increase or decrease) in response to many different stimuli and, in many cases, these genes may play important mechanistic roles by which the plant copes with different situations, for example, the genes involved in the plant immune system. Following this logic, a number of projects have developed approaches towards identifying the genes regulated by important biotic and abiotic factors affecting cacao growth, such as copper toxicity,[40] pathogen interactions[41,45,49,53] and oxidative stress.[50] A large number of potentially useful candidate genes have been identified in these studies, though their precise functions remain unknown.

Another approach to functional genomics is adapting knowledge from other plant species to help identify and characterize cacao genes (*i.e.*, a translational biology approach). Genomic and genetic resources in model plant species such as *Arabidopsis thaliana*, maize and others provide powerful tools for cacao scientists to apply translational biology to accelerate functional genomic analysis. Using this approach, Teixeira *et al.* have identified two genes encoding proteins that are similar to proteins from other plant species known to be critical for the immune system.[39] One such gene was up-regulated in response to witches' broom disease, suggesting a role for PR-1RKs during cacao defense responses. Likewise, an earlier report characterizing the cacao gene *TcNPR1* demonstrated that it functions similarly to the *AtNPR1* gene of *Arabidopsis*, as a central regulator of the plant's immune system.[3,53] A related gene, *TcNPR3*, was identified and it acts together with *NPR1*, but to repress the plant defense response.[56] Gregorio *et al.* identified a gene (*ABI-4*) thought to be involved in regulating plant response to the hormone abscisic acid, a component that is important during seed development and drought responses.[43] Britto *et al.* isolated a gene encoding a glucanase and demonstrated that the protein can inhibit the growth of witches' broom mycelia.[44] A study by Feltus *et al.* resulted in the identification of several genes potentially responsible for three *T. cacao* traits: black pod disease resistance, bean shape index and pod weight.[16] As a whole, the field of cacao functional genomics is moving rapidly and, as new approaches and techniques are developed, it will further accelerate.

4.3.4   Genomics of Cacao Health and Nutrition

Although a vast body of literature on various health and nutritional effects of chocolate has accumulated, much less is known about the genes that control the synthesis of the molecules involved. Using knowledge from model organisms such as *Arabidopsis*, the cacao genome sequence was searched to discover several hundred genes predicted to be involved in the major molecular groups found in cocoa: the flavonoids, lipids, terpenes and various seed storage proteins.[3] Genomic regions that are important in determining the cocoa butter melting point, seed coloration and cocoa astringency have been mapped.[38] A major challenge remains to determine which of these genes are most important for determining the concentrations and profiles of these important cocoa molecular constituents, and to explore the genetic variants of these genes within the germplasm collections. Armed with such knowledge, breeders may combine specific gene combinations to create new cacao varieties tailored towards molecular profiles that will enhance their health and nutrition effects.

To reach this goal, recent work in the Penn State cacao laboratory has begun to unravel the flavonoid and lipid biosynthesis pathways in detail.[54,57]

In work performed by Yi Liu, three genes encoding proteins that act at the major branch points of metabolic pathways leading to proanthocyanidins and anthocyanidins were characterized. Furthermore, a gene encoding a myb-type transcription factor that is thought to regulate these genes was also characterized. Using these genes to screen for varieties that differ in expression levels, it may be possible to identify new genotypes with increased levels or types of proanthocyanidins. Similarly, Yufan Zhang's thesis work in the Penn State cacao laboratory is focused on lipid biosynthesis genes. In this work in progress, a gene encoding a fatty acid desaturase is under study. Preliminary findings suggest that this gene is critical to carrying out the key desaturation step resulting in the synthesis of 18:1 fatty acids, which are critical to determining cocoa butter's melting point. This knowledge may be useful in screening for variants in this gene with specific altered fatty acid profiles and physical properties that provide health benefits.

4.3.5 Pathogens

The tools of genomics have also been used to study the pests and diseases of cacao and the interactions between these and their host tissues.[41,55,58-63] The genome of witches' broom disease has been previously reported[64-66] and ongoing work will soon release the genomic sequences of frosty pod, as well as the most devastating of the black pod rot species. These findings will open up new ways towards a better understanding of the genetic diversity and evolution of these pathogens and offer help in devising nuanced approaches to breeding for resistance and designing other pathogen control strategies. Two reports detail the expression of genes in the witches' broom genome that seem significant regarding its infection of cacao.[58,59] These studies provide insight into the mechanisms of infection and the life cycle of this devastating disease, including the role of the alternative oxidation pathway and the role of PR-1-like proteins, both of which are hypothesized to contribute towards inhibiting the plant immune system during pathogen infection. The work of Bailey *et al.* focused on the physiological and molecular changes that occur during the shift from biotrophy to necrotrophy during the life cycle of *Moniliophthora roreri*, the frosty pod pathogen.[55] This work describes many changes in gene expression and metabolites during the infection life cycle, thereby providing clues as to the precise mechanisms involved.

Oro *et al.* used molecular markers to study the genetic diversity and geographical distribution of different strains of CSSV in Togo.[60] Their results demonstrated that at least three different strains of virus can be detected and suggested at least two different origins and two of the strains originated in Ghana. This information will be helpful for the better understanding of CSSV genetics and how different strains of this virus spread through West Africa. Similar approaches (developing molecular markers to study the evolution and distribution of cacao pests and diseases) were reported for *Phytophthora megakarya* and cocoa mirid bugs, both of which are important problems in West Africa.[61,63]

4.4 The Future of Cacao Genomics

From this review of recent literature, it is clear that sequencing the cacao genome is beginning to have a significant effect upon research towards improving cacao through its breeding and in constructing more effective disease control strategies. This impact will grow dramatically over the coming decade, with the full impact being felt in a 10–20-year time horizon, primarily due to the long time required to complete cacao breeding programs aimed at releasing new varieties. In the future, we can expect to see:

- Genome sequencing of all of the different cacao genetic types, providing much deeper insight into the genetic structure and variation within this species.
- Identifying the most important genes for yield, disease resistance and quality traits, and discovering variations of genes that are most useful for crop improvement. This will involve the application of quantitative trait loci mapping and genome-wide association studies and the development of very-high-resolution genetic maps.
- Increasingly common use of DNA marker technologies for accelerating breeding programs, the traceability of cocoa genetic origins and screening for diseases.
- Identifying new varieties of cacao through combining the best genetic variations regarding strong disease resistance, high productivity and excellent flavor potential and, in some cases, regarding specialty flavors or other traits for specific applications.
- Enabling a better understanding of the genetic variation and evolution of cacao pests and diseases, all of which will greatly help efforts to fight their devastating effects.

It is the hope of the cacao research community that these advances will contribute towards a strengthened cocoa value chain and will benefit all of the stakeholders from the consumers on down, especially improving the livelihoods of the cacao farmers, on whose hard work and strong backs the entire cocoa value chain depends.

References

1. F. Collins, *Academy of Achievement web site*, 1998, Interview.
2. The Arabidopsis Genome Sequencing Initiative, *Nature*, 2000, **408**, 796–815.
3. X. Argout, J. Salse, J. M. Aury, M. J. Guiltinan, G. Droc, J. Gouzy, M. Allegre, C. Chaparro, T. Legavre, S. N. Maximova, M. Abrouk, F. Murat, O. Fouet, J. Poulain, M. Ruiz, Y. Roguet, M. Rodier-Goud, J. F. Barbosa-Neto, F. Sabot, D. Kudrna, J. S. S. Ammiraju, S. C. Schuster, J. E. Carlson, E. Sallet, T. Schiex, A. Dievart, M. Kramer, L. Gelley, Z. Shi, A. Berard, C. Viot, M. Boccara, A. M. Risterucci, V. Guignon, X. Sabau,

M. J. Axtell, Z. R. Ma, Y. F. Zhang, S. Brown, M. Bourge, W. Golser, X. A. Song, D. Clement, R. Rivallan, M. Tahi, J. M. Akaza, B. Pitollat, K. Gramacho, A. D'Hont, D. Brunel, D. Infante, I. Kebe, P. Costet, R. Wing, W. R. McCombie, E. Guiderdoni, F. Quetier, O. Panaud, P. Wincker, S. Bocs and C. Lanaud, *Nat. Genet.*, 2011, **43**, 101–108.
4. J. C. Motamayor, K. Mockaitis, J. Schmutz, N. Haiminen, D. Livingstone, 3rd, O. Cornejo, S. D. Findley, P. Zheng, F. Utro, S. Royaert, C. Saski, J. Jenkins, R. Podicheti, M. Zhao, B. E. Scheffler, J. C. Stack, F. A. Feltus, G. M. Mustiga, F. Amores, W. Phillips, J. P. Marelli, G. D. May, H. Shapiro, J. Ma, C. D. Bustamante, R. J. Schnell, D. Main, D. Gilbert, L. Parida and D. N. Kuhn, *Genome Biol.*, 2013, **14**, R53.
5. F. Micheli, M. Guiltinan, K. Gramacho, M. Wilkinson, A. Figueira, J. Cascardo, S. Maximova and C. Lanaud, in *Advances in Botanical Research*, ed. J.-C. Kader and M. Delseny, Academic Press, Burlington, 2010, vol. 55, pp. 119–177.
6. S. N. Maximova, T. C. Lock and M. J. Guiltinan, in *A Compendium of Transgenic Crop Plants*, ed. C. Kole and T. Hall, Plantation Crops, Ornamentals and Turf Grasses, Blackwell Publishing, Oxford, UK, 2008, vol. 8, pp. 85–98.
7. M. Guiltinan, J. Verica, D. Zhang and A. Figueira, in *Genomics of Tropical Crop Plants*, ed. P. Moore and R. Ming, Springer, New York, 2008.
8. M. Guiltinan, in *Biotechnology in Agriculture and Forestry: Transgenic Crops V*, ed. E. Pua and M. Davey, Springer-Verlag, Berlin Heidelbelg, 2007, vol. V, pp. 497–518.
9. L. Pray and K. Zhaurova, *Nature Education*, 2008, **1**, 169.
10. E. S. L. Santos, C. B. M. Cerqueira-Silva, G. M. Mori, D. Ahnert, R. X. Correa and A. P. Souza, *Biol. Plant.*, 2012, **56**, 789–792.
11. D. N. Kuhn, D. Livingstone, D. Main, P. Zheng, C. Saski, F. A. Feltus, K. Mockaitis, A. D. Farmer, G. D. May, R. J. Schnell and J. C. Motamayor, *Tree Genet. Genomes*, 2012, **8**, 97–111.
12. N. Kane, S. Sveinsson, H. Dempewolf, J. Y. Yang, D. P. Zhang, J. M. M. Engels and Q. Cronk, *Am. J. Bot.*, 2012, **99**, 320–329.
13. M. Allegre, X. Argout, M. Boccara, O. Fouet, Y. Roguet, A. l. Bérard, J. M. Thévenin, A. l. Chauveau, R. Rivallan, D. Clement, B. Courtois, K. Gramacho, A. Boland-Augé, M. Tahi, P. Umaharan, D. Brunel and C. Lanaud, *DNA Res.*, 2012, **19**, 23–35.
14. C. A. Saski, F. A. Feltus, M. E. Staton, B. P. Blackmon, S. P. Ficklin, D. N. Kuhn, R. J. Schnell, H. Shapiro and J. C. Motamayor, *BMC Genomics*, 2011, **12**, 413.
15. D. S. Livingstone, J. C. Motamayor, R. J. Schnell, K. Cariaga, B. Freeman, A. W. Meerow, J. S. Brown and D. N. Kuhn, *Mol. Breed.*, 2011, **27**, 93–106.
16. F. A. Feltus, C. A. Saski, K. Mockaitis, N. Haiminen, L. Parida, Z. Smith, J. Ford, M. E. Staton, S. P. Ficklin, B. P. Blackmon, C. H. Cheng, R. J. Schnell, D. N. Kuhn and J. C. Motamayor, *BMC Genomics*, 2011, **12**, 379.

17. J. Y. Yang, M. Scascitelli, L. A. Motilal, S. Sveinsson, J. M. M. Engels, N. C. Kane, H. Dempewolf, D. P. Zhang, K. Maharaj and Q. C. B. Cronk, *Tree Genet. Genomes*, 2013, **9**, 829–840.
18. B. Trognitz, E. Cros, S. Assemat, F. Davrieux, N. Forestier-Chiron, E. Ayestas, A. Kuant, X. Scheldeman and M. Hermann, *PLoS One*, 2013, **8**, e54079.
19. K. Ji, D. P. Zhang, L. A. Motilal, M. Boccara, P. Lachenaud and L. W. Meinhardt, *Genet. Resour. Crop Evol.*, 2013, **60**, 441–453.
20. E. J. Boza, B. M. Irish, A. W. Meerow, C. L. Tondo, O. A. Rodriguez, M. Ventura-Lopez, J. A. Gomez, J. M. Moore, D. P. Zhang, J. C. Motamayor and R. J. Schnell, *Genet. Resour. Crop Evol.*, 2013, **60**, 605–619.
21. D. P. Zhang, W. J. Martinez, E. S. Johnson, E. Somarriba, W. Phillips-Mora, C. Astorga, S. Mischke and L. W. Meinhardt, *Genet. Resour. Crop Evol.*, 2012, **59**, 239–252.
22. E. Thomas, M. van Zonneveld, J. Loo, T. Hodgkin, G. Galluzzi and J. van Etten, *PLoS One*, 2012, **7**, e47676.
23. J. M. Thevenin, V. Rossi, M. Ducamp, F. Doare, V. Condina and P. Lachenaud, *PLoS One*, 2012, **7**, e40915.
24. R. G. L. Solorzano, O. Fouet, A. Lemainque, S. Pavek, M. Boccara, X. Argout, F. Amores, B. Courtois, A. M. Risterucci and C. Lanaud, *PLoS One*, 2012, **7**, e48438.
25. L. A. Motilal, D. P. Zhang, P. Umaharan, M. Boccara, S. Mischke, A. Sankar and L. W. Meinhardt, *Plant Genet. Resour.*, 2012, **10**, 232–241.
26. E. Aragon, C. Rivera, H. Korpelainen, A. Rojas, P. Elomaa and J. P. T. Valkonen, *Plant Genet. Resour.*, 2012, **10**, 254–257.
27. B. Trognitz, X. Scheldeman, K. Hansel-Hohl, A. Kuant, H. Grebe and M. Hermann, *PLoS One*, 2011, **6**, e16056.
28. C. R. S. Silva, P. S. B. Albuquerque, F. R. Ervedosa, J. W. S. Mota, A. Figueira and A. M. Sebbenn, *Heredity*, 2011, **106**, 973–985.
29. L. A. Motilal, D. P. Zhang, P. Umaharan, S. Mischke, S. Pinney and L. W. Meinhardt, *Plant Genet. Resour.*, 2011, **9**, 430–438.
30. J. C. Motamayor, P. Lachenaud, E. M. J. W. da Silva, R. Loor, D. N. Kuhn, J. S. Brown and R. J. Schnell, *PLoS One*, 2008, **3**, e3311.
31. F. K. Padi, O. Domfeh, J. Takrama and S. Opoku, *Crop Prot.*, 2013, **51**, 24–31.
32. S. D. V. M. Silva, L. R. M. Pinto, B. F. de Oliveira, V. O. Damaceno, J. L. Pires and C. T. D. Dias, *Trop. Plant Pathol.*, 2012, **37**, 191–195.
33. D. Nyadanu, R. Akromah, B. Adomako, C. Kwoseh, S. T. Lowor, H. Dzahini-Obiatey, A. Y. Akrofi and M. K. Assuah, *Euphytica*, 2012, **188**, 253–264.
34. D. S. Livingstone, B. Freeman, J. C. Motamayor, R. J. Schnell, S. Royaert, J. Takrama, A. W. Meerow and D. N. Kuhn, *Mol. Breed.*, 2012, **30**, 33–52.
35. U. V. Lopes, W. R. Monteiro, J. L. Pires, D. Clement, M. M. Yamada and K. P. Gramacho, *Crop. Breed. Appl. Biotechnol.*, 2011, **11**, 73–81.

36. S. Royaert, W. Phillips-Mora, A. M. A. Leal, K. Cariaga, J. S. Brown, D. N. Kuhn, R. J. Schnell and J. C. Motamayor, *Tree Genet. Genomes*, 2011, **7**, 1159–1168.
37. W. Phillips-Mora, A. Arciniegas-Leal, A. Mata-Quirós and J. Motamayor-Arias *Technical series. Technical manual/CATIE*, 2013, vol. 105, p. 68.
38. C. Lanaud, O. Fouet, D. Clément, M. Boccara, A. M. Risterucci, S. Surujdeo-Maharaj, T. Legavre and X. Argout, *Mol. Breed.*, 2009, **24**, 361–374.
39. P. J. Teixeira, G. G. Costa, G. L. Fiorin, G. A. Pereira and J. M. Mondego, *Mol. Plant Pathol.*, 2013, **14**, 602–609.
40. V. L. Souza, A. A. de Almeida, S. S. J. de, P. A. Mangabeira, R. M. de Jesus, C. P. Pirovani, D. Ahnert, V. C. Baligar and L. L. Loguercio, *Environ. Sci. Pollut. Res. Int.*, 2013, 1217–1230.
41. R. M. F. Santos, D. Clement, L. S. L. Lemos, T. Legravre, C. Lanaud, R. J. Schnell, J. L. Pires, U. V. Lopes, F. Micheli and K. P. Gramacho, *Tree Genet. Genomes*, 2013, **9**, 117–127.
42. A. M. Noah, N. Niemenak, S. Sunderhaus, C. Haase, D. N. Omokolo, T. Winkelmann and H. P. Braun, *J. Proteomics*, 2013, **78**, 123–133.
43. J. Gregorio, A. F. Hernández-Bernal, E. Cordoba and P. León, *Mol. Plant*, 2013, 422–436.
44. D. S. Britto, C. P. Pirovani, B. S. Andrade, T. P. Dos Santos, C. Pungartnik, J. C. Cascardo, F. Micheli and A. S. Gesteira, *Mol. Biol. Rep.*, 2013, 5417–5427.
45. S. A. Naganeeswaran, E. A. Subbian and M. Ramaswamy, *Bioinformation*, 2012, **8**, 65–69.
46. L. C. Mejia, M. J. Guiltinan, Z. Shi, L. Landherr and S. N. Maximova, *Acs Symp. Ser.*, 2012, **1095**, 379–395.
47. B. V. de Oliveira, G. S. Teixeira, O. Reis, J. G. Barau, P. J. P. L. Teixeira, M. C. S. do Rio, R. R. Domingues, L. W. Meinhardt, A. F. P. Leme, J. Rincones and G. A. G. Pereira, *Fungal Genet. Biol.*, 2012, **49**, 922–932.
48. D. V. G. de Andrade, A. Goes-Neto, M. Comar and A. G. Taranto, *Int. J. Quantum Chem.*, 2012, **112**, 3164–3168.
49. B. T. da Hora, J. D. Poloni, M. A. Lopes, C. V. Dias, K. P. Gramacho, I. Schuster, X. Sabau, J. C. D. Cascardo, S. M. Z. Di Mauro, A. D. Gesteira, D. Bonatto and F. Micheli, *Mol. BioSyst.*, 2012, **8**, 1507–1519.
50. B. C. Rehem, A. A. F. Almeida, I. C. Santos, F. P. Gomes, C. P. Pirovani, P. A. O. Mangabeira, R. X. Correa, M. M. Yamada and R. R. Valle, *Photosynthetica*, 2011, **49**, 127–139.
51. T. T. Pinheiro, C. G. Litholdo, M. L. Sereno, G. A. Leal, P. S. B. Albuquerque and A. Figueira, *Genet. Mol. Res.*, 2011, **10**, 3291–3305.
52. O. Fouet, M. Allegre, X. Argout, M. Jeanneau, A. Lemainque, S. Pavek, A. Boland, A. M. Risterucci, G. Loor, M. Tahi, X. Sabau, B. Courtois and C. Lanaud, *Tree Genet. Genomes*, 2011, **7**, 799–817.
53. Z. Shi, S. Maximova, Y. Lui, J. Verica and M. Guiltinan, *BMC Plant Biol.*, 2010, **10**, 248.

54. Y. Liu, Z. Shi, S. Maximova, M. Payne and M. J. Guiltinan, *BMC Plant Biol.*, 2013, **13**, 202.
55. B. A. Bailey, J. Crozier, R. C. Sicher, M. D. Strem, R. Melnick, M. F. Carazzolle, G. G. L. Costa, G. A. G. Pereira, D. Zhang, S. Maximova, M. Guiltinan and L. Meinhardt, *Physiol. Mol. Plant Pathol.*, 2013, **81**, 84–96.
56. Z. Shi, S. Maximova, Y. Lui, J. Verica and M. Guiltinan, *BMC Plant Biol.*, 2010, **10**, 248.
57. Y. Liu, Ph.D. Thesis, Molecular analysis of genes involved in the synthesis of proanthocyanidins in *Theobroma cacao*, The Pennsylvania State University, 2010.
58. D. P. T. Thomazella, P. J. P. L. Teixeira, H. C. Oliveira, E. E. Saviani, J. Rincones, I. M. Toni, O. Reis, O. Garcia, L. W. Meinhardt, I. Salgado and G. A. G. Pereira, *New Phytol.*, 2012, **194**, 1025–1034.
59. P. J. P. L. Teixeira, D. P. T. Thomazella, R. O. Vidal, P. F. V. do Prado, O. Reis, R. M. Baroni, S. F. Franco, P. Mieczkowski, G. A. G. Pereira and J. M. C. Mondego, *PLoS One*, 2012, **7**, e45929.
60. F. Oro, E. Mississo, M. Okassa, C. Guilhaumon, C. Fenouillet, C. Cilas and E. Muller, *Arch. Virol.*, 2012, **157**, 509–514.
61. C. V. Mfegue, C. Herail, H. Adreit, M. Mbenoun, Z. Techou, M. Ten Hoopen, D. Tharreau and M. Ducamp, *Am. J. Bot.*, 2012, **99**, E353–E356.
62. G. G. L. Costa, O. G. Cabrera, R. A. Tiburcio, F. J. Medrano, M. F. Carazzolle, D. P. T. Thomazella, S. C. Schuster, J. E. Carlson, M. J. Guiltinan, B. A. Bailey, P. Mieczkowski, G. A. G. Pereira and L. W. Meinhardt, *Fungal Biol.*, 2012, **116**, 551–562.
63. R. Babin, C. Fenouillet, T. Legavre, L. Blondin, C. Calatayud, A. M. Risterucci and M. P. Chapuis, *Int. J. Mol. Sci.*, 2012, **13**, 4412–4417.
64. G. G. Costa, O. G. Cabrera, R. A. Tiburcio, F. J. Medrano, M. F. Carazzolle, D. P. T. Thomazella, S. C. Schuster, J. E. Carlson, M. J. Guiltinan, B. A. Bailey, P. Mieczkowski, G. A. Pereira and L. W. Meinhardt, *Fungal Biol.*, 2012, **116**, 551–562.
65. J. M. Mondego, M. F. Carazzolle, G. G. Costa, E. F. Formighieri, L. P. Parizzi, J. Rincones, C. Cotomacci, D. M. Carraro, A. F. Cunha, H. Carrer, R. O. Vidal, R. C. Estrela, O. Garcia, D. P. Thomazella, B. V. de Oliveira, A. B. Pires, M. C. Rio, M. R. Araujo, M. H. de Moraes, L. A. Castro, K. P. Gramacho, M. S. Goncalves, J. P. Neto, A. G. Neto, L. V. Barbosa, M. J. Guiltinan, B. A. Bailey, L. W. Meinhardt, J. C. Cascardo and G. A. Pereira, *BMC Genomics*, 2008, **9**, 548.
66. E. F. Formighieri, R. A. Tiburcio, E. D. Armas, F. J. Medrano, H. Shimo, N. Carels, A. Goes-Neto, C. Cotomacci, M. F. Carazzolle, N. Sardinha-Pinto, D. P. Thomazella, J. Rincones, L. Digiampietri, D. M. Carraro, A. M. Azeredo-Espin, S. F. Reis, A. C. Deckmann, K. Gramacho, M. S. Goncalves, J. P. Moura Neto, L. V. Barbosa, L. W. Meinhardt, J. C. Cascardo and G. A. Pereira, *Mycol. Res.*, 2008, **112**, 1136–1152.

CHAPTER 5

Nutritional and Physiological Aspects of Chocolate

MICHELLE A. BRIGGS,*[a] YUJIN LEE,[b] JENNIFER FLEMING,[b] CHRISTINA SPONSKY[b] AND PENNY M. KRIS-ETHERTON[b]

[a] Department of Biology, Lycoming College, 700 College Place, Williamsport, PA 17701, USA; [b] Department of Nutritional Sciences, Penn State University, 319 Chandlee Lab, University Park, PA 16802, USA
*Email: briggs@lycoming.edu

5.1 Introduction

Healthy dietary practices are the foundation for preventing chronic diseases. Impressive advances have been made in identifying dietary patterns that decrease risks associated with the leading chronic diseases globally. A major theme of current nutrition research is to identify strategies that further enhance the efficacy of healthy dietary patterns. One food in the limelight is chocolate because evidence has shown multiple health benefits of cacao and chocolate products. This chapter provides an update on the many important nutrients and bioactive constituents in chocolate that confer health benefits. It also discusses how post-harvest processing affects the composition of chocolate products. The unifying objective of the current research thrust is to develop the next generation of chocolate products that further enhance health and teach consumers how to incorporate them into a healthy diet.

Chocolate and Health: Chemistry, Nutrition and Therapy
Edited by Philip K. Wilson and W. Jeffrey Hurst
© The Royal Society of Chemistry 2015
Published by the Royal Society of Chemistry, www.rsc.org

5.2 Nutrient Composition from Cacao to Chocolate

5.2.1 History of Cacao Consumption

The Latin name for cacoa—*Theobroma*—literally means "food of the gods". This valuable crop played an important role in many ancient South American cultures where the Mayas, Incas and Aztecs cultivated the cacao fruit-producing tree, *Theobroma cacao*.[1] There are over 22 species of the genus *Theobroma*, but varieties have been classified into three groups: criollo, forastero and trinitario.[2] The more commonly used bean is forastero, with criollo falling second. In its earliest forms, the Mayans used cacao beans as the basis of a drink known as *xocolatl*, a bitter beverage made from roasted cacao beans, water and spices. Chocolate's medicinal qualities were discovered early on, and it was used for the treatment of coughs, fever and even discomfort during pregnancy.[3] Like the Mayans, the Aztecs also consumed large quantities of *xocolatl*. It was described as a luxury drink since it was enjoyed mostly by the elite upper class.

In 1520, Hernán Cortés tasted chocolate prepared by the Aztecs and learned how to use the bitter bean to make a new and very popular beverage by adding cane sugar and honey.[4] He brought this "hot chocolate" drink back to Spain where the origin and preparation method remained a secret for nearly a century. Consequent to the decline of Spain as a global power, the secret behind making cocoa became more commonly known across Europe. By the mid-1600s, chocolate in drink form gained widespread popularity, and it was praised as a delicious, healthy "food" enjoyed by the wealthy. One enterprising Frenchman opened the first hot chocolate shop in London, and by the mid-1700s, these "chocolate houses" were common elsewhere in England, too. With the development of new production methods as part of the Industrial Revolution, chocolate became mass produced, and its popularity spread quickly.

In the early 19th century, Dutch chocolate-maker Coenraad Van Houten invented the cocoa press, a machine designed to separate the cocoa butter from the beans, leaving behind the defatted fine powder (*i.e.*, cocoa powder). This powder was then treated with alkaline salts to reduce bitterness and to enhance cocoa solubility. The final product had a darker color, and the resulting beverage had a milder taste and a smoother consistency, further increasing cocoa's popularity. This process, known as "Dutching", is still used today and is what differentiates natural cocoa from Dutch-processed cocoa. Natural cocoa retains all the heart-healthy flavanols, whereas Dutch processing with alkali destroys some of these bioactives, thereby reducing some of chocolate's health benefits. Interestingly, many of today's leading chocolate businesses started with the aim of persuading the poor to give up alcohol in favor of the healthier chocolate beverage.

The chocolate bar was first developed by Joseph Fry in 1847.[5] Similar to the cocoa beverage, the first chocolate bars were made of bittersweet chocolate. In the mid-1860s, the Swiss pharmacist Henri Nestlé began experimenting with various combinations of cow's milk, wheat flour and sugar in an attempt to develop a stable, concentrated source of infant nutrition for mothers who were unable to breast feed. His goal was to help combat the problem of infant mortality due to malnutrition. Following the invention of powdered milk, Nestlé teamed up with chocolate-maker Daniel Peter and developed milk chocolate, which today is preferred by 80% of the world's chocolate consumers.

Chocolate today is an affordable indulgence with health benefits that has maintained its popularity even during difficult economic times. The International Cocoa Organization reported in 2010 that the intake of chocolate had increased 2.8% from the previous year, resulting in a record 5.4 million tons consumed.[6] Growing awareness of the benefits associated with dark chocolate has influenced consumer purchasing, with 35% of consumers reporting a preference for dark chocolate in 2012, up from 33% in 2011. It appears that understanding the health benefits of dark chocolate may be increasing its popularity as individuals look for indulgent foods that can serve multiple functions and promote health.

5.2.2 Nutrient Composition of Cacao

Nutritionally, cocoa contains many biologically active substances that have positive effects on human health. The most widely studied include the flavonoids (catechins and the isomer epicatechin [EPI]), theobromine and magnesium. In the diet, most theobromine is derived from chocolate and cocoa, but it is also present in tea. Theobromine is not degraded during cocoa processing and can be used as a marker of cocoa content. Interestingly, clinical studies that compared theobromine-containing cocoa products to theobromine-free control products repeatedly showed significant beneficial effects on cholesterol, specifically high-density lipoprotein (HDL) cholesterol.[7] In contrast, studies examining products that controlled for theobromine content generally did not affect HDL cholesterol.[8] These results suggest that theobromine in cocoa, rather than flavonoids, may be involved with or even responsible for the HDL cholesterol benefit. However, additional studies are needed to better understand its precise biological effects. Based on the US Department of Agriculture's National Nutrient Database, cocoa also contains significant amounts of magnesium (2–4 mg g^{-1} dry powder), an essential nutrient that is typically underconsumed by most Americans. While the amount of magnesium provided is dependent on the type of chocolate, a 40 g portion of 70% cocoa dark chocolate contains ≈ 40 mg of magnesium, enough to make a modest ($\sim 10\%$) contribution to the recommended daily allowance (300–400 mg magnesium per day in adults).

Not as well-known are the additional micronutrients that chocolate provides. Dark chocolate specifically is ranked among the top dietary sources of copper and zinc and is considered a good source of iron, phosphorus and potassium. Table 5.1 provides a summary of the key nutrients found in the

Table 5.1 Macro- and micronutrients of cocoa powder, chocolate syrup and dark, milk and white chocolate per 100 g serving.[a]

Product	Energy (kcal)	CHO (g)	Sugar (g)	Protein (g)	Fat (g)	Ca (mg)	Fe (mg)	Mg (mg)	P (mg)	K (mg)	Na (mg)	Zn (mg)	Cu (mg)
Dietary Reference Intake					1,000	8–18	300–400	700	4,700	1,500	11	0.9	
Cocoa Powder:													
100 g	228	57.9	1.8	19.6	13.7	128	13.9	499	734	1524	21	6.8	3.8
1 tbsp	12	3	<0.1	<1	0.5	7	0.8	27	40	82	1	0.4	0.2
Chocolate Syrup:													
100 g	279	65.1	50	2.1	1.1	14	2.1	65	129	224	72	0.7	0.5
2 tbsp	109	25	19	<1	<1	5	0.8	25	50	87	28	0.3	0.2
Dark Chocolate (45–59% cacao solids)	546	61.1	48	4.9	31.4	56	8.0	146	206	559	7	0.6	1.0
Dark Chocolate (60–69% cacao solids)	579	52.4	37	6.1	38.3	62	6.3	176	260	567	10	0.8	1.0
Dark Chocolate (75–85% cacao solids)	598	45.9	24	7.8	42.6	73	11.9	228	308	715	20	3.3	1.8
Milk Chocolate	535	59.4	52	7.7	29.7	189	2.4	63	208	372	79	2.3	.5
White Chocolate	539	59.3	59	5.9	32.1	199	0.2	12	176	286	90	0.7	0.1

[a] Data obtained from the United States Department of Agriculture (USDA) Agricultural Research Services Database. Released December 2011.

various types of chocolate. The 2010 Dietary Guidelines for Americans identified four nutrients of concern: dietary fiber, vitamin D, calcium and potassium, none of which are consumed in adequate amounts.[9] Chocolate products are a good source of dietary potassium. The adequate intake (AI) for potassium is set at 4700 mg for all adults; however, current dietary intake of potassium by all groups in the USA is considerably lower than the AI. In recent surveys, the median intake of potassium by adults was approximately 2800–3300 mg.[10] While a diet rich in fruits and vegetables promotes adequate potassium intake, dark chocolate can help achieve optimal potassium levels that may not otherwise be met. For example, a medium banana provides 422 mg of potassium, but dipping it in 22 g of dark chocolate (100 kcal) provides an additional 140 mg of potassium. Dietary potassium is linked to reductions in blood pressure. In a pooled meta-analysis of 13 clinical trials, results demonstrated a significant blood pressure-lowering effect of cocoa chocolate (mean blood pressure change ± standard error: systolic blood pressure: -3.2 ± 1.9 mmHg, $P=0.001$; diastolic blood pressure: -2.0 ± 1.3 mmHg, $P=0.003$).[11] In addition, subgroup analyses demonstrated greater effects for hypertensive versus normotensive individuals. Daily flavanol doses ranged from 30 to 1000 mg in the treatment groups, and interventions ran for 2–18 weeks.

Globally, iron deficiency is one of the most common nutritional disorders, primarily affecting young women and children.[12] A one ounce serving of dark chocolate contains 19% of the daily value (DV) for iron, which is greater than the amount found in one ounce of beef (3% of DV). While the non-heme iron present in the cacao bean is less absorbed than heme iron, absorption can be enhanced by the consumption of food sources of vitamin C, which are sometimes consumed alongside chocolate (*e.g.*, strawberries and chocolate or chocolate-covered raspberries).

EPI, the main flavanol in cocoa, has been studied extensively. Since 2007, the majority of research emanating from human and nonhuman intervention studies has evaluated the effects of cocoa products, or individual chemical components of cocoa, regarding the risk of cardiovascular disease (CVD) using surrogate biomarkers. These studies demonstrated beneficial effects on endothelial function, blood pressure and cholesterol levels.[13] Endothelial function and nitric oxide (NO) synthesis in particular have been the focus of much of this research. The predominant mechanistic hypothesis is that cocoa, and specifically EPI, stimulates NO activity, inhibits arginase and inhibits NADPH oxidase, thereby leading to lower levels of superoxide and hence higher levels of NO. Though not the only mechanism involved, a substantial increase in NO synthesis may account for flow-mediated vessel dilation and lower blood pressure following chocolate and cocoa intervention treatments.[14]

The flavonoids also act as antioxidants, which are compounds that reduce free radicals and stabilize membranes. The antioxidants in chocolate are found in the cacao solids; therefore, dark chocolate tends to have higher amounts of flavonoids than milk chocolate.[15] Specifically, Vinson *et al.*

reported that dark chocolate provides 951 mg of cocoa flavonoids in a 40 g serving compared to 395 mg in 40 g of milk chocolate.[16] The high antioxidant content of dark chocolate provides evidence to further support cocoa as a biologically active ingredient with potential benefits on biomarkers related to CVD.

While dark chocolate contains more nonfat cocoa solids than milk chocolate, white chocolate contains none. As the percentage of nonfat cocoa solids increases, the flavonoid content increases, while the proportion of sugar, and hence carbohydrate (CHO), decreases. Thus, the macronutrient composition of chocolate products varies significantly (Table 5.1), as does their flavonoid content, which would be expected to affect the magnitude of CVD benefits of chocolate consumption. Although epidemiological evidence has shown that consumption of 50–100 g per week of chocolate reduces CVD risk, further clarification is needed regarding which components of cacao and other ingredients in chocolate benefit CVD risk.[17] Furthermore, epidemiological data are not able to establish the type of chocolate consumed (*i.e.*, dark, milk or white), the dose and the possible effect of other ingredients such as fruit or nuts that may be important factors in the overall analysis of the evidence, in addition to the total diet and various other CVD risk factors.

Sucrose is the predominant type of CHO in chocolate, with small amounts of lactose also present when powdered milk is added as an ingredient.[18] Most commercial milk chocolate contains 43–58% CHO from sugar.[19] The main purpose of adding sucrose is to enhance sweetness, mask the natural bitterness of chocolate and increase its palatability.[20] The protein content of chocolate ranges from 4% (white chocolate) to 35% (cocoa powder) per 100 g serving.[21] Arginine, glutamine and leucine comprise the greatest proportion of amino acids. Glutamine and leucine are abundant constituents of plant and animal proteins, whereas the content of arginine in foods varies widely. These three amino acids are involved in the signaling pathways that promote protein synthesis and possibly inhibit protein degradation in intestinal epithelial cells.[22]

Among single amino acids, L-arginine, a semi-essential amino acid, is the natural substrate for NO synthase and is responsible for the production of the endothelium-derived relaxing factor NO, which is involved in a wide variety of regulatory mechanisms of the cardiovascular system. Several mechanisms may be responsible for the beneficial effect of L-arginine on blood pressure. As a substrate for NO synthase, L-arginine may exhibit antihypertensive effects by augmenting the production of NO in the endothelium and improving its bioavailability in vascular smooth muscle cells, all of which are essential to maintaining vascular homeostasis.[23] A meta-analysis by Bai *et al.* suggests that oral L-arginine supplementation is effective at improving endothelial cell function in individuals with endothelial dysfunction.[24] Results from clinical intervention trials suggest that cocoa-containing arginine and dietary flavanols exert vascular protective effects, improve NO metabolism and endothelial function and are associated with a reduced risk of CVD.[25]

CHO and fat are inversely related in chocolate products; that is, as more cocoa solids are added, fat increases, while CHO (*i.e.*, sugar) decreases. Thus, total energy and percent CHO are also inversely related. Per 100 g serving, the CHO content is lowest in dark chocolate, while total energy is highest (75–85% cacao solids; 30% CHO, 598 kcal), when compared to both white (44% CHO, 539 kcal) and milk chocolate (44%, 535 kcal).[26]

The total fat content varies based on type of chocolate. White chocolate contains 54% fat per 100 g, milk chocolate 50% fat per 100 g and dark chocolate (75–85% cacao solids) 64% fat per 100 g. The fatty acid profile of chocolate is comprised mainly of saturated fat (stearic acid; C18:0, 35% and palmitic acid; C16:0, 25%) and monosaturated fat (oleic acid; C18:1, 35%), with only minimal polyunsaturated fat (linoleic acid; C18:2, 3%).[27] Unlike other saturated fatty acids, stearic acid, which accounts for half of the saturated fat content in chocolate, has a neutral effect on total cholesterol (TC) and low-density lipoprotein cholesterol (LDL-C) levels.[28] In addition, a recent meta-analysis of ten clinical trials summarizing the effects of cocoa products/dark chocolate (20–100 g) on serum lipids concluded that interventions (2–12 weeks) with dark chocolate/cocoa products significantly reduced LDL-C and TC levels. The post-intervention difference in mean LDL-C concentration (95% confidence interval) comparing active treatment with placebo was -5.90 mg dl^{-1} (-10.47, -1.32 mg dl^{-1}) and -6.23 mg dl^{-1} (-11.60, -0.85 mg dl^{-1}) for LDL-C and TC, respectively.[29] These results are consistent with a similar review conducted on eight studies, which reported that interventions with dark chocolate/cocoa products led to LDL-C reductions by 5.87 mg dl^{-1} (-11.13, -0.61 mg dl^{-1}) and TC reductions by 5.82 mg dl^{-1} (-12.39, 0.76 mg dl^{-1}) compared with placebo (white chocolate or nutrient-matched control).[30] Doses of polyphenols in the studies ranged from 30 to 963 mg per day, and the treatment duration varied from 2 to 18 weeks.

5.3 Incorporating Cacao into the Diet

The 2010 Dietary Guidelines Advisory Committee (which provided expertise and recommendations for the Dietary Guidelines for Americans, 2010) concluded that incorporating moderate amounts of natural cocoa and dark chocolate as part of a healthy, balanced diet can provide cardiovascular health benefits.[31] Improvements in cardiovascular health have been reported with one to two tablespoons of cocoa per day (10 kcal per tbsp) as an ingredient in beverages, meals or snacks, or 22 g (100 kcal) of dark chocolate with at least 50% cacao.[32]

There is growing evidence to support the role of insulin resistance and endothelial function as independent predictors of CVD risk. Hooper *et al.* conducted the first systematic review to comprehensively assess the overall effects and validity of the available randomized control trials (RCT) data on chocolate or cocoa on a range of important CVD risk factors, including insulin resistance and flow mediated dilation (FMD).[33] In their analyses, they observed improvements in homeostasis model assessment of insulin resistance (HOMA-IR) or FMD after twice-daily consumption of cocoa beverages containing 19, 22 or 54 g of cocoa per day, 46 or 100 g of dark chocolate per day or 48 g of chocolate plus 18 g of cocoa per day. By comparison, intakes of chocolate and cocoa in prospective studies (*e.g.*, 7.5 g of chocolate per day in the upper quintile of the European Prospective Investigation into Cancer and Nutrition study[34] and 4.2 g of cocoa per day in the Zutphen Elderly Study[35]) are lower in comparison to the chocolate/cocoa flavan-3-ol doses studied in the RCTs. Thus, larger and longer-duration trials with optimally designed treatments and controls are required to reach a consensus regarding the recommended quantity of chocolate/cocoa for health. Implicit to any recommendation made in the future is that energy balance must be achieved. Even though greater benefits are achievable with higher intakes, doing so will adversely affect body weight if the diet is not isocalorically balanced. For example, a 50 g serving of chocolate per day provides $\sim$280 kcal ($\sim$13% of daily energy intake) and should be used to replace foods of similar energy content and lower nutritional value in order to maintain energy balance and improve nutrient adequacy. Indeed, inclusion of chocolate can help achieve nutrient adequacy for micronutrients in chocolate, but it can also compromise nutrient adequacy for others if large amounts are incorporated into the diet (both isocalorically and in excess of calorie needs).

In addition to its micronutrients, chocolate products are also a source of added sugars. The amount of sugar varies greatly according chocolate type, with white chocolate containing the greatest amount (59 g of sugar per 100 g of white chocolate). As illustrated in Table 5.1, dark chocolate, containing the greatest percentage of cacao solids (75–85%), has the lowest sugar content (24 g per 100 g). Consequently, dark chocolate often has a deeper, more bitter flavor than milk chocolate, which contains both added sugars and milk sugar (52 g per 100 g). The American Heart Association (AHA) recommends limiting the amount of added sugars consumed to no more than 50% of the daily discretionary calorie allowance. For most American women, this is no more than 100 calories per day, or about six teaspoons (25 g) of sugar. For men, it is 150 calories per day, or about nine teaspoons of sugar (38 g). Based on the evidence in support of the nutritional benefits of chocolate as well as the AHA recommendations for added sugar, 0.75–1.75 oz of dark chocolate per day may be included in the diet for promoting optimal health. For women, 0.75 oz (21 g) of dark chocolate (>50% cacao solids) would provide only 100 calories and 10 g of sugar ($\sim$2 tsp). A 1.75 oz (50 g) serving of dark chocolate would provide 270 calories and 24 g of sugar ($\sim$5.5 tsp).

5.4 Physiological Functions of Cacao's Bioactive Compounds

Cacao reportedly has more than 500 different chemical constituents.[36] Although not all of these chemicals occur in physiologically significant concentrations, synergism among minor constituents may also significantly impact physiology.[37] Further, given the complex array of phytochemicals in chocolate, some negative interactions may be created due to opposing effects. For example, Mitchell et al. compared the effects of different cocoa methylxanthines at levels typical of chocolate products and found that 700 mg of theobromine significantly lowered systolic blood pressure one hour after ingestion, whereas 120 mg of caffeine increased diastolic blood pressure.[38] When given in a combined dose, however, the effects on blood pressure were cancelled out. This response might be due to the competitive action of these compounds for similar adenosine receptors in the brain, although relative to caffeine, theobromine apparently has a lower affinity for these receptors.[39] Other potential synergistic effects among chocolate constituents are noted in a recent review of chocolate's psychoactive constituents.[40]

5.4.1 Alkaloids

Alkaloids constitute 2.5–3.5% of cocoa powder.[41] Whereas the majority of cocoa alkaloids are the methylxanthines theobromine and caffeine (with negligible amounts of theophylline), processed cocoa products also contain four different tetrahydro-β-carbolines (THβC's). These apparently form during fermentation and/or heating, and can reach concentrations of several $\mu g\ g^{-1}$ in chocolate, which are comparable to those found in some alcoholic beverages.[42] Like many alkaloids, THβC's affect the central nervous system (CNS), although it is not clear whether their concentrations in chocolate would elicit a physiologically significant effect. In addition to inhibiting monoamine oxidase (MAO),[43] THβC's also interfere with serotonin cycling[44] and benzodiazepine receptors.[45] These actions could partially explain the improved mood many report after consuming chocolate.

Theobromine is the dominant alkaloid in chocolate, reaching typical ratios of one caffeine for every eight theobromine molecules.[46] Dark chocolate contains 25–35 mg of caffeine and 200–300 mg of theobromine per 40 g serving.[47] When consumed, these alkaloids are thought to elicit physiological responses by the non-selective binding of adenosine receptors,[48] although theobromine has a lower affinity for adenosine receptors than caffeine.[49] These methylxanthines appear to cause the majority of the psychopharmacological effects associated with cocoa products. This claim derives from a study in which subjects provided with a capsule containing 19 mg of caffeine and 250 mg of theobromine reported similar improvements in mood and cognitive function as compared to subjects provided with cocoa powder-containing identical levels of methylxanthines.[50] Indeed, these alkaloids may be partially responsible for why we like chocolate. Smit and Blackburn administered a capsule containing the above alkaloid levels while also consuming a novel fruit drink. Subjects were asked to score their opinion of that drink. Those who consuming the combined theobromine and caffeine in comparison to a placebo group significantly increased their opinions of the novel drink over several exposures.[51]

Although theobromine and caffeine are both regarded as non-selective adenosine antagonists, they elicit remarkably different effects when examined both singularly and in combination. Relative to mood, when theobromine was administered to healthy subjects at doses of 250, 500 and 1,000 mg, subjects reported that high doses elicited negative moods.[52] Similarly, a separate study found that 700 mg of theobromine decreased subjects' calmness, although 120 mg of caffeine increased both alertness and contentedness. However, a combined dose of 700 mg of theobromine and 120 mg of caffeine acted similarly to caffeine with respect to mood effects.[53] Based on the results on mood and blood pressure noted above, Mitchell *et al.*[53] concluded that caffeine has a greater influence on the CNS, whereas theobromine most likely produces its greatest effects on peripheral physiology.

The influence of caffeine on mood has been observed in many other studies. Ruxton reviewed 41 double-blind, placebo-controlled studies that investigated the effects of caffeine.[54] The majority reported that moderate doses of caffeine led to a more positive mood state, in addition to improving the participants' recall, reaction time and alertness. However, higher caffeine doses notably cause anxiety in some individuals.[55] Caffeine's effect on adenosine receptors also promotes inflammation at physiologically relevant concentrations.[56] Although high caffeine concentrations may be anti-inflammatory due to the inhibition of phosphodiesterase, the threshold for this activity is not achieved in humans under typical chocolate consumption.[57]

Caffeine's influence on adenosine receptors triggers a cascade of events in the brain as many other receptors and neurotransmitters are, in turn, affected.[58] In brief, caffeine's influence on adenosine receptors has been shown to increase turnover rates of serotonin; serotonin is often thought of as an anti-panic neurotransmitter.[59] Caffeine also increases turnover rates of dopamine and noradrenaline, neurotransmitters typically associated with stimulation.[60] Researchers have also shown that caffeine can increase cortical acetylcholine, perhaps partially explaining caffeine's psychostimulant effects.[61] Caffeine also alters GABA function[62] and elevates levels of glutamate.[63]

In humans, theobromine has been shown to influence a variety of physiological functions due, in part, to its effects on adenosine receptors.[64] For example, theobromine increases heart rate in a dose-dependent manner,[65] an issue related to decreased adenosine levels. Other studies found that a 10 mg kg^{-1} dose of theobromine increased bronchodilation in asthmatic youths.[66] Theobromine has also been shown to suppress capsaicin-induced cough in humans. Indeed, theobromine's anti-tussive effectiveness was greater than that of the widely used codeine. This anti-tussive property is apparently due to vagus nerve suppression, which partially reflects theobromine's influence on adenosine receptors.[67]

5.4.2 Opioids

Cocoa products also appear to influence endogenous opioid production.[68] The most commonly accepted mechanism for increasing opioid production involves the hedonic factor of chocolate. Foods that humans find pleasant tasting (*e.g.*, sweet or palatable) trigger endorphin release.[69] Increasing these endogenous opioids leads to a mild euphoria and most likely a mild analgesia,[70] thus decreasing pain sensitivity.

5.4.3 Biogenic Amines

Chocolate contains several amines including phenylethylamine (PEA, the "endogenous amphetamine"), tryptamine, clovamide, tyramine and serotonin. Although the levels of these compounds absorbed may be insufficient to create physiological responses, chocolate has the highest PEA levels among tested foods.[71] PEA is known to significantly influence other neurotransmitters, such as dopamine, noradrenalin and serotonin.[72] It is believed that PEA, when consumed in chocolate, is unlikely to cause pharmacological effects due to its rapid metabolism by MAO-B, although there is evidence that biogenic amines may affect those who are deficient in MAO.[73] These effects could include blushing, headaches and increased blood pressure.[74] PEA might also have some effect in individuals due to a synergistic association with compounds such as salsolinol and β-carbolides, both of which are chocolate alkaloids that inhibit MAO-B.[75] These interactions may be responsible for amplifying the effects of biogenic amines. Another cocoa amide of interest is clovamide, a neuroprotective compound whose function is apparently related to its antioxidant ability.[76]

5.4.4 Endocannabinoids

Anandamide, a compound that binds competitively to the brain's cannabinoid receptors, affects a variety of functions in the body that include controlling cognition,[77] suppressing pain[78] and protecting neurons,[79] whereas lower anandamide levels are associated with the cognitive dysfunction seen in dementia.[80] Although chocolate contains anandamide, it is typically a very low content (0.05 $\mu g\ g^{-1}$). Thus, it is unlikely that chocolate's anandamide level could actively trigger a physiological response.[81] However, two N-acylethanolamines are present in chocolate at 10^3–10^4 times higher concentrations.[82] One of these compounds, N-linoleoylethanolamine, interferes with anandamide hydrolysis even at 5 μM concentrations.[83] Given the high levels of these compounds in chocolate and their low effective dose, future research may show that consuming these compounds increases endogenous anandamide.

5.4.5 Polyphenolics: Flavanols

Fermented cacao beans contain approximately 6% phenolic compounds, with the most common being the monomeric flavanols EPI and catechin, and their oligomers the procyanidins.[84] These flavanols are also chocolate's main antioxidants.[85] On a per-gram basis, the antioxidant capacity of cocoa powder is significantly greater than blueberry, cranberry or pomegranate powders.[86] Antioxidants protect cell membranes, proteins and DNA from reactive oxygen species (ROS); uncontrolled ROS can lead to degenerative diseases such as heart disease, cancer, arthritis and neurodegenerative diseases resulting in dementia.[87] While polyphenolics function *in vitro* as antioxidants by inactivating free radicals and reducing their oxidative potential, high levels of polyphenolics may be toxic.[88] However, the levels of polyphenolics typically absorbed from food do not appear to be high enough to contribute significantly to total antioxidant capacity.[89]

Rather than acting directly as antioxidants, it appears more likely that polyphenolic compounds such as chocolate flavanols or their degradation products upregulate endogenous antioxidant enzymes, such as superoxide dismutase, catalase or glutathione peroxidase.[90] Upregulation of these enzymes may occur because polyphenolics generate ROS. Another mechanism for upregulating endogenous enzymes may involve catechins binding to DNA and RNA, thus allowing catechins to regulate gene transcription.[91] This process also occurs with green tea catechins. Cocoa flavonoids exhibit a range of physiological effects, including some that reinforce the outcomes seen from other cocoa compounds, although they work *via* different pathways. For example, many studies show that cocoa flavanols improve neurocognitive function,[92] as do the methylxanthines.[93] Flavanols also enhance mood and calm feelings,[94] as was also found with low doses of caffeine.[95]

A brief list of other biological effects of ingesting chocolate flavanols includes lowering blood pressure[96] and marginally improving cholesterol.[97] Flavonoids cross the blood–brain barrier,[98] supporting observations that cocoa flavanols appear to improve mental processes. For example, cocoa flavanols improve blood flow to cerebral gray matter, even in young subjects.[99] Cocoa flavanols also appear to slow the progression of some forms of dementia,[100] most likely because they inhibit the apoptosis triggered by amyloid-β protein.[101] Indeed, brain-related associations with cocoa flavanols reflect the currently accepted methods explaining the benefits of cocoa. These proposed functions include promoting neuron survival by inhibiting apoptosis and enhancing cerebral blood flow.[100,102] Increased blood flow is most likely due to improved endothelial function.[103]

5.4.6 Oxalates

While many of cacao's constituents have been linked to health benefits, not all are beneficial. For example, oxalic acid constitutes 0.3–0.5% of cocoa powder.[104] Chocolate is generally considered a high oxalate food since its ingestion is associated with high levels of urinary oxalate.[105] When ingested, oxalate can bind to minerals such as calcium, thus decreasing its absorption from food.[106] Once absorbed, oxalate can precipitate in the kidneys, forming kidney stones.[107]

5.5 Impact of Processing on the Nutrient/Bioactive Compound Content of Cacao Beans

5.5.1 Cacao Processing

Cacao beans are processed *via* a series of steps in manufacturing chocolate products. Fermentation, roasting and milling are fundamental in producing palatable chocolate products.

After harvesting the cacao beans, a wide range of yeasts dominate the fermentation process during the first 24 hours, then lactic acid bacteria replace the yeasts, whereupon the pectinaceous pulp surrounding the cacao beans is converted into ethanol. Chemical and physical changes occur in cacao beans during this process.[108] For example, when cacao beans are harvested, only (−)-EPI and (+)-catechin are present. After the fermentation process, (−)-catechin is formed due to the epimerization of (−)-EPIs.[109]

The fermented cacao beans are roasted for varying times (10–40 minutes) and temperatures (100–190 °C). During this roasting, sugars and amino acids are reduced, whereby amines and aldehydes are formed.[110] Torres-Moreno *et al.* found that the roasting time strongly correlates with the chocolate's palatability. Consumer acceptability, based on sensory attributes (*e.g.*, color, odor, flavor and texture), was lower with longer roasting times (38.5 minutes) compared with shorter roasting times (30.5 and 34.5 minutes) for Ghanaian samples.[111] Following the roasting process, cacao beans are milled to a fine grain size. The goal of milling is to reach a homogeneous particle size of 30 microns. Like roasting, particle size and conformation strongly influence the flavor and mouth-feel of the final product.[112]

Cacao powder is treated with alkali, known as the Dutch process, by which free organic acids are neutralized and bitterness is reduced. This process also darkens the color of the cacao powder and increases the dispersability of the cacao powder, thereby allowing cacao powder to be used for various products, most notably cacao-containing beverages.[113]

5.5.2 Processing Changes in Nutrient and Bioactive Compounds

Catechins (37%), anthocyanins (4%) and proanthocyanidins (58%) are the three main groups of polyphenols found in cocoa.[114] Concentrations of these chemicals vary greatly, depending on the levels of processing. As discussed above, although cacao processing is critical to increasing the palatability of cacao beans, some nutrients and bioactive compounds are lost during the process.

The loss of EPIs begins before the cacao beans become fully ripe. Payne *et al.* found that unripe cacao beans had 29% higher levels of EPIs compared with fully ripe beans.[115] They suggested that biological oxidation and/or polymerization of monomers to procyanidins might be responsible for this loss during the ripening process. When the beans are fermented, substantial amounts of (−)-EPIs and (+)-catechins are lost as (−)-catechins are generated. Hurst et al. suggested that the heat of fermentation may be responsible for the formation of this enantiomer (i.e., [−]-catechins).[116] Moreover, Donovan et al. ranked the bioavailability of the three bioactive compounds as: (−)-EPIs > (+)-catechins > (−)-catechins.[117] Fermentation appears to generate the poor bioavailability of the catechin, (−)-catechin. Moreover, it contributes significantly to the loss of bioactive compounds and the decreased levels of crude fiber and phosphorus.[118]

Roasting beans causes further changes due to epimerization of the remaining EPI, resulting in decreased EPI and increased catechin levels. However, chemical changes do not always simply reflect shifts from one category of polyphenolic compounds to another. Oliviero et al. studied the effects of roasting on total antioxidant activity of catechin. They found a 60% decrease of antioxidant activity of catechin after roasting at 180 °C for 30 minutes.[119]

Other processes affecting chocolate's constituents include the Dutch processing that removes bitterness. While this increases palatability, it also significantly reduces antioxidant capacity,[120] primarily due to the removal of beneficial flavanols.[121] Miller et al. investigated the impact of Dutch processing on the flavanol content of cocoa powders. They grouped alkalized cacao powders into three categories: lightly treated (pH 6.50–7.20), medium treated (pH 7.21–7.60) and heavily treated (pH 7.61 and higher). Natural cocoa powder was found to contain total flavanol contents ranging from 22.86 to 40.25 mg g^{-1}. After the light, medium and heavy Dutch processing, the total flavanol content decreased from 8.76 to 24.65 mg g^{-1}, 3.93 to 14.00 mg g^{-1} and 1.33 to 6.05 mg g^{-1}, respectively.[122]

Phenolic compounds like the flavanols are not the only bioactives that change during processing; microbial fermentation and heating increase alkaloids such as the THβC's. THβC's are known to have antioxidative and neuroactive properties.[123] Buckholtz and Boggan found that THβC's inhibited monoamine uptake and the monoamine oxidase enzymes responsible for oxidizing neurotransmitters.[124] However, a relatively low concentration of THβC's is present in chocolate (0.21 mg of THβC's/30 g of chocolate).[125] Further research is warranted to elucidate the likely roles performed by THβC's in chocolate.

5.6 Summary

The multitude of chocolate products in the marketplace varies widely in composition with the products' nutrient profiles and bioactive contents. All chocolate products provide energy, macronutrients, micronutrients and bioactives, many of which are important for achieving nutrient adequacy and conferring health benefits. Many factors affect the nutrient and bioactive content of chocolate products, including the post-harvest processing methods utilized. Although much has been learned about the nutritional benefits derived from cacao and chocolate, a need persists to develop chocolate products that deliver maximum health benefits. Moreover, consumer education is needed in order to learn ways of better incorporating chocolate products into diets such that their optimal health benefits may be realized.

References

1. K. Bruinsma and D. L. Taren, *J. Am. Diet. Assoc.*, 1999, **99**, 1249–1256.
2. P. Hebbar, H. C. Bittenbender and D. O'Doherty, Farm and Forestry Production and Marketing Profile for Cacao (*Theobroma cacao*), ed. C. R. Elevitch, in *Specialty Crops for Pacific Island Agroforestry*, Permanent Agriculture Resources (PAR), Holualoa, Hawai'i, http://agroforestry.net/scps.
3. S. Aaron and M. Bearden, *Chocolate: A Healthy Passion*, Prometheus Books, New York, 2008.
4. K. Bruinsma and D. L. Taren, *J. Am. Diet. Assoc.*, 1999, **99**, 1249–1256; S. D. Coe & M. D. Coe, *The True History of Chocolate*, Thames and Hudson, London, 1999.
5. L. E. Grivetti and H. Y. Shapiro, *Chocolate: History, Culture, and Heritage*, John Wiley & Sons, Inc., Hoboken, NJ, 2011.
6. International Cocoa Organization, (2012). The World Cocoa Economy: Past and Present. Retrieved from website: http://www.icco.org/about-us/international-cocoa-agreements/cat_view/30-related-documents/45-statistics-other-statistics.html.
7. S. Baba, N. Osakabe, Y. Kato, *et al.*, *Am. J. Clin. Nutr.*, 2007, **85**, 709–717; N. Khan, M. Monagas, C. Andres-Lacueva, *et al.*, *Nutr., Metab. Cardiovasc. Dis.*, 2012, **22**, 1046–1053; D. D. Mellor, T. Sathyapalan, E. S. Kilpatrick, *et al.*, *Diabetic Med.*, 2010, **27**, 1318–1321; J. Mursu, S. Voutilainen, T. Nurmi, *et al.*, *Free Radical Biol. Med.*, 2004, **37**, 1351–1359.
8. K. Davison, A. M. Coate, J. D. Buckley, *et al.*, *Int. J. Obes.*, 2008, **32**, 1289–1296; M. B. Engler, M. M. Engler, C. Y. Chen, *et al.*, *J. Am. Coll. Nutr.*, 2004, **23**, 197–204.
9. USDA Nutrient Database, Release 11 December, 2011.
10. National Research Council, *Dietary Reference Intakes for Water, Potassium, Sodium, Chloride, and Sulfate*, The National Academies Press, Washington, DC, 2005.
11. K. Ried, T. Sullivan, P. Fakler, *et al.*, *BioMed Central*, 2010, **28**, 39.
12. World Health Organization, 2013. Accessed October 2013. http://www.who.int/nutrition/topics/ida/en/.
13. S. Ellam and G. Williamson, *Ann. Rev. Nutr.*, 2013, **33**, 105–128.
14. I. Ramirez-Sanchez, L. Maya, G. Ceballos, *et al.*, *Hypertension*, 2010, **55**, 1398–1405; T. Brossette, C. Hundsdoerfer, K. D. Kroencke, *et al.*, *Eur. J. Nutr.*, 2011, **50**, 595–599.
15. F. M. Steinberg, M. M. Bearden and C. L. Keen, *J. Am. Diet. Assoc.*, 2003, **103**, 215–223.

16. J. A. Vinson, J. Proch and L. Zubik, *J. Agric. Food Chem.*, 1999, **47**, 4821–4824.
17. P. J. Mink, C. G. Scrafford, L. M. Barraj, *et al.*, *Am. J. Clin. Nutr.*, 2007, **85**, 895–909; I. Janszky, K. J. Mukamal, R. Ljung, *et al.*, *J. Int. Med.*, 2009, **266**, 248–257; L. Djousse, P. N. Hopkins, K. E. North, *et al.*, *Clin. Nutr.*, 2011, **30**, 182–187; B. Buijsse, E. J. Feskens, F. J. Kok, *et al.*, *Arch. Int. Med.*, 2006, **166**, 411–417.
18. S. T. Beckett, *The Science of Chocolate*, Royal Society of Chemistry, Cambridge, 2008.
19. M. Prawira and S. A. Barringer, *J. Food Process. Preserv.*, 2009, **33**, 571–589.
20. M. Prawira and S. A. Barringer, *J. Food Process. Preserv.*, 2009, **33**, 571–589.
21. *USDA Nutrient Database*, Release 11 December, 2011.
22. J. M. Rhoads and G. Wu, *J. Amino Acids*, 2009, **37**, 111–122.
23. J. Y. Dong, L. Q. Qin, Z. Zhang, *et al.*, *Am. Heart J.*, 2011, **162**, 959–965.
24. Y. Bai, L. Sun, T. Yang, *et al.*, *Am. J. Clin. Nutr.*, 2009, **89**, 77–84.
25. V. Habauzit and C. Morand, *Ther. Adv. Chronic Dis.*, 2012, **3**, 87–106.
26. USDA Nutrient Database, Release 11 December, 2011.
27. S. Aaron and M. Bearden, *Chocolate: A Healthy Passion*, Prometheus Books, New York, 2008; F. M. Steinberg, M. M. Bearden and C. L. Keen, *J. Am. Diet. Assoc.*, 2003, **103**, 215–223.
28. S. Aaron and M. Bearden, *Chocolate: A Healthy Passion*, Prometheus Books, New York, 2008.
29. O. A. Tokede, J. M. Gaziano, L. Djousse, *et al.*, *Eur. J. Clin. Nutr.*, 2011, **65**, 879–886.
30. J. Lei, L. Xuan, B. Yong Yi, *et al.*, *Am. J. Clin. Nutr.*, 2010, **92**, 218–225.
31. DGAC Report. Report of the Dietary Guidelines Advisory Committee on the Dietary Guidelines for Americans, 2010. United States Department of Agriculture, United States Department of Health and Human Service, Washington, DC, 2010.
32. K. D. Monahan, R. P. Feehan, A. R. Kunselman, *et al.*, *J. Appl. Physiol.*, 2011, **111**, 1568–1574; S. Desch, J. Schmidt, D. Kobler, *et al.*, *Am. J. Hypertens.*, 2010, **23**, 97–103.
33. L. Hooper, C. Kay, A. Abdelhamid, *et al.*, *Am. J. Clin. Nutr.*, 2012, **95**, 740–751.
34. B. Buijsse, C. Weikert, D. Drogan, *et al.*, *Eur. Heart J.*, 2010, **31**, 1616–1623.
35. B. Buijsse, E. J. Feskens, F. J. Kok, *et al.*, *Arch. Intern. Med.*, 2006, **166**, 411–417.
36. W. J. Hurst, S. M. Tarka, T. G. Powis, *et al.*, *Nature*, 2002, **418**, 289–290.
37. K. Bruinsma and D. L. Taren, *J. Am. Diet. Assoc.*, 1999, **99**, 1249–1256; H. J. Smit, in *Methylxanthines: Handbook of Experimental Pharmacology*, ed. B. B. Fredholm, Springer, Berlin Heidelberg, 2011, 200, pp. 201–234, p. 225; W. D. Crews Jr., D. W. Harrison and K. P. Gregory, *et al.*, in *Chocolate in Health and Nutrition: Nutrition and Health*, ed.

R. R. Watson, V. R. Preedy and S. Zibadi, Springer, Berlin Heidelberg, 2013, 7, pp. 369–379, p. 377.
38. E. S. Mitchell, M. Slettenaar, N. van der Meer, et al., Physiol. Behav., 2011, **104**, 816–822.
39. B. B. Fredholm and K. Lindström, Eur. J. Pharmacol., 1999, **380**, 197–202.
40. N. Rodrigues-Silva, in Chocolate in Health and Nutrition: Nutrition and Health, ed. R. R. Watson, V. R. Preedy and S. Zibadi, Springer, Berling Heidelberg, 2013, 7, pp. 421–435, p. 425.
41. J. L. Apgar and S. M. Tarka, Jr., in Caffeine, ed. G. A. Spiller, CRC Press, 1998, pp. 163–192, pp. 172–175.
42. T. Herraiz, J. Agric. Food Chem., 2000, **48**, 4900–4904.
43. N. S. Buckholtz and W. O. Boggan, Biochem. Pharmacol., 1977, **26**, 1991–1996.
44. R. A. Glennon, M. Dukat, B. Grella, et al., Drug Alcohol Depend., 2000, **60**, 21–132.
45. C. Braestrup, M. Nielsen and C. E. Olsen, Proc. Natl. Acad. Sci., 1980, **77**, 2288–2292.
46. J. L. Apgar and S. M. Tarka, Jr., in Caffeine, ed. G. A. Spiller, CRC Press, 1998, pp. 163–192, pp. 172–175.
47. M. J. Baggott, E. Childs, A. B. Hart, et al., Psychopharmacol, 2013, **228**(109–118), 109.
48. H. J. Smit, Methylxanthines, in Handbook of Experimental Pharmacology, ed. B. B. Fredholm, Springer-Verlag, Berlin Heidelberg, 2011, 200, pp. 201–234, p. 206.
49. B. B. Fredholm and K. Lindström, Eur. J. Pharmacol., 1999, **380**, 197–202.
50. H. J. Smit, E. A. Gaffan and P. J. Rogers, Psychopharmacol, 2004, **176**, 412–419.
51. H. J. Smit and R. J. Blackburn, Psychopharmacol, 2005, **181**, 101–106.
52. M. J. Baggott, E. Childs, A. B. Hart, et al., Psychopharmacol, 2013, **228**(109–118), 109.
53. E. S. Mitchell, M. Slettenaar, N. vd Meer, et al., Physiol. Behav., 2011, **104**, 816–822.
54. C. H. S. Ruxton, Nutr. Bull., 2008, **33**, 15–25.
55. E. Childs, C. Hohoff, J. Deckert, et al., Neuropsychopharmacol, 2008, **33**, 2791–2800.
56. A. Ohta, D. Lukashev, E. K. Jackson, et al., J. Immunol., 2007, **179**, 7431–7438.
57. A. Ohta and M. Sitkovsky, Methylxanthines, in Handbook of Experimental Pharmacology, ed. B. B. Fredholm, Springer-Verlag, Berlin Heidelberg, 2011, 200, pp. 469–481, p. 470.
58. B. B. Fredholm, K. Bättig, J. Holmén, et al., Pharmacol. Rev., 1999, **51**, 83–133.
59. J. D. Coplan, J. M. Gorman and D. F. Klein, Neuropsychopharmacol, 1992, **6**, 189–200.

60. B. B. Fredholm, K. Bättig, J. Holmén, *et al.*, *Pharmacol. Rev.*, 1999, **51**, 83–133.
61. A. J. Carter, W. T. O'Connor, M. J. Carter, *et al.*, *J. Pharmacol. Exper. Ther.*, 1995, **273**, 637–642.
62. F. Lopez, L. G. Miller, D. J. Greenblatt, *et al.*, *Eur. J. Pharmacol., Mol. Pharmacol. Sect.*, 1989, **172**, 453–459.
63. M. Solinas, S. Ferré, Z. B. You, *et al.*, *J. Neurosci.*, 2002, **22**, 6321–6324.
64. H. J. Smit, Methylxanthines, in *Handbook of Experimental Pharmacology*, ed. B. B. Fredholm, Springer-Verlag Berlin Heidelberg, 2011, 200, pp. 201–234, p. 206.
65. M. J. Baggott, E. Childs, A. B. Hart, *et al.*, *Psychopharmacol*, 2013, **228**(109–118), 109.
66. F. Simons, A. B. Becker, K. J. Simons, *et al.*, *J. Allergy Clin. Immunol.*, 1985, **76**, 703–707.
67. O. S. Usmani, M. G. Belvisi, H. J. Patel, *et al.*, *FASEB J.*, 2005, **19**, 231–233.
68. K. M. Eggleston and T. White, in *Chocolate in Health and Nutrition: Nutrition and Health*, ed. R. R. Watson, V. R. Preedy and S. Zibadi, Springer, Berling Heidelberg, 2013, 7, pp. 437–447, pp. 439–440.
69. G. Parker, I. Parker and H. Brotchie, *J. Affective Disord.*, 2006, **92**, 149–159.
70. F. N. Segato, C. Castro-Souza, E. N. Segato, *et al.*, *Braz. J. Med. Biol. Res.*, 1997, **30**, 981–984.
71. N. Rodrigues-Silva, in *Chocolate in Health and Nutrition: Nutrition and Health*, ed. R. R. Watson, V. R. Preedy and S. Zibadi, Springer, Berlin Heidelberg, 2013, 7, pp. 421–435, p. 425.
72. I. A. Paterson, A. V. Juorio and A. A. Boulton, *J. Neurochem.*, 1990, **55**, 1827–1837.
73. H. J. Smit, Methylxanthines, in *Handbook of Experimental Pharmacology*, ed. B. B. Fredholm, Springer-Verlag, Berlin Heidelberg, 2011, pp. 201–234, p. 206.
74. A Bertazzo, S. Comai, F. Mangiarini, *et al.*, in *Chocolate in Health and Nutrition: Nutrition and Health*, ed. R. R. Watson, V. R. Preedy and S. Zibadi, Springer, Berlin Heidelberg, 2013, 7, pp. 105–117.
75. H. J. Smit, Methylxanthines, in *Handbook of Experimental Pharmacology*, ed. B. B. Fredholm, Springer-Verlag, Berlin Heidelberg, 2011, pp. 201–234, p. 206.
76. S. Fallarini, G. Miglio, T. Paoletti, *et al.*, *Br. J. Pharmacol.*, 2009, **157**, 1072–1084.
77. K.-M. Jung, G. Astarita, S. Yasar, *et al.*, *Neurobiol. Aging*, 2012, **33**, 1522–1532.
78. J. R. Clapper, G. Moreno-Sanz, R. Russo, *et al.*, *Nat. Neurosci.*, 2010, **13**, 265–1270.
79. E. Eljaschewitsch, A. Witting, C. Mawrin, *et al.*, *Neuron*, 2006, **49**, 67–79.
80. K.-M. Jung, G. Astarita, S. Yasar, *et al.*, *Neurobiol. Aging*, 2012, **33**, 1522–1532.

81. H. J. Smit, Methylxanthines, in *Handbook of Experimental Pharmacology*, ed. B. B. Fredholm, Springer-Verlag, Berlin Heidelberg, 2011, pp. 201–234, p. 206.
82. V. Di Marzo, N. Sepe, L. De Petrocellis, *et al.*, *Nature*, 1998, **396**, 636.
83. N. Rodrigues-Silva, in *Chocolate in Health and Nutrition: Nutrition and Health*, ed. R. R. Watson, V. R. Preedy and S. Zibadi, Springer, Berlin Heidelberg, 2013, 7, pp. 421–435, p. 425.
84. A. Bertazzo, S. Comai, F. Mangiarini, *et al.*, in *Chocolate in Health and Nutrition: Nutrition and Health*, ed. R. R. Watson, V. R. Preedy and S. Zibadi, Springer, Berlin Heidelberg, 2013, 7, 105–117.
85. L. Gu, S. E. House, X. Wu, *et al.*, *J. Agric. Food Chem.*, 2006, **54**, 4057–4061.
86. S. J. Crozier, A. G. Preston, J. W. Hurst, *et al.*, *Chem. Central J.*, 2011, **5**, 3–4.
87. M. Valko, C. J. Rhodes, J. Moncol, *et al.*, *Chemi.-Biol. Interact.*, 2006, **160**, 1–40.
88. D. E. Stevenson and R. D. Hurst, *Cell. Mol. Life Sci.*, 2007, **64**(2900–2916), 2903.
89. M. N. Clifford, *Plant Med.*, 2004, **70**, 1103–1114.
90. D. E. Stevenson and R. D. Hurst, *Cell. Mol. Life Sci.*, 2007, **64**(2900–2916), 2903.
91. T. Kuzuhara, Y. Sei, K. Yamaguchi, *et al.*, *J. Biol. Chem.*, 2006, **281**, 17446–17456.
92. W. D. Crews Jr., D. W. Harrison, K. P. Gregory, *et al.*, in *Chocolate in Health and Nutrition: Nutrition and Health*, ed. R. R. Watson, V. R. Preedy and S. Zibadi, Springer, Berlin Heidelberg, 2013, 7, pp. 369–379, p. 377.
93. H. J. Smit, E. A. Gaffan and P. J. Rogers, *Psychopharmacol*, 2004, **176**, 412–419.
94. M. P. Pase, A. B. Scholey, A. Pipingas, *et al.*, *J. Psychopharmacol.*, 2013, **27**, 451–458.
95. E. S. Mitchell, M. Slettenaar, N. vd Meer, *et al.*, *Physiol. Behav.*, 2011, **104**, 816–822.
96. S. Desch, J. Schmidt, D. Kobler, *et al.*, *Am. J. Hypertens.*, 2010, **23**, 97–103.
97. L. Hooper, C. Kay, A. Abdelhamid, *et al.*, *Am. J. Clin. Nutr.*, 2012, **95**, 740–751.
98. D. Vauzour, K. Vafeiadou, A. Rodriguez-Mateos, *et al.*, *Genes Nutr.*, 2008, **3**, 115–126; A. Faria, D. Pestana, D. Teixeira, *et al.*, *Food Funct.*, 2011, **2**, 39–44.
99. S. T. Francis, K. Head, P. G. Morris, *et al.*, *J. Cardiovasc. Pharmacol.*, 2006, **47**, S215–S220.
100. N. Sokolov, M. A. Pavlova, S. Klosterhalfen, *et al.*, *Neurosci. Biobehav. Rev.*, 2013, **37**(10, Part 2), 2445–2453.
101. H. J. Heo and C. Y. Lee, *J. Agric. Food Chem.*, 2005, **53**, 1445–1448.

102. D. Vauzour, K. Vafeiadou, A. Rodriguez-Mateos, et al., Genes Nutr., 2008, **3**, 115–126; A. Faria, D. Pestana, D. Teixeira, et al., Food Funct., 2011, **2**, 39–44; W. D. Crews Jr., D. W. Harrison, K. P. Gregory, et al., in Chocolate in Health and Nutrition: Nutrition and Health, ed. R. R. Watson, V. R. Preedy and S. Zibadi, Springer, Berlin Heidelberg, 2013, 7, pp. 369–379, p. 377; A. Nehlig, Br. J. Clin. Pharmacol., 2013, **75**, 716–727.
103. M. B. Engler, M. M. Engler, C. Y. Chen, et al., J. Am. Coll. Nutr., 2004, **23**, 197–204.
104. A Bertazzo, S. Comai, F. Mangiarini, et al., in Chocolate in Health and Nutrition: Nutrition and Health, ed. R. R. Watson, V. R. Preedy and S. Zibadi, Springer, Berlin Heidelberg 2013, 7, 105–117.
105. L. K. Massey, H. Roman-Smith and R. A. Sutton, J. Am. Diet. Assoc., 1993, **93**, 901–906.
106. T. Schroder, L. Vanhanen and G. P. Savage, J. Food Compos. Anal., 2011, **24**, 916–922; A Bertazzo, S. Comai, F. Mangiarini, et al., in Chocolate in Health and Nutrition: Nutrition and Health, ed. R. R. Watson, V. R. Preedy and S. Zibadi, Springer, Berlin Heidelberg, 2013, 7, pp. 105–117.
107. L. K. Massey, H. Roman-Smith and R. A. Sutton, J. Am. Diet. Assoc., 1993, **93**, 901–906; T. Schroder, L. Vanhanen and G. P. Savage, J. Food Compos. Anal., 2011, **24**, 916–922.
108. R. F. Schwan and A. E. Wheals, Crit. Rev. Food Sci. Nutr., 2004, **44**, 205–221.
109. M. J. Payne, W. J. Hurst, K. B. Miller, et al., J. Agric. Food Chem., 2010, **58**, 10518–10527.
110. M. Granvogl, S. Bugan and P. Schieberle, J. Agric. Food Chem., 2006, **54**, 1730–1739.
111. M. Torres-Moreno, A. Tarrega, E. Costell, et al., J. Sci. Food Agric., 2012, **92**, 404–411.
112. A. McShea, E. Ramiro-Puig and S. B. Munro, Nutr. Rev., 2008, **66**, 630–641.
113. T. Stark and T. Hofmann, J. Agric. Food Chem., 2006, **54**, 9510–9521.
114. C. L. Hii, C. L. Law, S. Suzannah, et al., Asian J. Food Agro-Ind., 2009, **2**, 702–722.
115. M. J. Payne, W. J. Hurst, K. B. Miller, et al., J. Agric. Food Chem., 2010, **58**, 10518–10527.
116. W. J. Hurst, S. H. Krake, S. C. Bergmeier, et al., Chem. Cent. J., 2011, **5**, 53–62.
117. J. L. Donovan, V. Crespy, M. Oliveira, et al., Free Radical Res., 2006, **40**, 1029–1034.
118. C. Y. Aremu, M. A. Agiang and J. O. Ayatse, Plant Food Hum. Nutr., 1995, **48**, 217–223.
119. T. Oliviero, E. Capuano, B. Cämmerer, et al., J. Agric. Food Chem., 2009, **57**, 147–152.

120. L. Gu, S. E. House, X. Wu, et al., *J. Agric. Food Chem.*, 2006, **54**, 4057–4061.
121. N. D. Fisher and N. K. Hollenberg, *J. Hypertens.*, 2005, **23**, 1453–1459; M. J. Payne, W. J. Hurst, K. B. Miller, et al., *J. Agric. Food Chem.*, 2010, **58**, 10518–10527.
122. K. B. Miller, W. J. Hurst, M. J. Payne, et al., *J. Agric. Food Chem.*, 2008, **56**, 8527–8533.
123. T. Herraiz and J. Galisteo, *J. Agric. Food Chem.*, 2003, **51**, 7156–7161.
124. N. S. Buckholtz and W. O. Boggan, *Biochem. Pharmacol.*, 1977, **26**, 1991–1996.
125. T. Herraiz, *J. Agric. Food Chem.*, 2000, **48**, 4900–4904.

CHAPTER 6

Chocolate, the Digestive Tract and Diabetes

JOHN W. FINLEY,* GABRIELLA CRESPO AND ZUYIN LI

School of Nutrition and Food Sciences, Louisiana State University, 201 Animal and Food Sciences Laboratories Building, Baton Rouge, Louisiana 70803, USA
*Email: jfinley@agcenter.lsu.edu

6.1 Introduction to Cocoa Phenolics

The raw cocoa bean is a rich source of potentially bioactive simple phenolic and polyphenolic compounds. Cocoa phenolic compounds are excellent *in vitro* antioxidants, and cocoa has been shown to reduce oxidative stress related conditions *in vivo*. Although many *in vivo* effects are well documented, chocolate may provide more than simple antioxidant effects. Antioxidant activity, however, has been used as a stable marker for changes regarding the potential health benefits associated with cocoa.

Major changes in antioxidant activity have been shown as a result of cocoa bean processing together with changes in the distribution of phenolic and polyphenolic compounds within the product. For example, total anthocyanins and total flavonoids were compared before and after fermentation. Phenolic compounds were reduced by 1.8 times and the anthocyanins, after fermentation, were nearly undetectable. The same study demonstrated that the fermented cocoa beans improved antioxidant and free radical scavenging activity beyond that of raw beans, reducing the total phenolic content by 14%, and "Dutching" reduced the phenolic content an additional 64%. The greatest losses were found in procyanidins B1 and B2 and epigallocatechin, whereas caffeic acid derivatives showed the greatest stability throughout

---

Chocolate and Health: Chemistry, Nutrition and Therapy
Edited by Philip K. Wilson and W. Jeffrey Hurst
© The Royal Society of Chemistry 2015
Published by the Royal Society of Chemistry, www.rsc.org

processing.[1] Stahl et al. further investigated the changes in total polyphenolics, flavan-3-ols and procyanidins along with antioxidant activity in foods prepared from cocoa powder.[2] Measurements were made in chocolate frosting, a hot cocoa beverage, chocolate cookies and chocolate cake produced with cocoa powder. Antioxidant activity, total phenolic, flavanol monomer and procyanidin recoveries ranged from 86% to over 100% in the chocolate frosting, hot cocoa drink and chocolate cookies. Chocolate cake had lower recoveries, ranging from 5% for epicatechin to 54% of the original antioxidant activity. The greater losses of antioxidant activity, total phenolics, flavanol monomers and procyanidins noted in baked chocolate cakes were associated with the increased pH caused by the baking powder used in the cake recipes.

A survey was conducted to determine the procyanidin levels (flavan-3-ol oligomers and polymers) in commercially available products sold in the USA. Percent fat, percent nonfat cocoa solids, oxygen radical absorbance capacity (ORAC), total polyphenols, epicatechin, catechin, total monomers and flavan-3-ol oligomers and polymers (procyanidins) were determined in three or four top-selling products within six categories: natural cocoa powder, unsweetened baking chocolate, dark chocolate, semisweet baking chips, milk chocolate and chocolate syrup. The catechin content of the products follow in decreasing order: cocoa powder > baking chocolate > dark chocolate = baking chips > milk chocolate > chocolate syrup. The monomer and oligomer profiles within product categories showed products containing high monomers with decreasing levels of oligomers and products in which the level of dimers is equal to or greater than the monomers. Results show a strong correlation ($R^2 = 0.834$) of epicatechin to the level of nonfat cocoa solids. Principal component analysis showed that the products group discretely into five classes: (1) cocoa powder, (2) baking chocolate, (3) dark chocolate and semisweet chips, (4) milk chocolates and (5) chocolate syrup. The results also indicated that most factors group closely together regarding antioxidant activity, total polyphenols and the flavan-3-ol measures, with the exception of catechin and percent fat in the product. Catechin distribution appears to be different from the other flavan-3-ol measures. The epicatechin to catechin ratio demonstrated that there is a >5-fold variation in this measure across the studied products. The cocoa products tested range from cocoa powder with 227.34 ± 17.23 mg of procyanidins per serving to 25.75 ± 9.91 mg of procyanidins per serving for chocolate syrup.[3] Studies of the antioxidant activity, total polyphenols, flavan-3-ol monomers and procyanidin levels in milk and dark chocolate, in cocoa powder and in cocoa beans showed that antioxidant activity and flavan-3-ol levels are stable over typical shelf lives of one year under controlled storage and over 2 years in ambient storage in the laboratory.[4]

6.2 Cocoa, Cocoa Liquor, Cocoa Powder and Chocolate

Cocoa powder typically contains between 12 and 18% polyphenolics. The bioavailability of these components depends on interactions of the phenolics with other food constituents in their food matrix. Cocoa phenolic compounds have been associated with antioxidant functions, reduced blood pressure *via* the induction of nitric oxide-dependent vasodilation in human and flow-mediated dilation of the brachial artery, improvements of endothelial function, vasoprotection, prevention of cancer and blood pressure-lowering effects, improvements of blood flow regulation, prevention of cardiovascular disease, prevention of diabetes, improvements of insulin sensitivity, decreases of platelet activation and function and treatment of glaucoma, as well as a modulation of immune function and inflammation. Cocoa and chocolate may also exert a protective effect on low-density lipoprotein (LDL) oxidization, which has been correlated with a reduced risk of developing atherosclerosis.

Inhibition of DNA synthesis, modulation of reactive oxygen species production, regulation of cell cycle arrest and modulation of survival/proliferation pathways are the most important mechanisms that can exert chemopreventive effects on different organ-specific cancers. Cocoa polyphenols can directly influence different points of the apoptotic process and/or the expression of regulatory proteins. These activities have been associated with the high concentrations of flavanols, catechins and procyanidins, all of which exhibit strong antioxidative activities. To function *in vivo*, the active components must be absorbed and transported to the target tissue in a biologically active form. Epicatechin appears to be absorbed much more efficiently than catechin. Catechins have a bioavailability of approximately 25–30% based on the quantity of ingested catechins found in urine. These flavonoids have potent antioxidant activity and account for the antiplatelet actions observed following consumption of cocoa or chocolate. They are present in blood as metabolites such as glucuronides and/or as sulfate conjugates and/or methylated forms. In acute feeding studies, flavanol-rich cocoa and chocolate increased plasma antioxidant capacity and reduced platelet reactivity. Some studies propose that 150 mg of flavonoids are needed to trigger a rapid antioxidant effect and changes in prostacyclin. Dose–response evidence demonstrates an antioxidant effect at the level of approximately 500 mg flavonoids. Antioxidants in cocoa can prevent the oxidization of LDL-cholesterol, which is related to protection from cardiovascular disease. For example, (−)-epicatechin and other cocoa flavan-3-ols have shown significant cardiovascular protection effects. They have been reported to decrease the plasma concentrations of proinflammatory cysteinyl leukotrienes. They also inhibit LDL oxidation, reduce thrombosis, improve endothelial function and reduce inflammation. The origin from which cocoa beans are derived together with the methods of their processing also influence the antioxidant polyphenols in cocoa products and the bioavailability of the active components.[5]

6.3 Bioavailability

The first human bioavailability trial of polyphenols from chocolate found that epicatechin was absorbed when consumed in dark chocolate.

The epicatechin in plasma was present as conjugates with either glucuronide or sulfates. The maximum concentration was reached about two hours after consumption, and small amounts remained after 8 hours.[6] A number of factors impact the bioavailability of the cocoa polyphenolic compounds in humans. Absorption rates are affected by the short half-life *in vivo*, and subsequent rapid excretions lower the overall bioavailability. The bioavailability of polyphenols is generally inverse to the molecular weight of the bioactive molecule. Generally, polymers larger than dimers are essentially not absorbed; however, when they reach the colon and interact with the microbiome, they may be broken down into smaller, more readily absorbable molecules. Monomeric epicatechin or O-methylated forms are conjugated as glucuronides and sulfates.[7]

The biological value of flavanols or procyanidins is dependent upon whether they can be absorbed, their tissue and cellular distributions after absorption and their chemical forms. Unlike other polyphenolics, the flavanols are not present as glycosides. Flavanols in cocoa are present as aglycones, oligomers or gallic acid esters. The flavanols and procyanidins are stable during gastric digestion[8] and absorbed from the jejunum epithelial layer, where they are rapidly methylated, sulfated or conjugated.[9]

The rest of the metabolites are rapidly excreted. In human volunteers, epicatechin metabolites (sulfates, glucuronides and sulfoglucuronides) reached maximum serum levels two hours after consumption of cocoa or chocolate. Epicatechin metabolites in the urine after chocolate or cocoa consumption were measured at $29.8 \pm 5.3\%$ and $25.3 \pm 8.1\%$ of total intake, respectively. This finding clearly demonstrates that epicatechins from chocolate and cocoa are partly absorbed and found to be present as a component of various conjugates in plasma and that they are rapidly excreted in urine.[10] Donovan *et al.* and Bell *et al.* found similar results with red wine and dealcoholized red wine.[11] All of these studies demonstrate that the epicatechin that is absorbed is quickly converted into conjugates and rapidly excreted.

Since the flavanols are only partially absorbed, some portion of the flavanols reaches the lower gastrointestinal tract, where they are metabolized by the microflora. The gut microflora can form simple phenolics, which can contribute to the bioactivity of the cocoa.[12] The microbial metabolites of the flavanols are *m*-hydroxyphenylpropionic acid, ferulic acid, 3,4-dihydroxyphenylacetic acid, *m*-hydroxyphenylacetic acid, vanillic acid and *m*-hydroxybenzoic acid.[13]

Information on the metabolism of the procyanidins is limited. The dimer B2 [epicatechin-(4-8)-epicatechin] and dimer B5 [epicatechin-($4\alpha \rightarrow 6$)-epicatechin] have been detected in the nanomolar range in the plasma of humans and rats,[14] whereas other procyanidins, such as dimer B3 [catechin-($4\alpha \rightarrow 8$)-catechin and trimer C2 [catechin-($4\alpha \rightarrow 8$)-catechin-($4\alpha \rightarrow 8$)-catechin], could not be detected.[15] It is important to note that the oligomers found in the plasma are those consisting of epicatechin and are not catechin subunits.[16]

The bioavailability of the cocoa polyphenols is influenced by the delivery matrix and other components in the diet. Table 6.1 summarizes clinical

Table 6.1 Clinical trials exploring the health benefits of cocoa.

Intervention	Assay Marker	Subjects/Duration	Outcome	Reference
Semi-sweet chocolate Doses 27, 53 and 80 g	AOX	20 adults (20–56 years) Single dose	Dose-dependent increase in plasma epicatechin NS trend in plasma AOX activity Decrease on thiobarbituric acid reactive substances (TBARS)	1
18–75 g procyanidin-rich cocoa powder in 330 ml water	Platelet	30 adults (30–50 years) 10 per group Single dose	Suppression of platelet activation Aspirin-like hemostasis at 6 hours	2
105 g (80 g from chocolate semi-sweet baking bits	AOX	13 healthy adults (26–39 years) Single dose	After 2 hours there was a 12-fold increase in plasma epicatechin Increase in plasma AOX activity Decrease in TBARS	3
12 g cocoa powder	AOX	15 healthy men (32.5 ± 6.4 years) 3 times per day for 2 weeks	Increased LDL oxidation time No change in lipids or AOX Higher excretion of epicatechin metabolites	4
22 g cocoa powder 16 g dark chocolate	AOX	23 healthy adults (21–62 years)	Increased LDL oxidation time No change in plasma lipids Increased excretion of epicatechin metabolites	5
18–75 g cocoa powder in 300 ml water with sugar with and without aspirin	Platelet	16 healthy adults (22–49 years) One dose	After 6 hours cocoa inhibited epinephrine-stimulated platelet activation	6
36–90 g dark chocolate 30–95 g cocoa powder in drinks	Inflammation	25 healthy adults (20–60 years) One drink per day for 6 weeks	Lower LDL oxidation No effect on inflammation or AOX capacity	7
25 g semi-sweet chocolate chips	Inflammation	18 healthy adults One dose	Increase in plasma epicatechin at 2 hours Increase in prostacyclin:leukotriene ratio Reduction in platelet-related hemostasis	8

Table 6.1 (Continued)

Intervention	Assay Marker	Subjects/Duration	Outcome	Reference
100 g dark chocolate	Vascular	13 elderly adults (55–64 years) with mild hypertension	Lower systolic and diastolic BP	9
Cocoa flavanol/procyanidin tablets	Platelet	28 adults (31–51 years) 28 days	Lower platelet aggregation No change in oxidation/anti-oxidation markers	10
High-polyphenol cocoa drink	Vascular	27 healthy adults (27–72 years) 4 days	Increase in plasma catechin and epicatechin Improved peripheral vasodilation at 4 days Response after 90 minutes not chronic	11
Four 230-ml doses/day 100 ml high-polyphenol cocoa drink	FMD	20 adults all with one coronary heart disease (CHD) risk factor (41 ± 14 years) Single dose	NO bioactivity and arterial FMD increase	12
100 g dark chocolate with and without 200 ml milk	AOX	12 healthy adults (25–35 years) Single dose	Dark chocolate increased plasma AOX and plasma epicatechin Milk attenuated effects Milk chocolate was even less effective	13
75 g dark or high-phenolic dark chocolate	AOX	45 healthy adults (19–49 years) 3 weeks	Increase in HDL cholesterol Decreased lipid peroxidation No change in AOX capacity	14
46 g high-phenolic dark chocolate	FMD	21 healthy adults (21–55 years) 14 days	Improved endothelium-dependent FMD No change in BP, oxidative markers or blood lipids	15
High-polyphenol cocoa drink (100 ml)	Anti-inflammation (ANTIINFL)	20 healthy males (20–40 years) Single dose	Higher plasma epicatechin F2 isoprostanes improved at 2 and 4 hours	16
100 g dark chocolate	Insulin BP	15 healthy adults (34 ± 7.6 years) Single dose	Higher insulin sensitivity Lower insulin resistance Lower systolic BP	17

Intervention	Marker	Subjects/Duration	Results	Ref
75 g dark chocolate or high-phenolic dark chocolate	Lipids	45 healthy adults (19–49 years) 3 weeks	Both dark chocolates increased HDL cholesterol and lipid peroxidation decreased (but also with white chocolate control) No change in plasma antioxidant capacity	14
46 g day^{-1} high-phenolic dark chocolate	FMD	21 healthy adults 2 weeks	Improved endothelium-dependent FMD, no change in BP, oxidative markers or blood lipids Higher plasma epicatechin	1
100 ml high-polyphenol cocoa drink	Inflammation	20 healthy males (20–40 years) Single dose	F2 isoprostanes improved at 2 and 4 hours after exercise	18
Dark chocolate, 100 g	Insulin/BP	15 healthy adults (34 ± 7.6 years) Single dose	Insulin sensitivity higher and insulin resistance lower Lower systolic BP	17
Flavonoid-rich drink at 0.25, 0.375 and 0.5 g kg^{-1} body weight	AOX	8 healthy males (26 ± 2 years) Single dose	Reduction in the rate of free radical-induced hemolysis	19
105 g day^{-1} milk chocolate	BP BP	28 healthy males (18–20 years) under exercise stress for 14 days	Decrease in diastolic and mean blood pressure, plasma cholesterol, LDL, malondialdehyde, urate and lactate dehydrogenase activity, increase in vitamin E:cholesterol ratio No change in plasma epicatechin but samples were fasting	20
100 g dark chocolate	FMD	20 never-treated adults with essential hypertension (44 ± 8 years) 15 days	Increase in resting and hyperemic brachial artery diameter Increase in FMD at 60 minutes Aortic augmentation index decreased No significant change in malondialdehyde, total antioxidant capacity and pulse wave velocity	21

Table 6.1 (Continued)

Intervention	Assay Marker	Subjects/Duration	Outcome	Reference
High-polyphenol cocoa drink, 100 ml	FMD	11 adult smokers (average 31 years) Single dose	Insulin sensitivity improved, lower systolic and diastolic BP and LDL and improved FMD	22
300 ml high-polyphenol cocoa drink		16 healthy males (25–32 years) Single dose	Acute elevations in levels of circulating NO species, an enhanced FMD response of conduit arteries and an augmented microcirculation	23
40 g dark chocolate	FMD	20 male smokers (age not given) Single dose	Improved FMD after 2 hour lasting for 8 hours Reduction in platelet function Increased plasma total antioxidant status	24
High-polyphenol cocoa drink 4×230 ml day^{-1}	FMD	4–6 days	NO synthesis after cocoa was suppressed in older volunteers FMD was enhanced in both groups but more so in the older group Pulse wave amplitude was enhanced in both groups, with acute rises with cocoa ingestion being more robust in older subjects No change in BP	25
22 g cocoa powder and 16 g dark chocolate (in a muffin)	AOX	4 (30–49 years) normolipidemic subjects (pilot trial)	Dark chocolate increased resistance of LDL and VLDL to oxidation, whilst cocoa butter alone decreased resistance Noted after examination of dietary data that chocolate is third-highest contributor of antioxidants to the American diet	26

Treatment	Measure	Subjects / Duration	Results	Ref
41 g day^{-1} of high-polyphenol dark chocolate either with or without almonds 60 g day^{-1}	Lipids	49 women with cholesterol 4.1–7.8 mmol l^{-1} (22–65 years) 6 weeks	Dark chocolate decreased triacyl glycerides (TAG) by 21%, 19% when eaten with almonds, 13% with almonds alone and 11% with no intervention Circulating intercellular adhesion molecule was reduced with dark chocolate alone	18
High-flavanol cocoa drink 100 ml	FMD	6 male smokers with smoking-related endothelial dysfunction (11 total) (22–32 years)	Daily continual FMD increases at baseline (fasted) and a sustained FMD augmentation at 2 hours post-ingestion	27
3 doses per day	FMD	1 week	A dose-dependent effect was also seen with FMD and nitrate Biomarkers for oxidative stress unaffected	
40 g dark chocolate	FMD FMD	25 male smokers 2 hours	Improved FMD Reduced platelet function Increased AOX	24
1. Solid dark chocolate bar/placebo	FMD	45 healthy adults	Improved endothelial function	28
2. Sugar-free hot cocoa/sugared hot cocoa	BP		Decreased BP	
High-flavanol beverage	FMD	21 overweight adults 2 hours	Increased FMD Attenuated increase in BP during exercise	29
100 g flavonoid-rich dark chocolate	FMD	17 young healthy volunteers 3 hours	Increased FMD Decreased aortic augmentation index	21
100 g white, dark or milk chocolate		30 healthy adults 4 hours	Inhibited collagen-induced platelet aggregation	30
Flavanol-rich cocoa drink	FMD	11 males with smoking-related endothelial dysfunction 7 days	Increased FMD Increased circulating nitrite	31
3 × 306 mg flavanol/day 100 g dark chocolate		13 healthy adults 14 days	Decreased systolic and diastolic BP	9

Table 6.1 (*Continued*)

Intervention	Assay Marker	Subjects/Duration	Outcome	Reference
100 g dark chocolate		15 healthy adults 15 days	Decreased BP Improved insulin sensitivity	17
100 g dark chocolate		20 adults with hypertension 15 days	Decreased BP, homeostatic model assessment insulin resistance (HOMA-IR), isolated systolic hypertension (ISO) and LDL cholesterol Increased FMD	17
100 g flavanol-rich dark chocolate		19 adults with hypertension and impaired glucose tolerance 15 days	Ameliorated insulin sensitivity and beta cell function Decreased BP Increased FMD	32
Flavanol-rich cocoa		41 medicated diabetics 30 days	Increased FMD	31
High-flavonoid dark chocolate bars		21 healthy adults 14 days	Increased FMD	12
Flavanol-rich cocoa drink		20 subjects with hypertension 14 days	Increased brachial artery diameter Increased insulin sensitivity	15
Cocoa powder with skimmed milk		42 high-risk adults 28 days	Lowered VLA-4, CD40 and CD36 in monocytes and P-selectin	33

Cocoa powder with skimmed milk	42 high-risk adults 28 days	Improved lipid metabolism	34
Flavanol-rich chocolate	20 adults with congestive heart failure	Increased FMD	35
Cocoa beverage	32 men and 20 post-menopausal women 6 weeks	Decreased BP	36
Sugared and sugar-free cocoa beverage	44 overweight adults 6 weeks	Improved endothelial function	37
High-flavanol cocoa beverage	32 post-menopausal hypercholesterolemic women 6 weeks	Increased brachial artery flow	38
Flavanol-rich chocolate bar and cocoa beverage	40 adults with coronary artery disease	Unaltered vascular function No change in vascular function	39
High-flavanol drink and exercise	49 overweight and obese adults	Increased FMD Reduced insulin resistance Reduced BP	40
6.3 g dark chocolate	44 upper-range pre-hypertensives	Increased NO Decreased BP	41

BP, blood pressure; FMD, flow-mediated dilation; HDL: High-density lipoprotein; LDL: Low-density lipoprotein; NO, Nitric oxide; NS: Nonsignificant; VLDL: Very-low-density lipoprotein.

trials with cocoa in a range of delivery systems. None of these studies monitored plasma polyphenolics and metabolites over time, so it is very difficult to accurately assess uptake. However, the range of delivery systems included supplements, flavanol-rich cocoa, dark chocolate, confections and cocoa-containing drinks. Many of the trials used single doses, whereas some spanned several weeks of consumption. Table 6.1 focuses on outcomes rather than bioavailability. Rusconi and Conti concluded that human intake of polyphenols is in the order of 1 g per day,[17] but there is enormous individual variation, with some Spanish diets reaching over 3 g per day.[18] The bioavailability is, in part, driven by the chemical structure of the polyphenol, including the basic structure (benzene or flavone derivatives), basic molecular size, the extent of glycosylation, conjugation with other phenolics, polymerization and complexing with other food ingredients such as proteins.[19] Catechins and procyanidins do not form glycosides, thus molecular size remains the most likely determining factor. The low bioavailability of cocoa polyphenolics is a result of a low C_{max} in plasma, a short half-life and the fact that they are rapidly excreted.[20] Monomeric procyanidins, on the other hand, are absorbed 10–100-times better than procyanidins and other polymeric forms. The major proportion of flavanols in cocoa products exist as oligomeric procyanidins.[21] Chocolate has been reported to contain predominantly (−)-epicatechin and (−)-catechin, with only small amounts of (+)-epicatechin and (+)-catechin. The (+) form appears to be much better absorbed than the (−) form.[22] The stereochemistry of the structures cause different degrees of hydrophobicity, which likely account for the differences in bioavailability (Figure 6.1).[23]

From the point of ingestion, cocoa components are altered throughout the digestive tract. In the mouth, the flavanols bind to proline-rich proteins, which contributes to the astringent response. Both flavanols and procyanidins survive gastric transport, with flavanol oligomers reaching the small intestine.[24] The flavanol oligomers are hydrolyzed to monomers or dimers in the acidic environment of the stomach, which, in turn, enhances their absorption in the small intestine. Flavanols exist predominantly in a conjugated form in mesenteric circulation and are absorbed from the jejunal lumen at the epithelial cell layer, where they are methylated and glucuronated.[25]

Epicatechin metabolites (glucuronide and 3-O-methylglucuronide) have been shown to cross the blood–brain barrier and potentially act at a cerebral level.[26] These flavonoids are thought to provide neuroprotection in aging.

Serafini et al. reported that the absorption of epicatechin from chocolate is dramatically lower when consumed with milk or as milk chocolate, which prevents absorption.[27] The hypothesis was that milk proteins bind to cocoa polyphenols, which, in turn, prevents their absorption in the gastrointestinal tract. Subsequent studies have failed to demonstrate a reduction in epicatechin bioavailability when cocoa was consumed with milk.[28] Individuals likely exhibit a wide variation in their ability to absorb polyphenols, which may explain the differences observed in absorption and ultimately the

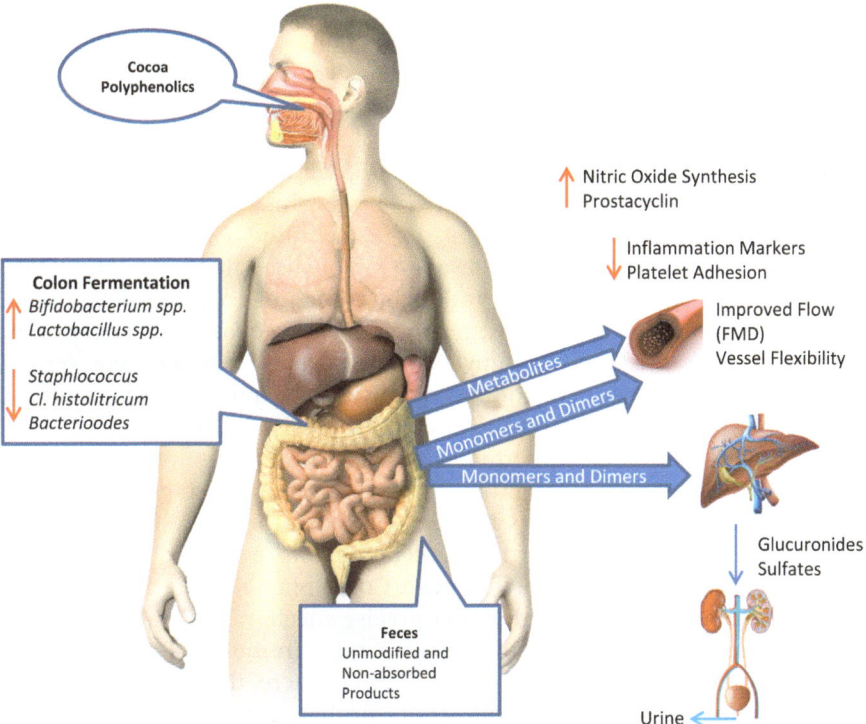

Figure 6.1 Cocoa components provide benefits throughout the gastrointestinal tract.

biological effects as summarized in Table 6.1. In a more recent study, catechin and epicatechin levels were measured in blood from adults who consumed 2 g of chocolate polyphenols plus sugar and cocoa butter in 200 ml of water with or without 2.45 g of milk solids in the drink. No differences were observed in the average polyphenol concentration in blood samples. Milk protein caused a slight increase in concentration at the early time points and a decrease at the later time points; however, no physiological significance was determined.[29]

In other recent work, the effect of milk has been measured in terms of colonic microbial metabolism of the nonabsorbed flavanol fraction that reaches the colon, which is then metabolized by the colonic microbiota into various phenolic acids. Of the 15 metabolites assessed, the excretion of nine different phenolic acids was affected by the intake of milk. The urinary concentration of 3,4-dihydroxyphenylacetic, protocatechuic, 4-hydroxybenzoic, 4-hydroxyhippuric, hippuric, caffeic and ferulic acids diminished after the intake of cocoa with milk, whereas urinary concentrations of vanillic and phenylacetic acids increased. It was concluded that milk partially affects the formation of microbial phenolic acids derived from the colonic degradation of procyanidins and other compounds present in cocoa powder.[30]

The human colon carcinoma cell line (Caco-2) model was used to investigate the permeability of two flavanol-C-glucosides and five dimeric and trimeric flavan-3-ols, namely procyanidins. The study measured the transport of pure single compounds at a range of concentrations from the apical to the basolateral side of the cell monolayer. The results showed slight permeations of procyanidins A2, B2 and B5 and cinnamtannin B1, as well as (−)-catechin-6-C-glucoside and (−)-catechin-8-C-glucoside of about 0.02–0.2% after 24 hours. No transport of cyanidin C1 was observed.[31]

Vitaglione investigated the bioavailability of cocoa flavanols and phenolic acids from a cocoa-nut cream and from cocoa-nut cream enriched with a 1.5% (w/w) cocoa polyphenol extract in both the free form and encapsulated with a gastric-resistant high-amylose maize starch.[32] It was reported that epicatechin was absorbed from cocoa-nut cream and that consumption of cocoa polyphenols increased the circulating levels of phenolic acids.

It is of paramount importance to understand the interrelationships between the host and the microbiome in order to understand nutrition and metabolism. The interactions between the host, the microbiome and food can have important implications in dietary habits, improved health and the potential delay or prevention of disease. Cocoa phenolics are associated with many health benefits; however, the precise mechanisms underlying these benefits are still being unveiled. Polyphenols are not well absorbed, thus they reach the lower gastrointestinal tract where they interact with the microbiome. In the microbiome, the polyphenolics are extensively metabolized by gut bacteria into a complex series of end-products. The resulting symbiotic relationship can influence the host and the distribution of dominant organisms in the microbiome.

The human lower digestive tract is populated by some 100 million bacterial cells. There is a wide variation of active communities within the microbiome that are influenced by initial inoculation in early life as well as by the lifestyle of the host.[33] The diversity of the organisms constituting the microbiome facilitates our capacity to consume and thrive on widely diverse diets.[34]

As substantial portions of cocoa polyphenolics are not absorbed, this raises the question of what happens when these compounds reach the lower gastrointestinal tract where they can be altered by gut microbiota. The bioavailability of flavonoids is influenced by intestinal metabolism, and the flavonoids that reach the lower gastrointestinal tract can be further modified by microbial deconjugation and degradation of the flavanols and flavanol glycosides. The initial investigation of the influence of microbiota was conducted using microbiota isolated from the cecal lumen of pigs. The study was aimed as assessing the microbial degradation of quercetin-3-O-beta-D-rutinoside 1, quercetin-3-O-beta-D-glucopyranoside 2, quercetin-4′-O-beta-D-glucopyranoside 3, quercetin-3-O-beta-D-galactopyranoside 4, quercetin-3-O-beta-D-rhamnopyranoside 5, quercetin-3-O-[alpha-l-dirhamnopyranosyl-(1 → 2)-(1 → 6)-beta-D-glucopyranoside 6, kaempferol-3-O-[alpha-l-dirhamnopyranosyl-(1 → 2)-(1 → 6)-beta-D-glucopyranoside 7, apigenin 8, apigenin-8-C-glucoside (vitexin)

9 and feruloyl-O-beta-D-glucopyranoside 10 (100 µM), representing flavonoids with different aglycones, sugar moieties and types of glycosidic bonds. The flavonol O-glycosides were almost completely metabolized by the intestinal microbiota within 4 hours depending on the sugar moiety and the type of glycosidic bond. The liberated aglycones were completely metabolized within 8 hours. Phenolic compounds such as 3,4-dihydroxyphenylacetic acid 12,4-hydroxyphenylacetic acid 13 and phloroglucinol were the primary degradation products.[35]

Investigating changes in flavan-3-ols was performed in a similar study with pigs in which it was observed that the flavan-3-ols were almost completely metabolized by the intestinal microbiota within 4–8 hours. No difference was observed for catechin enantiomers. In addition to monomeric flavonoids, procyanidins are also metabolized by the intestinal microbiota. The resulting hydroxylated phenolcarboxylic acids were similar for all of the tested polyphenolics.[36] The small hydroxylated phenolic degradation products have been shown to be antioxidants and thus could account for some of the antioxidant activity observed in Table 6.1.

Polyphenols exert a wide range of beneficial health effects. Most work has focused on the effects of microbiota on polyphenols. Polyphenol-rich environments exert effects on the gut microbiota distribution of dominant organisms. When microbiota were exposed to 150 or 1000 mg l^{-1} of (+)-catechin, a greater growth of the beneficial bacteria *Eubacterium rectale – Clostridium coccoides*, and at the lowest concentration, *Lactobacillus* spp. and *Bifidobacterium* spp. growth were stimulated. Addition of (+)-catechin inoculation at 150 mg l^{-1} induced *Escherichia coli*, whereas the addition of a 1000-mg l^{-1} concentration decreased *Clostridium histolyticum*. Addition of 150 mg/l of (−)-epicatechin stimulated the growth of *E. rectale – C. coccoides*.[37] When female Wistar rats were fed with 10% (w/w) cocoa in their diet, significantly lower levels of *Bacteroides*, *Staphylococcus* and *Clostridium* genera were found at the end of the intervention. The reduction in *Clostridium* spp. correlated with weight loss.[38]

In a human intervention study, subjects consumed cocoa containing either 29 mg flavanol or 494 mg flavanol in a beverage for 4 weeks. The high-flavanol group exhibited a significant increase in bifidobacteria after the 4-week intervention. At the same time, there was a significant decrease in *C. histolyticum*, a group that includes the human pathogen *Clostridium perfringens*. It was suggested that the change in microflora was responsible for a reduction in C-reactive protein (CRP).[39] Stoupi *et al*. determined the interactions of (−)-epicatechin (2,3-*cis* stereochemistry) and its dimer of pure procyanidin B2 with microbiota from the human microbiome.[40] Less than 10% of procyanidin B2 was converted to epicatechin by scission of the interflavan bond. Five phenolic acid catabolites from (−)-epicatechin were unique to procyanidin B2 and ten phenolic acid catabolites were common to both substrates. The dominant catabolites were 5-(3′-hydroxy phenyl) valeric acid, 3-(3′-hydroxyphenyl) propionic acid and phenyl acetic acid. The procyanidin dimer was degraded more rapidly than the (−)-epicatechin.

Small amounts of catabolites of procyanidin were found to retain the flavanol A-ring and the C4–C8 interflavan bond.

Modern instrumentation provides tools to measure metabolic products in considerable detail. In one key study, subjects received 40 g day^{-1} of cocoa powder with 500 ml of skimmed milk (cocoa with skimmed milk intervention) or 500 ml day^{-1} of skimmed milk (skimmed milk intervention) for 4-weeks. Urine (24 hour) samples were collected at baseline and after each intervention and were analyzed by HPLC-hybrid quadrupole TOF in negative and positive ionization modes followed by multivariate analysis. The results show major differences between the cocoa with skimmed milk intervention and skimmed milk intervention and baseline periods. Thirty-nine compounds were associated with cocoa intake, including alkaloid metabolites, polyphenol host and gut microbial metabolites (hydroxyphenylvalerolactones and hydroxyphenylvaleric acids), diketopiperazines and N-phenylpropenoyl-L-amino acids.[41]

6.4 Diabetes, Metabolic Syndrome and Glucose

Several epidemiological studies have demonstrated that polyphenols from cocoa, green tea and citrus are associated with metabolic and/or cardiovascular morbidity and mortality. Metabolic syndrome is defined by the International Diabetes Foundation as a group of conditions that includes central obesity and at least two of the following: low HDL cholesterol, high triglycerides, high systolic blood pressure or diastolic blood pressure and increased fasting glucose.[42] Approximately 34% of the US population has metabolic syndrome according to the 2003–2006 National Health and Nutrition Examination Survey.[43] Metabolic syndrome patients are more prone to diabetes, cardiovascular disease and certain cancers.[44] Chronic inflammation commonly occurs in conjunction with metabolic syndrome and represents a link between obesity and obesity-related pathologies, including dyslipidemia, hypertension and insulin resistance. Visceral fat deposition becomes dysfunctional, resulting in increased chronic inflammation, which, in turn, leads to the metabolic syndrome-associated pathological consequences.[45] Adipose tissue acts as an endocrine organ mediating pro-inflammatory cytokines, chemokines and adipokines including tumor necrosis factor-alpha, interleukins, monocyte chemo attractant protein-1 and adiponectin. Collectively, these chemokines are involved with insulin resistance and the pathophysiology of dyslipidemia and hypertension associated with metabolic syndrome.[46]

Gu and Lambert have reviewed the current literature that suggests that the initiation of adipose tissue inflammation begins with excessive intake of macronutrients, thereby leading to hypertrophy or necrosis of adipocytes.[47] These changes can be modulated by cocoa.[48] The changes in the adipocytes induce the recruitment, proliferation and activation of macrophages.[49] Macrophages are recruited to the adipose tissue as a consequence of excess energy and the increase in arachidonic acid levels from the metabolism of high levels of omega-6 fatty acids in the diet.[50] Stressed adipocytes recruit macrophages, which results in the increased production of inflammatory cytokines. Obesity disrupts the dynamic role of the adipocytes in energy homeostasis, thereby resulting in inflammation and changes in leptin and adiponectin signaling. Obesity causes secondary changes, including disruption of insulin signaling and lipid deregulation.[51] The inflammation induced in the high-fat diet requires 12/15-lipoxygenase, which has also been shown to induce insulin resistance in mice.[52] The eicosanoids produced in these reactions constitute a diverse class of bioactive lipid mediators produced from arachidonic acid. These eicosanoids play critical roles in cell signaling and the inflammatory aspects of numerous diseases, including diabetes. Eicosanoid metabolites are produced in RAW264.7 macrophage cells in response to Toll-like receptor 4 signaling and have been demonstrated to induce transcripts coding for the enzymes involved in the eicosanoid metabolite biosynthetic pathways. An integrated genomic, proteomic and metabolomic analysis of the eicosanoid biochemical pathway has been reported by Sabido et al.[53]

Insulin resistance arises from the inability of insulin to regulate carbohydrate metabolism in peripheral tissues. Human population studies and animal research have established correlative as well as causative links between chronic inflammation and insulin resistance. Xu et al. demonstrated that many inflammation and macrophage-specific genes are dramatically upregulated in white adipose tissue in mouse models of genetic and high-fat diet-induced obesity.[54] The upregulation is progressively increased in the white adipose of obese mice and precedes increases in circulating insulin levels. These data suggest that macrophages in white adipose tissue play an active role in morbid obesity and that macrophage-related inflammatory activities may contribute to the pathogenesis of obesity-induced insulin resistance.

Cocoa and cocoa products have the potential to help control metabolic syndrome in multiple ways, including reducing obesity and controlling inflammation, dyslipidemia, hypertension and insulin resistance.[55] Growing evidence from both observational and experimental studies suggests that the consumption of cocoa and cocoa products may reduce inflammatory biomarkers including serum levels of CRP and pro-inflammatory cytokines.[56] In the only epidemiological study of cocoa consumption on inflammation, a J-shaped relationship was found between dark chocolate consumption and serum CRP, with moderate consumers ($\leq$20 g of dark chocolate every third day) having significantly lower serum CRP concentrations than either non-consumers or those high consuming $\geq$20 g chocolate per day.[57]

Evidence increasingly supports a beneficial role of cocoa and cocoa products in the prevention and treatment of metabolic syndrome. Gu and Lambert have reviewed the clinical, animal and cell culture studies that have explored the potential pathways by which cocoa may reduce this disorder.[58] The benefits can be explained by multiple proposed mechanisms, many of which are discussed elsewhere in this chapter. In addition to their antioxidant activity, the flavanols appear to help regulate cytokine secretion and expression, modulate eicosanoid metabolism, inhibit NF-kappaB activation and modulate nitric oxide-mediated effects. Cocoa flavanols have potential anti-inflammatory actions *in vitro* and *in vivo*; however, current data drawn from human subjects remain inconclusive. Several animal studies have demonstrated that cocoa has a protective role in obesity and diabetes, though results regarding metabolic syndrome-related inflammation are limited. *In vitro* studies provide some mechanistic context, but they are inconclusive. Although the available data seem to support the anti-inflammatory effects of cocoa, further studies are needed with regard to the dose–response relationship, as well as to discern the underlying mechanisms of action.

Grassi tested the effects of flavanol-rich dark chocolate on endothelial function, insulin sensitivity, beta-cell function and blood pressure in hypertensive patients with impaired glucose tolerance (IGT).[59] The study included 19 hypertensive subjects with IGT. Flavanol-rich dark chocolate ameliorated insulin sensitivity and beta-cell function, decreased blood pressure and increased flow-mediated dilation in IGT hypertensive patients. The results suggest that flavanol-rich, low-energy cocoa food products may have a positive impact on cardiovascular disease risk factors.

Hanhineva *et al.* reviewed the influence of polyphenols on carbohydrate metabolism.[60] In an extensive review, they reported the multiple effects of polyphenols on carbohydrate metabolism, including slowed glucose uptake and increased insulin secretion. The possible mechanisms include inhibition of carbohydrate digestion and glucose absorption in the intestine, stimulation of insulin secretion from the pancreatic beta-cells, modulation of glucose release from the liver, activation of insulin receptors and glucose uptake in the insulin-sensitive tissues and modulation of intracellular signaling pathways and gene expression. Cordero-Herrera *et al.* studied the modulation of insulin signaling by (−)-epicatechin and a cocoa phenolic extract on hepatic HepG2 cells.[61] They analyzed the key proteins of the insulin pathways, including insulin receptor, insulin receptor substrates 1 and 2, PI3K/AKT and 5-AMP-activated protein kinase, the levels of the glucose transporter GLUT-2 and the hepatic glucose production. Their data demonstrated that (−)-epicatechin and a cocoa phenolic extract strengthened insulin signaling by activating key proteins of that pathway and regulating glucose production through an 5-AMP-activated protein kinase and AMPK modulation in HepG2 cells.

Martin *et al.* demonstrated that cocoa phenolics protected beta-cells against oxidative stress in a model system using t-butylhydroxyperoxide (y-BOOH) to stress Ins-1E pancreatic beta-cells.[62] Pre-treatment of cells with cocoa phenolics reduced the t-BOOH-induced reactive oxygen species and carbonyl groups. Antioxidant defenses were returned to adequate levels after treatment with cocoa extracts. Ins-1E pancreatic beta-cells treated with cocoa powder extract showed a remarkable recovery of cell viability damaged by t-BOOH, indicating that the integrity of surviving machineries in cocoa powder extract (CPE)-treated cells was notably protected against the oxidative insult.

6.5 Cocoa and the Prebiotic Environment

Significant portions of chocolate reach the lower gastrointestinal tract where they are metabolized by the microflora in the microbiome. Cocoa products are metabolized by the microbiota, but they also have the capacity to alter the distribution of dominant bacteria in the microbiome, thus altering the effects on the host. The influence of changes in diet suggests that dietary modification could influence long-term health outcomes.[63] In one clinical trial, 30 humans who were classified as either high or low anxiety consumed 40 g day^{-1} of dark chocolate/day for 14 days. Urine and blood plasma were measured for global changes in metabolites at the beginning of the study, after 7 days and after 14 days. High-anxiety subjects showed a distinct metabolic profile indicative of different energy homeostasis (lactate, citrate, succinate, *trans*-aconitate, urea and proline), hormonal metabolism (adrenaline, DOPA and 3-methoxy-tyrosine) and gut microbial activity (methylamines, *p*-cresol sulfate and hippurate). After 2 weeks of consumption of dark chocolate, there was a significant reduction in urinary excretion of the stress hormone cortisol and catecholamines and partially normalized stress-related differences in energy metabolism (glycine, citrate, *trans*-aconitate, proline and beta-alanine) and gut microbial activities (hippurate and *p*-cresol sulfate). It was concluded that 2 weeks is a sufficient period of time to modify human gut microbial metabolism.

Flavanols are poorly absorbed in the small intestine and, as a result, the majority of consumed flavanols are metabolized by the gut microbiota. Tzounis *et al.* conducted a randomized, double-blind, crossover controlled intervention study with human volunteers who were randomly assigned to either a high-cocoa flavanol group (494 mg of cocoa flavanols/day) or a low-cocoa flavanol group (23 mg of cocoa flavanols/day) for 4 weeks.[64] At the end of each leg of the study, fecal samples were analyzed and bacterial numbers were measured in addition to other biochemical and physiological markers. When subjects consumed the high-flavanol drink, a significant increase was found in the bifidobacterial ($P<0.01$) and lactobacilli ($P<0.001$) populations and significant decreases in Clostridia counts ($P<0.001$). Subjects on high-flavanol diets also exhibited significant reductions in plasma triacylglycerol ($P<0.05$) and CRP ($P<0.05$) concentrations. Furthermore, changes in CRP concentrations were linked to changes in lactobacilli counts ($P<0.05$, $R^2 = -0.33$ for the model). The results imply that potential prebiotic effects occur in diets containing flavanol-rich foods. When water-insoluble cocoa fractions are predigested in an *in vitro* system emulating human digestion and the non-digested residue is fermented by colonic bacteria, there is a significant increase in bifidobacteria and lactobacilli. In the same system, the flavanols were converted to high concentrations of phenolic acids, particularly 3-hydroxyphenylpropionic acid.[65]

The monomeric (+)-catechin and (−)-epicatechin in cocoa have been shown to promote or inhibit the growth of selected intestinal bacteria. Tzounis *et al.* found that (+)-catechin induced an inhibitory effect in the growth of the *C. histolyticum* group in batch cultures, and that both (+)-catechin and (−)-epicatechin enhanced the growth rate of the beneficial bacteria group, *E. rectale–C. coccoides*.[66] They also observed increases in both *Lactobacillus* spp. and *Bifidobacterium* spp. following (+)-catechin exposure. There was a small increase in *E. coli* after (+)-catechin incubation.[67] Animal studies have shown similar results: when young rats were fed diets containing cocoa polyphenol-enriched powders for 6 weeks or older rats were fed for 4 weeks, a significant decrease in the proportions of *Bacteroides*, *Staphylococcus* genus and the *C. histolyticum* sub-group were observed.[68] The effect of high-flavanol cocoa on gut microbiota has been assessed in a human intervention study. After 4 weeks of ingestion of a high-cocoa flavanol beverage containing 494 mg of flavanols, the growth of *Lactobacillus* spp. and *Bifidobacterium* spp. increased compared to diets with beverages containing little flavanol.[69] These findings suggest that cocoa polyphenols can influence the growth of select gut microbiota and, therefore, potentially impact the immune system.

Cocoa polyphenols are metabolized in the colon, generating various phenolic acids, including phenylpropionic, phenylacetic and benzoic acid derivatives.[70] These smaller metabolites are more readily absorbed into the bloodstream from the colon, thereby providing additional sources of potentially bioactive compounds.[71] Once absorbed, the microbial metabolites from flavanols are mainly metabolized in the liver by phase-II enzymes as hepatic-conjugated derivatives that are subsequently eliminated in urine.[72] In particular, the presence of 5-(3′,4′,5′-trihydroxyphenyl)-gamma-valerolactone and 5-(3′,4′-dihydroxyphenyl)-gamma-valerolactone in the urine are considered to be potential biomarkers of flavan-3-ol consumption after cocoa intake.[73]

6.6 Cocoa, Blood Lipids and Diabetes

Cocoa ingestion influences the immune system, including the innate inflammatory response and the systemic and intestinal adaptive immune response. When any food is ingested, the level and activity of epithelial cells, immune cells and bacteria are affected. The gut-associated lymphoid tissue is the most extensive part of the human immune system. As such, it must distinguish between pathogens, food allergens and innocuous antigens in food.

The relationship between fat consumption, the influence on blood lipid profiles and cardiovascular disease continues to be controversial. The question arises as to whether there is a cause and effect or an association in regard to atherosclerosis diseases. Dietary lipids affect postprandial circulating lipids, particularly HDL and LDL cholesterol levels and total serum triacylglycerols. The greatest variability in diets is reflected in the ratios of saturated, monounsaturated and polyunsaturated fatty acids. The general view is that saturated fatty acids tend to raise serum cholesterol and that polyunsaturated fatty acids reduce serum cholesterol, whereas monounsaturated fatty acids remain neutral. Strong evidence suggests that replacing saturated fatty acids with monounsaturated fatty acids reduces total serum cholesterol.[74] Good agreement exists supporting the view that saturated fatty acids increase serum cholesterol; however, there is no convincing evidence that high consumption of saturated fatty acids is correlated with coronary heart disease or associated side effects.[75]

It seems fair to conclude that moderate consumption of cocoa butter is not likely to be a health risk, despite the notable levels of saturated fat. Based on the saturated fatty acid content of cocoa butter, one would anticipate that it would increase LDL cholesterol, whereas with oleic acid, one would expect a decrease in LDL cholesterol. Overall, the consumption of chocolate appears to reduce LDL cholesterol, with the effect being attributed to the polyphenols. Other polyphenols including tea catechins, genistein, diadzein, naringenin, hesperetin and red wine polyphenols have been shown to reduce LDL cholesterol.[76] These polyphenols inhibit cholesterol absorption from the digestive tract, inhibit LDL biosynthesis, suppress hepatic secretion of apolipoprotein B-100 and increase LDL receptors in the liver.[77]

Several studies from 2001 to the present have been designed to assess the influence of chocolate on blood lipids. The studies consistently show that long-term exposure to chocolate, particularly polyphenol-rich chocolate, results in enhanced HDL cholesterol levels.[78] When increasing levels of polyphenols were consumed, there was a concomitant increase in HDL cholesterol.[79] This finding leads to the conclusion that the polyphenols in chocolate are primarily responsible for these effects.

Some of the most profound health benefits of cocoa are associated with effects on metabolic syndrome, diabetes and the control of the side effects of these conditions. Latif reviewed the broad benefits of cocoa.[80] Recent findings on the biologically active phenolic compounds in cocoa has stimulated further research into its effects on aging, oxidative stress, blood pressure regulation and atherosclerosis, all of which are related to diabetes. Because of the earlier work in phenolic antioxidant activity, cocoa drew particular interest due to its antioxidant potential. Cocoa is one of the richest food sources of a range of phenolic compounds with antioxidant potential. Many studies have reported contradictory results and methodological issues have been raised, thereby making it difficult to assess the true extent of cocoa's benefits.

Early epidemiological evidence suggesting beneficial effects of chocolate were derived from a study of the Kuna Indian population living on the islands of Panama. This population exhibits a low prevalence of atherosclerosis, Type 2 diabetes and hypertension. This group notably consumes a concentrated cocoa beverage on a daily basis. After they moved to urban areas, their cocoa consumption significantly dropped and their lowered risk of atherosclerosis, Type 2 diabetes and hypertension disappeared.[81] The Kuna's drink flavanol-rich cocoa as their main beverage, contributing more than 900 mg day^{-1} of flavanoids and thus routinely consume a more flavonoid-rich diet than any other known population. Deaths directly related to diabetes mellitus were much more common in the mainland (24.1 ± 0.74) than in the San Blas (6.6 ± 1.94).[82] A longitudinal study found cocoa intake to be inversely related to blood pressure. After multivariate adjustment, mean systolic blood pressure was 3.8 mmHg lower in the highest cocoa intake group as compared with the lowest intake group. Similar results were found in a perspective analysis that associated a higher cocoa intake with a reduction in cardiovascular disease and in all-cause mortality.[83]

Epidemiological studies have reported that chocolate consumption is inversely correlated with cardiovascular disease. Epidemiological studies citing the relationship between plant polyphenols and cardiovascular risk were recently reviewed by Ros and Hu.[84] Reductions in cardiovascular incidents were demonstrated in a meta-analysis of five prospective studies including 60 455 subjects with a history of coronary heart disease.[85] Larson *et al.* reported a reduction in the incidence of stroke in five other prospective studies based upon patients with histories of coronary heart disease.[86] Improvements in lipid profiles were observed in meta-analyses performed on ten studies by Tokede *et al.* and on 21 studies by Hooper *et al.*[87] Zomer reported dark chocolate consumption to be beneficial for reducing cardiovascular risk in subjects with metabolic syndrome.[88] Hooper *et al.* also reported improved glycemic control in both healthy subjects and subjects who were at risk of cardiovascular disease. Van Dam *et al.* reported that cohort studies suggested that cocoa flavan-3-ols showed benefits by lowering cardiovascular risk and improving insulin sensitivity.[89] In a broader study, Hanhineva *et al.* summarized the positive effects of polyphenols on glucose homeostasis in a large number of *in vitro* and animal models, which are supported by epidemiological evidence on polyphenol-rich diets.[90] They confirmed that polyphenol consumption may improve insulin resistance in metabolic syndrome and eventually Type 2 diabetes; however, more human trials using well-defined diets with clinically relevant end-points together with systems biology profiling technologies are needed.

6.7 Cocoa, Anti-oxidation and Inflammation

As discussed elsewhere in this volume, polyphenolics as a broad group exhibit anti-oxidant and anti-inflammatory activities *in vitro* and *in vivo*. Sies *et al.* have extensively reviewed the role of antioxidants in post-prandial oxidative stress induced by triglyceridemia and/or hyperglycemia.[91] Excess nutrient loads result in the production of reactive oxygen species and the development of oxidized LDLs. These reactive oxygen species lead to endothelial dysfunction, a frequent consequence of diabetes. Cocoa is a rich source of flavanols and other polyphenol antioxidants that can help attenuate the development of these effects. Flavonoids, including catechin, epicatechin and procyanidins, are the predominant contributors to cocoa's antioxidant activity. The tricyclic structure of the flavonoids provides the basis for their antioxidant effects, including the scavenging reactive oxygen species, chelation of Fe^{2+} and Cu^+ and upregulation of antioxidant defenses. Epicatechin in cocoa is primarily responsible for the impact on vascular endothelium through the increased production of nitric oxide. The anti-inflammatory effects of cocoa polyphenols modulate the activity of NF-kappaB. Cocoa consumption can stimulate changes in redox-sensitive signaling pathways involved in gene expression and the immune response.[92]

Pancreatic beta-cells are prone to oxidative stress, compromising their ability to generate insulin. Martin investigated the chemo-protective effect of a cocoa phenolic extract against a model for oxidative stress induced by t-BOOH.[93] The pancreatic Ins-1E cells treated with 5–20 µg ml^{-1} of cocoa phenolic extract exhibited no cell damage. Treatment of the cells with 50 µM of t-BOOH for 2 hours increased reactive oxygen species and carbonyl group content and reduced the glutathione level. Pre-treatment of cells with cocoa phenolic extract prevented the t-BOOH-induced reactive oxygen species production and reduced carbonyl group formation. Pancreatic Ins-1E cells treated with cocoa phenolics showed a remarkable recovery of cell viability after damage by t-BOOH. More recently, Martin et al. have shown that pretreatment of pancreatic cells with epicatechin prevented the t-BOOH-induced reactive oxygen species generation, carbonyl group formation, p-JNK expression and cell death, and also recovered insulin secretion.[94]

The cardiovascular benefits of cocoa for diabetic and non-diabetic subjects were initially attributed to the antioxidant and anti-inflammatory activities of the polyphenols. In vitro as well as cell culture data demonstrated that cocoa polyphenols exhibit anti-oxidant, anti-inflammatory and anti-atherogenic activity. The targets appear to be NF-kappaB, endothelial nitric oxide synthase and angiotensin-converting enzyme. The cocoa polyphenol concentrations utilized in many cell culture studies are orders of magnitude higher than levels reached in vivo. Plasma concentrations in vivo are in the nanomolar range.[95] The health effects derived from cocoa polyphenols have been attributed to the activity of the flavan-3-ols. The bioavailability of the flavan-3-ols is dependent on their metabolism and that of the procyanidins that reach the colon. Khan et al. studied the bioavailability of cocoa polyphenols and inflammatory biomarkers of atherosclerosis in patients at high risk of cardiovascular disease.[96] In a review, Lecour and Lamont suggested that polyphenols can exert their cardioprotective effects through the activation of several powerful prosurvival cellular pathways that involve metabolic intermediates, microRNAs, sirtuins and mediators of the recently described reperfusion injury salvage kinases and survivor-activating factor enhancement pathways.[97] Several studies have demonstrated that the consumption of flavanol-rich chocolate and cocoa products can decrease blood pressure and promote vasodilation. Overwhelming evidence indicates that the flavanols in cocoa and chocolate increase the levels of the vasodilator nitric oxide, which contributes to the vascular effects of these foods.

Of particular concern for diabetics are studies that show that endothelial function is impaired during hyperglycemia. Dark chocolate increases flow-mediated dilation in healthy and hypertensive subjects who exhibited normal and impaired intolerance. Grassi et al. investigated the influence of flavanol-rich dark chocolate administration on (1) flow-mediated dilation and wave reflections and (2) blood pressure, endothelin-1 and oxidative stress, before and after an oral glucose tolerance test.[98] They compared the influence of either white chocolate or dark chocolate ingestion on flow-mediated dilation, wave reflections, endothelin-1 and 8-iso-PGF(2α). After white chocolate ingestion, flow-mediated dilation was reduced after the oral glucose tolerance test from 7.88 ± 0.68 to 6.07 ± 0.76 ($P = 0.027$) and 6.74 ± 0.51 ($P = 0.046$) at 1 and 2 hours after the glucose load, respectively. Similarly, after white chocolate but not after dark chocolate, wave reflections, blood pressure and endothelin-1 and 8-iso-PGF(2α) levels increased after the glucose tolerance test. The oral glucose tolerance test causes acute, transient impairment of endothelial function and oxidative stress, which is attenuated by flavanol-rich dark chocolate. These results suggest that cocoa flavanols may contribute to vascular health by reducing the postprandial impairment of arterial function associated with the pathogenesis of atherosclerosis.

In a 16-week intervention study with Type 2 diabetics, Mellor *et al.* reported that consumption of high-polyphenol chocolate was effective at improving the atherosclerotic cholesterol profile in patients with diabetes by increasing HDL cholesterol and improving the cholesterol:HDL ratio without affecting weight, inflammatory markers, insulin resistance or glycemic control.[99] Mellor *et al.* further investigated the effects of high-polyphenol chocolate upon endothelial function and oxidative stress in Type 2 diabetes mellitus patients during acute transient hyperglycemia.[100] A 75-g oral glucose load was used to induce hyperglycemia, which was administered to participants 60 minutes after they had ingested either low- (control) or high-polyphenol chocolate. Endothelial function was assessed by both functional (reactive hyperemia peripheral artery tonometry, EndoPAT-2000) and serum markers (including intercellular adhesion molecule 1, P-selectin and P-selectin glycoprotein ligand 1). Urinary 15-F2t-isoprostane adjusted for creatinine was monitored as an oxidative stress marker. Measurements were made at baseline and 2 hours post-ingestion of the glucose load. Prior consumption of high-polyphenol chocolate before a glucose load improved endothelial function (1.7 ± 0.1 *vs.* $2.3 \pm 0.1\%$, $P = 0.01$), whereas prior consumption of control chocolate resulted in a significant increase in intercellular adhesion molecule 1 (321.1 ± 7.6 *vs.* 373.6 ± 10.5 ng ml^{-1}, $P = 0.04$) and 15-F2t-isoprostane (116.8 ± 5.7 *vs.* 207.1 ± 5.7 mg mol^{-1}, $P = 0.02$). The authors concluded that high-polyphenol chocolate protected against acute hyperglycemia-induced endothelial dysfunction and oxidative stress in individuals with Type 2 diabetes mellitus.

In a recent review, van Dam *et al.* discussed the effects of flavonoid intake on the development of Type 2 diabetes and cardiovascular diseases.[101] The availability of comprehensive databases that include polyphenols in a wide range of foods has facilitated the evaluation of more comprehensive assessments of flavonoid consumption in epidemiological studies. Results from both cohort studies and randomized trials suggest that anthocyanidins from berries and flavan-3-ols from green tea and cocoa may lower the risk of Type 2 diabetes and cardiovascular diseases. Meta-analyses of randomized trials indicate the beneficial effects of flavan-3-ol-rich cocoa on endothelial function and insulin sensitivity.

Previous studies have reported the health benefits of cocoa polyphenols in reducing the risk of cardiovascular diseases, and cell culture studies have suggested that cocoa polyphenolics may help prevent oxidative damage to pancreatic cells. Still, few reports have shown direct effects of cocoa in preventing the development of diabetes mellitus. Ruzaidi et al. tested cocoa liquor and cocoa powder streptozotocin (STZ)-induced diabetic rats.[102] Male Sprague–Dawley rats were divided into diabetic control, diabetic cocoa extract (CE) and diabetic glibenclamide groups. Cocoa phenolic-rich extracts normalized the body weight loss caused by STZ. In the 20-mg CE-pretreated group, a 143% increase in plasma glucose levels was noted, compared with a 226% increase in diabetic control rats. The results suggest that cocoa phenolic extracts from roasted cocoa beans could prevent the development of diabetes induced by STZ injection in rats.

In clinical trials, Grassi et al. observed a reduction in insulin resistance after the ingestion of flavanol-rich chocolate in both healthy and hypertensive subjects.[103] After chocolate ingestion, an increase in insulin sensitivity and a decrease in blood pressure were found. Other work has shown that flavanol-rich chocolate ingestion produces a positive effect on oral glucose tolerance tests in hypertensive adults.[104]

Flavanols potentially protect against insulin resistance. Stote et al. investigated the influence of cocoa flavanols over a range of consumption levels on improving biomarkers of glucose regulation, inflammation and hemostasis in obese adults who were at risk for insulin resistance.[105] Green tea and cocoa flavanols were compared for their ability to modulate these biomarkers. In a randomized crossover design, 20 adults consumed a controlled diet for 5 days along with four cocoa beverages containing from 30–900 mg of flavanol per day, or tea matched to a cocoa beverage for monomeric flavanol content. Cocoa beverages produced no significant changes in glucose, insulin or total area under the concentration–time curve (AUC) for glucose or total insulin AUC. As the dose of cocoa flavanols increased, total 8-isoprostane concentrations were lowered (linear contrast, $P=0.02$), as were CRP concentrations (linear contrast, $P=0.01$). The relationship between cocoa flavanol levels and interleukin-6 (IL-6) concentrations was quadratic. There were no significant effects on measured indices of glucose regulation, nor on those of total 8-isoprostane, CRP and IL-6 concentrations when cocoa and green tea were compared. It was concluded that the short-term intake of cocoa and green tea flavanols does not appear to improve glucose metabolism, but it does affect selected markers of one or more measures of oxidative stress, inflammation or hemostasis in obese adults who are at risk for insulin resistance.

Studies on the influence of gut microbiota on poorly digested food components provide new understanding of the potential metabolic effects of the microbial decomposition products. Polyphenols are purported to exert a wide range of health benefits. The gut microbiota can modify polyphenolic compounds, resulting in smaller molecules such as phenolic acids, which are more easily absorbed. Also, the phenolic compounds can modify the microbiota distribution in the gut. The direct impact of these interactions in diabetes remains to be explained. Etxeberria *et al.* have recently reviewed the impact of polyphenols on gut microbiota composition, and Moco *et al.* have reviewed the metabolomics of food polyphenolics by the gut microbiome.[106]

In conclusion, it appears that cocoa polyphenolic compounds offer cellular protection against inflammation and oxidative stress. There is clear evidence that cocoa offers protection against hypertension by improving nitric oxide status. However, cocoa's direct impact on glucose metabolism and insulin resistance remains in question. It is also possible that metabolites from the colonic fermentation of polyphenols will provide greater insight into the protective nature of cocoa polyphenolics.

References

1. S. Mazor Jolić, *et al.*, *Int. J. Food Sci. Technol.*, 2011, **46**, 1793–1800.
2. L. Stahl, *et al.*, *J. Food Sci.*, 2009, **74**, C456–C461.
3. K. B. Miller, *et al.*, *J. Agric. Food Chem.*, 2006, **54**, 4062–4068.
4. W. J. Hurst, M. J. Payne, K. B. Miller and D. A. Stuart, *J. Agric. Food Chem.*, 2009, **57**, 9547–9550.
5. M. Lacroix, in *Polyphenols in cocoa: influence of processes on their composition and biological activities*, Nova Science Publishers, Inc., 2010, pp. 183–197.
6. M. Richelle, I. Tavazzi, M. Enslen and E. A. Offord, *Eur. J. Clin. Nutr.*, 1999, **53**, 22–26.
7. J. P. E. Spencer, *J. Nutr.*, 2003, **133**, 3255S–3261S; K. A. Cooper, J. L. Donovan, A. L. Waterhouse and G. Williamson, *Br. J. Nutr.*, 2008, **99**, 1–11.
8. J. P. Spencer, *et al.*, *Free Radical Biol. Med.*, 2001, **31**, 1139–1146.
9. J. P. E. Spencer, *J. Nutr.*, 2003, **133**, 3255S–3261S; M. P. Gonthier, *et al.*, *Free Radical Biol. Med.*, 2003, **35**, 837–844.
10. S. Baba, *et al.*, *Free Radical Biol. Med.*, 2000, **33**, 635–641.
11. J. L. Donovan, *et al.*, *J. Nutr.*, 1999, **129**, 1662–1668; J. R. Bell, *et al.*, *Am. J. Clin. Nutr.*, 2000, **71**, 103–108.
12. A. Serra, *et al.*, *Eur. J. Nutr.*, 2013, **52**, 1029–1038; V. Fogliano, *et al.*, *Mol. Nutr. Food Res.*, 2011, **55**, S44–S55.
13. L. Y. Rios, *et al.*, *Am. J. Clin. Nutr.*, 2003, **77**, 912–918.
14. S. Baba, *et al.*, *J. Agric. Food Chem.*, 2001, **49**, 6050–6056; R. R. Holt, *et al.*, *Am. J. Clin. Nutr.*, 2002, **76**, 798–804; Q. Y. Zhu, *et al.*, *Clin. Dev. Immunol.*, 2005, **12**, 27–34; C. L. Keen, *et al.*, *Am. J. Clin. Nutr.*, 2005, **81**, 298S–303S.
15. M.-P Gonthier, *et al.*, *Free Radical Biol. Med.*, 2003, **35**, 837–844.
16. C. L. Keen, *et al.*, *Am. J. Clin. Nutr.*, 2005, **81**, 298S–303S.

17. M. Rusconi and A. Conti, *Pharmacol. Res.*, 2010, **61**, 5–13.
18. F. Saura-Calixto and I. Goni, *Crit. Rev. Food Sci. Nutr.*, 2009, **49**, 145–52.
19. L. Bravo, *Nutr. Rev.*, 1998, **56**, 317–333.
20. C. Manach, A. Mazur and A. Scalbert, *Curr. Opin. Lipidol.*, 2005, **16**(1), 77–84.
21. J. F. Hammerstone, *et al.*, *J. Agric. Food Chem.*, 1999, **47**, 490–496.
22. J. L. Donovan, *et al.*, *Free Radical Res.*, 2006, **40**, 1029–1034.
23. C. L. Keen, *et al.*, *Am. J. Clin. Nutr.*, 2005, **81**, 298S–303S.
24. L. Y. Rios, *et al.*, *Am. J. Clin. Nutr.*, 2003, **77**, 912–918.
25. J. L. Donovan, *et al.*, *Free Radical Res.*, 2006, **40**, 1029–1034; J. P. Spencer, *et al.*, *Free Radical Biol. Med.*, 2001, **31**, 1139–1146.
26. A. M. N. Faria, *Nutr. Aging*, 2012, **1**, 89–97.
27. M. Serafini, *et al.*, *Nature*, 2003, **424**, 1013.
28. E. A.-L. C. Roura, *et al.*, *Nutr. Metab.*, 2007, **51**, 493–498.
29. J. B. Keogh, J. McInerney and P. M. Clifton, *J. Food Sci.*, 2007, **72**, S230–S233.
30. M. Urpi-Sarda, *et al.*, *J. Agric. Food Chem.*, 2009, **57**, 10134–10142.
31. S. Hemmersbach, *et al.*, *J. Agric. Food Chem.*, 2013, **61**, 7932–7940.
32. P. Vitaglione, *Br. J. Nutr.*, 2013, **109**, 1832–1843.
33. J. Qin, *et al.*, *Nature*, 2010, **464**, 59–65; S. Moco, F.-P. J. Martin and S. Rezzi, *J. Proteome Res.*, 2012, **11**, 4781–4790.
34. B. D. Muegge, *et al.*, *Science*, 2011, **332**, 970–974.
35. E.-M. Hein, *et al.*, *J. Agric. Food Chem.*, 2008, **56**, 2281–2290.
36. G. van't Slot and H. U. Humpf, *J. Agric. Food Chem.*, 2009, **57**, 8041–8048.
37. X. Tzounis, *et al.*, *Br. J. Nutr.*, 2008, **99**, 782–792.
38. M. Massot-Cladera, *et al.*, *Arch. Biochem. Biophy.*, 2012, **527**, 105–112.
39. X. Tzounis, *et al.*, *Am. J. Clin. Nutr.*, 2011, **93**, 62–72.
40. S. Stoupi, *et al.*, *Mol. Nutr. Food Res.*, 2010, **54**, 747–759.
41. R. Llorach, *et al.*, *Mol. Nutr. Food Res.*, 2013, **57**, 962–973.
42. E. S. Ford, *Diabetes Care*, 2005, **28**, 2745–2749.
43. E. S. Ford, C. Li and G. Zhao, *J. Diabetes*, 2010, **2**, 180–193.
44. S. Devaraj, *et al.*, *Am. J. Clin. Pathol.*, 2008, **129**, 815–822.
45. J. P. Despres, *Can. J. Cardio.*, 2012, **28**, 642–652; S. de Ferranti and D. Mozaffarian, *Clin. Chem.*, 2008, **54**, 945–955; M. Laclaustra, D. Corella and J. M. Ordovas, *Nutr. Metab. Cardiovasc. Dis.*, 2007, **17**, 125–139.
46. E. E. Kershaw and J. S. Flier, *J. Clin. Endocrin. Metab.*, 2004, **89**, 2548–2556; K. E. Wellen and G. S. Hotamisligil, *J. Clin. Invest.*, 2003, **112**, 1785–1788.
47. Y. Gu and J. D. Lambert, *Mol. Nutr. Food Res.*, 2013, **57**, 948–961.
48. Y. Gu, S. Yu and J. D. Lambert, *Eur. J. Nutr.*, 2014, **53**, 149–158.
49. J. Haase, *et al.*, *Diabetologia*, 2014, **57**, 562–571.
50. S. P. Weisberg, *et al.*, *J. Clin. Invest.*, 2003, **112**, 1796–1808; D. D. Sears, *et al.*, *PLoS One*, 2009, **4**, e7250.

51. S. P. Weisberg, *et al.*, *J. Clin. Invest.*, 2003, **112**, 1796–1808; M. J. Khandekar, P. Cohen and B. M. Spiegelman, *Nat. Rev. Cancer*, 2011, **11**, 886–895.
52. D. D. Sears, *et al.*, *PLoS One*, 2009, **4**, e7250.
53. E. Sabido, *et al.*, *Mol. Cell. Proteomics*, 2012, **11**, M111.014746–M111.014749.
54. H. Xu, *et al.*, *J. Clin. Invest.*, 2003, **112**, 1821–1830.
55. Y. Gu, S. Yu and J. D. Lambert, *Eur. J. Nutr.*, 2014, **53**(1), 149–158.
56. Y. Gu, S. Yu and J. D. Lambert, *Eur. J. Nutr.*, 2014, **53**(1), 149–158.
57. Y. Gu, *et al.*, *J. Nutr. Biochem.*, 2014, **25**, 439–445.
58. Y. Gu and J. D. Lambert, *Mol. Nutr. Food Res.*, 2013, **57**, 948–961.
59. D. Grassi, *J. Nutr.*, 2008, **138**, 1671–1676.
60. K. Hanhineva, *et al.*, *Int. J. Mol. Sci.*, 2010, **11**, 1365–1402.
61. I. Cordero-Herrera, *et al.*, *Mol. Nutr. Food Res.*, 2013, **57**, 974–985.
62. M. A. Martin, *et al.*, *Nutrients*, 2013, **5**, 2955–2968.
63. F. P. Martin, *et al.*, *J. Proteome Res.*, 2009, **8**, 5568–5579.
64. X. Tzounis, *et al.*, *Am. J. Clin. Nutr.*, 2011, **93**, 62–72.
65. V. Fogliano, *et al.*, *Mol. Nutr. Food Res.*, 2011, **55**(Suppl. 1), S44–S55.
66. X. Tzounis, *et al.*, *Br. J. Nutr.*, 2008, **99**, 782–792; X. Tzounis, *et al.*, *Am. J. Clin. Nutr.*, 2011, **93**, 62–72.
67. X. Tzounis, *et al.*, *Br. J. Nutr.*, 2008, **99**, 782–792.
68. M. Massot-Cladera, *et al.*, *Arch. Biochem. Biophys.*, 2012, **527**, 105–112.
69. X. Tzounis, *et al.*, *Am. J. Clin. Nutr.*, 2011, **93**, 62–72.
70. Kohri, M. Suzuki and F. Nanjo, *J. Agric. Food Chem.*, 2003, **51**, 5561–5566; X. Tzounis, *et al.*, *Am. J. Clin. Nutr.*, 2011, **93**, 62–72; V. Fogliano, *Mol. Nutr. Food Res.*, 2011, **55**(Suppl. 1), S44–S55.
71. L. Y. Rios, *Am. J. Clin. Nutr.*, 2003, **77**, 912–918.
72. A. P. Neilson and M. G. Ferruzzi, *Annu. Rev. Food Sci. Techol.*, 2011, **2**, 125–151.
73. M. Urpi-Sarda, *et al.*, *J. Agric. Food Chem.*, 2009, **57**, 10134–10142.
74. A. Keys, *Acta Med. Scand.*, 1980, **207**, 153–160.
75. C. M. Skeaff and J. Miller, *Ann. Nutr. Metab.*, 2009, **55**, 173–201; C. M. Skeaff, *Eur. J. Clin. Nutr.*, 2009, **63**(Suppl. 2), S34–S49; C. M. Skeaff and R. Jackson, *J. Prim. Health Care*, 2011, **3**, 320–321.
76. T. Ikeda, *et al.*, *Diabete Metab.*, 1992, **18**, 465–467; S. Pal, *et al.*, *J. Nutr.*, 2003, **133**, 700–706.
77. C. Galli, Cocoa, Chocolate and Blood Lipids. in *Chocolate and Health*. 3013, Springer, Milan, pp. 127–135.
78. D. D. Mellor, *et al.*, *Diabetic Med.*, 2010, **27**, 1318–1321; C. E. O'Neil, V. L. Fulgoni, III and T. A. Nicklas, *Nutr. Res.*, 2011, **31**, 122–130.
79. Y. Wan, *Am. J. Clin. Nutr.*, 2001, **74**, 596–602.
80. R. Latif, *Neth. J. Med.*, 2013, **71**, 63–68.
81. M. L. McCullough, *et al.*, *J. Cardiovasc. Pharmacol.*, 2006, **47**(Suppl. 2), S103–S109; discussion 119–21.
82. V. Bayard, *et al.*, *Int. J. Med. Sci.*, 2007, **4**, 53–58.

83. B. Buijsse, *et al.*, *Arch. Intern. Med.*, 2006, **166**, 411–417; I. Janszky, *J. Intern. Med.*, 2009, **266**, 248–257.
84. E. Ros and F. B. Hu, *Circulation*, 2013, **128**, 553–565.
85. A. Buitrago-Lopez, *BMJ*, 2011, **343**, d4488.
86. S. C. Larsson, J. Virtamo and A. Wolk, *Neurology*, 2012, **79**, 1223–1229.
87. L. Hooper, *et al.*, *Am. J. Clin. Nutr.*, 2012, **95**, 740–751; O. A. Tokede, J. M. Gaziano and L. Djousse, *Eur. J. Clin. Nutr.*, 2011, **65**, 879–886.
88. E. Zomer, *BMJ*, 2012, **344**, e3657.
89. R. M. van Dam, N. Naidoo and R. Landberg, *Curr. Opin. Lipidol.*, 2013, **24**, 25–33.
90. K. Hanhineva, *Int. J. Mol. Sci.*, 2010, **11**, 1365–1402.
91. H. Sies, W. Stahl and A. Sevanian, *J. Nutr.*, 2005, **135**, 969–972.
92. D. L. Katz, K. Doughty and A. Ali, *Antioxid. Redox Signaling*, 2011, **15**, 2779–2811.
93. M. A. Martin, *Nutrients*, 2013, **5**, 2955–2968.
94. M. A. Martin, *et al.*, *Mol. Nutr. Food Res.*, 2014, **58**, 447–456.
95. G. Rimbach, *Int. J. Mol. Sci.*, 2009, **10**, 4290–4309.
96. K. Khan, *et al.*, *Nutr. Metab. Cardiovasc. Dis.*, 2012, **22**, 1046–1053.
97. S. Lecour and K. T. Lamont, *Mini-Rev. Med. Chem.*, 2011, **11**, 1191–1199.
98. D. Grassi, *et al.*, *Hypertension*, 2012, **60**, 827–832; D. Grassi, G. Desideri and C. Ferri, *Curr. Opin. Clin. Nutr. Metab. Care*, 2013, **16**, 662–668.
99. D. D. Mellor, *et al.*, *Diabetic Med.*, 2010, **27**, 1318–1321.
100. D. D. Mellor, *et al.*, *Diabetic Med.*, 2013, **30**, 478–483.
101. R. M. van Dam, N. Naidoo and R. Landberg, *Curr. Opin. Lipidol.*, 2013, **24**, 25–33.
102. A. M. M. Ruzaidi, *et al.*, *J. Sci. Food Agric.*, 2009, **88**, 1442–1447.
103. D. Grassi, *et al.*, *Am. J. Clin. Nutr.*, 2005, **81**, 611–614.
104. D. H. Ryan, *Int. J. Clin. Pract.*, 2003, **134**(Supplement), 28–35.
105. K. S. Stote, *et al.*, *Eur. J. Clin. Nutr.*, 2012, **66**, 1153–1159.
106. U. Etxeberria, *et al.*, *J. Agric. Food Chem.*, 2013, **61**, 9517–9533; S. Moco, F. P. Martin and S. Rezzi, *J. Proteome Res.*, 2012, **11**, 4781–4790.

CHAPTER 7

Chocolate and Cardiovascular Health

GABRIELA GUTIÉRREZ-SALMEÁN,[a] EDUARDO MEANEY,[a] GUILLERMO CEBALLOS-REYES[a] AND FRANCISCO VILLARREAL*[b]

[a] Sección de Estudios de Posgrado e Investigación, Escuela Superior de Medicina, Instituto Politécnico Nacional, Mexico City, Mexico;
[b] University of California San Diego School of Medicine, Department of Medicine, 9500 Gilman Dr., BSB4028 La Jolla, 92093-0613J San Diego, CA, USA
*Email: fvillarr@ucsd.edu

7.1 Introduction

Theobroma cacao is the tree from which cocoa—more accurately cacao—is obtained. As per ancient Mesoamerican civilization beliefs, the gods discovered cacao and used it as food, hence the meaning of its the botanical name, which translates as "food of the gods".

The Aztec (Nahuatl) term for chocolate was *cacahuatl*; the Maya name for hot water is *chocol ha*; and the word chocolate is probably a combination of these two. *Cacahuatl* or *xocolatl* refers to a bitter, strong, foaming beverage made from cacao that translates to "sour water".[1] What we now refer to as chocolate is an industrialized product elaborated from cacao seeds. The seeds are roasted, husked, ground and pressed into chocolate liquor. The liquor can be separated into two products: a solid part (cocoa powder) and a fatty fraction (cocoa butter). Different types of chocolates are made from

Chocolate and Health: Chemistry, Nutrition and Therapy
Edited by Philip K. Wilson and W. Jeffrey Hurst
© The Royal Society of Chemistry 2015
Published by the Royal Society of Chemistry, www.rsc.org

mixing varying proportions of cocoa liquor and butter, together with other ingredients such as vanilla or milk.

- White chocolate is prepared with at least 20% of cocoa butter, milk and sugar. It lacks cocoa liquor or powder.
- Dark chocolate, in the strict sense, is prepared by mixing at least half-and-half of cocoa liquor and butter. Some brands can reach up to 95% of cocoa liquor.
- Milk chocolate contains <40% cocoa liquor. As its name indicates, it also contains milk as well as added sugar.

Although the terms cacao, cocoa and chocolate are often used interchangeably, the bioactive substances are actually contained in the solid fraction. Thus, it is the consumption of cocoa powder or dark chocolate that exerts potential cardiovascular and metabolic effects. For convenience, we will refer to these as cocoa products (CPs).

7.2 Historical Remarks

The concept that cacao yields health benefits is not new. Even though its consumption as chocolate was popularized by Europeans, cacao was an important part of Mesoamerican native culture for several millennia. The Olmecs first systematically cultivated cacao. However, it is from the Mayans that we gain more information about its use and consumption.[2] The word cacao is a Mayan term (*kakaw*), borrowed from the Mixe-Zoqueaen language of the Olmec and Soconusco region.

The Mayan glyph (Figure 7.1) that is syllabically read as *kakaw* is the first written proof of the word "cacao" in history. The phoneme *ka* is represented as the shape of a fish, either in its complete form or as a comb (which represent the fish's fins and scales); moreover, the two points located between this figure indicate that the phoneme must be repeated twice. The last phoneme *wa*, is captured in the glyph's third component: the

Figure 7.1 Mayan glyph for cacao (*kakaw*).

figure of a corn whose syllabic value is derived from the Mayan *waaj*, which alludes to tamal (a traditional food made from corn dough). Given that in Mayan writing the last letter is not pronounced, the glyph is read *kakaw*.

Cacao possessed a high value among the Mayans as it was used not only as a key element in food and beverage preparation, but also as currency and for commercial exchange. In fact, *Ek Chuah* served as both the Mayan god of cacao and the merchants' benefactor.[3,4] Mesoamerican Indians also used cacao as part of their traditional medicine. For example, they drank it to alleviate dyspepsia and intestinal colic, diarrhea and dysentery, fever, asthenia, angina and convulsive episodes, amongst other uses. Cacao was also utilized in combination with other ingredients to treat diverse maladies. For instance, cacao ground together with dried corn beans and a flower called *tlacoxochitl* was consumed to mitigate fever. At other times, it was combined with cotton sap to treat gastrointestinal colic and discomfort or mixed with Mexican pepperleaf (*Piper auritum*, hoja santa) and vanilla flowers to allay productive cough.[2] Although some current perceptions of CPs associate them with a confectionery pleasure that might seemingly impair health, recent scientific evidence has emerged demonstrating the cardiovascular and metabolic benefits of moderate CP consumption.

7.3 Nutrition and Biochemistry as Related to Health and Medicine

One hundred grams of dark chocolate (*i.e.*, 70–80% cacao) provides $\sim$550 kilocalories or $\sim$27.5% of a daily recommended intake of 2000 kcal. Cacao seeds, as a raw product, are $\sim$30% fat, 7% protein, 60% carbohydrates and 3% water. Other compounds of note include:

- The lipid fraction of cacao seeds, which contains monounsaturated oleic acid and saturated stearic and palmitic fatty acids, as well as short-chain fatty acids whose content varies according to the type of cacao seed.[5] These are important components given that much has been discussed regarding the high fat content of dark chocolate. It is well known that consuming CPs does not raise cholesterol levels due to the fact that although palmitic acid increases cholesterolemia, oleic acid counters these effects and stearic acid does not impact the lipid profile (*i.e.*, it contributes a neutral effect). Thus, overall, CPs are not pro-atherogenic.[6]
- As only a negligible amount of sitosterol and sodium ($\approx$24 mg) is present in cacao, the consumption of CPs is not likely to increase the risk of hypertension.
- Cacao contains about 1% of the methylxanthine theobromine, which stimulates diuresis and has vasodilator properties.[7,8] Contrary to a once-common belief, cacao contains very little caffeine,[9] therefore its noted effects cannot be attributed to this component.

- Cacao beans contain a significant amount of bran; however, much of it can be lost during processing. Powdered cacao can contain up to 2 g of dietary (mostly insoluble) fiber per 100 g. CPs contain even less, and therefore do not contribute significantly to the recommended daily intake of dietary fiber.[10]
- Cacao is rich in polyphenols, particularly flavonoids (in monomeric or multimeric forms), with the flavanols epicatechin (EPI) and catechin being most abundant. Polyphenols are often considered as mere antioxidants, whereby the health benefits of CPs have frequently been attributed to their antioxidant potential. However, many studies have demonstrated that CPs or EPI can regulate important metabolic pathways, leading to a range of healthy effects.[11-16] As a consequence, CPs have the potential to trigger many beneficial effects on cardiometabolic diseases.

As it is evident that CPs contain multiple bioactive compounds, identifying a single agent responsible for mediating its beneficial effects is challenging. However, EPI has emerged as a prominent candidate since it can reproduce many of cacao's healthy cardiometabolic effects.

7.4 Cardiometabolic Effects of CPs

Cardiometabolic diseases represent the number one public health threat in developed countries and emerging nations. Diseases or clinical conditions such as coronary heart disease, stroke, dyslipidemia, hypertension and diabetes mellitus are listed among the more important epidemiological problems in modern society, each imposing a high clinical and economic burden. The prevention and/or treatment of these "civilization diseases" demands a holistic approach that should include behavior modification and dietary intervention supplemented when necessary by pharmacological approaches.

The concept that the regular consumption of natural products yields an improved health profile and can prevent some diseases such as atherosclerosis or cancer is certainly not new.[16] In recent decades, a sociocultural phenomenon that can be referred to as "antioxidant mania" has advocated for the consumption of natural "antioxidants" such as flavonoid-containing products to prevent and/or treat multiple conditions or diseases. Most of these claims lack scientific rigor and are merely advertising overstatements. In this regard, dark chocolate and other rich CPs have been labeled as natural foods that are full of healthy advantages, mainly because of their high flavonoid content.[17] Although flavonoids contained in CPs may exert salutary effects given their antioxidant potential, recent work indicates that EPI (the most important flavanol in cacao) may exert metabolic and cardiovascular protection *via* other mechanisms.

Kuna Indians, who live in the San Blas archipelago off the coast of Panama, consume a cacao beverage on a regular basis and have extremely low incidences of hypertension, cancer, diabetes or other cardiovascular diseases. When the Kuna migrated to mainland Panama, they had no access to this beverage and incurred the same incidence for these diseases as the rest of the mainland population.[18,19]

Although a large proportion of past medical reports concerning the healthy effects of CPs have focused on hypertension,[20] evidence is mounting regarding the effects of CPs on important variables such as oxidative stress, glycemia, insulin sensitivity, lipid profile and inflammation, amongst others, all of which influence cardiometabolic risk. In this chapter, the evidence regarding the effects of CPs on some of these variables will be reviewed and discussed.[21,22]

7.4.1 Systemic Arterial Hypertension

Systemic arterial hypertension (SAH) is one of the most prevalent cardiovascular diseases, affecting a considerable segment of the worldwide population. SAH appears to be the result of a mixture of genetic and environmental factors that disrupt basic blood pressure (BP) control mechanisms. Underlying the pathophysiology of the disease are factors including inflammation, oxidative stress, endothelial dysfunction, metabolic, humoral and hormonal derangements, insulin resistance and increased vascular stiffness, amongst others. Lifestyle modifications including a healthy diet, exercise and pharmacological treatment are useful to control BP, in that they reduce the risk of associated pathologies such as stroke, coronary disease and heart failure (HF).

Evidence indicates that the inclusion of CPs within the daily diet may contribute to controlling SAH. One of the first long-term studies substantiating an inverse relationship between CP consumption and elevated BP and cardiovascular mortality was the Zutphen Elderly Study, which followed 470 elderly men for 5 years.[23] Subjects who consumed CPs had systolic BP (SBP) at 3.7 mmHg and diastolic BP (DBP) at 2.1 mmHg lower than controls. Furthermore, subjects with the highest consumption of CPs had a SBP at 3.8 mmHg less than those with the lowest intake. A recent meta-analysis contrasted results from individuals consuming CPs containing 50–70% of cacao or other high-flavanol products against those who consume cocoa-free or low-flavanol products. A small but significant reduction in SBP and DBP was reported of -3.2 ± 1.9 mmHg and -2.0 ± 1.3 mmHg, respectively.[4] When different groups were analyzed, it was noted that the beneficial effect occurred only in hypertensive and pre-hypertensive subjects (-5.0 ± 3.0 mmHg; $p = 0.0009$ for SBP; and -2.7 ± 2.2 mmHg, $p = 0.01$ for DBP), but not in normotensive individuals. The range of flavanol consumption was between 30 and 1008 mg during observation periods ranging from 2 to 18 weeks.[24] These reductions in BP associated with CP consumption can be seen as marginal. However, lowering of SBP by just 2 mmHg leads to a 10% reduction in stroke and a 7% drop in coronary mortality.[25] A separate meta-analysis comprising 1106 subjects from 24 separate studies concluded that flavonoid-rich CPs on average decreased SBP by 1.63 mmHg, but that neither DBP nor heart rate were affected.

The antihypertensive mechanisms underlying CP consumption have not been fully elucidated, but flavanols likely exert multiple biological effects that can aid in BP control. CPs or EPI can reduce arginase activity, leading to a greater concentration of L-arginine, the main substrate of endothelial nitric oxide synthase (eNOS), which is the calcium-dependent endothelial enzyme that produces nitric oxide (NO). Interestingly, EPI can even stimulate NO synthesis under Ca^{2+}-free conditions and independently of the translocating effect of calmodulin on eNOS from the membrane to the cytoplasm.[26–28] NO is a potent physiological vasodilator that lowers arteriolar resistance and thus BP. In addition, NO can antagonize the actions of angiotensin II, a hormone that induces vasoconstriction, cellular hyperproliferation, nitroxidation and inflammation.[29] The hypothesis that the greater release of NO is the main antihypertensive mechanism of cacao or EPI is reinforced by the increase of NO urine metabolites, the augmentation of flow-mediated dilatation and the reduction of nitroxidative stress in subjects consuming CPs.[30] Moreover, the release of the endothelium-derived hyperpolarizing factor is also stimulated by flavanols, yielding greater prostacycline release and reduced synthesis of the vasoconstrictor endothelin.[30]

7.4.2 Obesity

The prevention and treatment of obesity, which is a major risk factor for cardiovascular diseases, are in need of alternative approaches, since drugs against obesity are scarce and of limited usefulness. Moreover, weight reduction through restrictive diets and the frequent practice of physical exercise face poor patient compliance.

As noted above, CPs can contain a significant amount of fat and, therefore, may be considered as an energy-dense foodstuff. Interestingly, clinical trials have failed to document significant weight gain when CPs are consumed in limited amounts. In fact, a recent study found that rats fed with a cacao-enriched chow reduced their body weight at the expense of body fat.[31,32] The authors hypothesized that flavanols may reduce body weight *via* the suppression of fatty acid synthesis and the enhancement of energy expenditure by mitochondria.[33] The ability of flavanols to stimulate oxidative metabolism *via* improvements in mitochondrial structure and function is perhaps one of the mechanisms that may best explain the weight loss and the improved metabolic profile (*i.e.*, the glucose and lipid profiles discussed below). Preclinical experiments demonstrate that flavanols, particularly EPI, increase the expression of the peroxisome PGC-1α, which in turn activates downstream pathways that stimulate mitochondriogenesis and increase cristae density.[32–44] PGC-1α also stimulates β-oxidation and GLUT-4, thus improving blood glucose and lipid levels.[42] PGC-1α can also increase the expression of uncoupling proteins such as UCP-1 that are responsible for decreasing the efficiency of energetic substrate oxidation (*i.e.*, glucose and fatty acids), thereby augmenting energy expenditure (*i.e.*, metabolic rate) and possibly leading to a loss of body weight.[40] Indeed, animal studies demonstrate a reduction of weight with the addition of EPI or flavonoid-rich products, a response associated with an increase in energy expenditure.[43]

Other mechanisms for the positive effect of CPs on obesity may encompass satiation. Some studies have reported that CP intake suppresses appetite (not hunger) and thus favors satiation (*i.e.*, promotes meal termination) and satiety between meals.[44] Such actions may result in weight loss as subjects decrease their overall caloric intake. It should be noted that while some studies report an inverse correlation between CP consumption and body mass index, others have found conflicting or negative results regarding the beneficial effects of such products on weight or abdominal girth.

7.4.3 Glycemia and Insulin Resistance

The epidemiological relationship between type 2 diabetes mellitus (T2DM) and abdominal obesity is firmly established.[45] A pathophysiological link between both conditions is insulin resistance. Contrary to common belief, CP consumption appears to yield beneficial effects in subjects with insulin resistance.[46] The consumption of CPs has been reported to reduce insulin resistance in hypertensive patients following 2 weeks of ingestion of 100 g day^{-1} of flavonoid-rich chocolate.[47,48] Furthermore, in a meta-analysis conducted by Shrime *et al.*,[22] the consumption of CPs was associated with a significant decrease in the insulin resistance index by 0.94 points and an increased insulin sensitivity index by 4.97 points. In contrast, Muniyappa *et al.*[49] did not detect improvements in insulin sensitivity or BP in hypertensive patients with insulin resistance. However, other investigators have found that the regular consumption of flavonoid-rich products favorably modifies metabolic syndrome risk factors, including dysglycemia, dyslipidemia and hypertension.[50]

Studies have also reported that flavonoids, including those from CPs, attenuate both postprandial and fasting hyperglycemia, together with a decrease in plasma insulin levels. Apparently, the effects are secondary to inhibiting pancreatic amylase activity, thereby decreasing the enteral absorption of dietary glucose.[34] Other studies have reported that flavonoid consumption decreases the expression of enteral transporters (*e.g.*, GLUT-2 and SGLT2), which also results in lower carbohydrate absorption.[34] CPs have also been proposed as insulin sensitizers and possibly as secretagogues through the regeneration of pancreatic β-cells.[36] Hence, the use of CPs may be considered for preventive and therapeutic approaches as not only may they attenuate and improve dysglycemia and insulin resistance, but they may also revert cell dysfunction in those patients with more advanced metabolic disruptions. In this regard, a study in Japan reported on the diminished risk associated with coffee, green tea (rich in catechins), other types of teas and chocolate for exacerbating diabetes.[37] However, to date, no well-controlled studies have been reported on the effects of CPs in patients with insulin resistance or T2DM, so its usefulness in this condition remains speculative.

7.4.4 Lipids and Atherosclerosis

CP consumption has demonstrated beneficial effects on blood lipid profiles. Specifically, CP consumption increases high-density lipoprotein (HDL) cholesterol, reduces low-density lipoprotein (LDL) and total cholesterol and attenuates lipid peroxidation.[16] As for cholesterol, CPs and flavanols increase ApoA while decreasing ApoB by upregulating sterol regulatory element binding proteins, as well as by increasing the scavenging activity of the LDL receptor. Although significant amounts of fatty acids are present in cacao, clinical studies have reported that CPs and/or EPI are able to reduce plasma triglyceride levels in either fasting or postprandial states[19] This effect may be due to the inhibition of pancreatic lipase and phospholipase 2, thereby decreasing the enteral absorption of dietary fats.[20] Another proposed mechanism is that flavonoids bind to dietary lipids and co-precipitate them, hence reducing their transport into the enterocyte.

Several short-term trials have shown that the consumption of CPs decreases LDL cholesterol by 5.8 mg dl^{-1} without any significant change in total and HDL cholesterol levels. However, in subjects with high cardiovascular risk, the consumption of CPs lowered total and LDL cholesterol by 8 and 7.7 mg dl^{-1}, respectively.[9] In contrast, other investigations indicate that subjects consuming milk chocolate *versus* high-content carbohydrate snacks raised their HDL cholesterol and decreased triglycerides without modifying LDL cholesterol.[6] A study in hypertensive subjects indicated that flavonoid-rich products consumed over 2 weeks caused a 12% reduction in total cholesterol.[24] Studies have also demonstrated reductions in lipid oxidation.[24] In research performed by Baba, daily intake of cocoa powder in humans over several weeks increased HDL levels and decreased LDL oxidation. Oxidized LDL levels were found to correlate negatively with urinary excretion of EPI metabolites.[25] However, other researchers have not documented changes in the lipid profile of hypertensive individuals consuming 50 g of chocolate with 70% cacao over 4 weeks.[22]

Altogether, the culmination of contradictory results and lack of long-term studies fail to provide sufficient evidence to support the conclusion that flavanol-rich CPs have a consistent salutary effect on lipid profiles. However, through several complementary mechanisms (enhancement of endothelial function,[26] antiplatelet aggregation,[27] lipid antioxidation[28] and reduction of BP and insulin resistance), CPs may exert suppressive effects on atherogenesis. Adding further support of this claim is evidence generated in studies using hypercholesterolemic rabbits that mimic human familial hypercholesterolemia,[51] whereby the chronic administration of cocoa powder was found to suppress the production of fatty streaks and intimal thickening in the aorta. Although no changes were observed in the lipid profile of these animals, thiobarbituric acid reactive substances were noted to decrease, indicating reduced lipid peroxidation. Therefore, this model suggests it was the decrease in oxidative stress rather than the effects on lipids that caused the suppression of atherosclerosis.[29] Additional support derives from *in vitro* studies using human aortic vascular smooth muscle cells,[52] in which it has been shown that the use of procyanidin-rich cocoa fractions and other cyanidin products inhibits the activation of matrix metalloproteinases, which are involved in vascular cell migration, a key early step in the atherosclerosis process and plaque rupture.[30] Monagas *et al.* studied the effects of chronic cocoa consumption on cellular and serum biomarkers related to atherosclerosis and found that the intake of polyphenol-rich cocoa modulated inflammatory mediators in patients who were at high risk of cardiovascular disease.[53] In spite of these and similar findings, no well-controlled long-term clinical studies regarding cocoa have focused on the primary or secondary prevention of atherosclerosis.

Specific at-risk populations who may also benefit from CP consumption include post-menopausal women and smokers. In a study of hypercholesterolic post-menopausal women who consumed cocoa beverages for 6 weeks, the lipid profile was found to improve (*i.e.*, total, HDL and LDL cholesterol levels). Another study reported that after 1 year of daily consumption of flavonoid-enriched chocolate, diabetic post-menopausal women reduced their overall cardiovascular risk by 10% through significant improvements in arterial stiffness. Elsewhere, smokers are known to have a high cardiovascular risk given that cigarette smoke impairs NO bioavailability and raises lipoprotein oxidation. A small pilot study indicated that administering CPs in the form of a 100 mL beverage significantly increased circulating NO levels in plasma, as well as improving flow-mediated vasodilation.

7.4.5 Heart Failure

Observational studies suggest that consuming CPs results in significant reductions in HF hazard ratios. Participants from the Swedish Mammography Cohort study who reportedly consumed no more than two weekly servings of chocolate demonstrated significantly lower HF and hospitalization rates. Interestingly, this benefit was not observed when the intake was one or more serving per day.

HF is associated with detrimental alterations in skeletal muscle structure and function, including major derangements in mitochondrial structure and in the integrity of the dystrophin-associated protein complex (DAPC), which is essential for maintaining sarcomere microstructure and function. Recently, Taub *et al.*[58] studied a group of five patients with stage II/III HF and T2DM that consumed CPs containing $\sim$100 mg of EPI for 3 months. The results demonstrated improved skeletal muscle protein and/or activity levels of signaling pathways involved in mitochondrial biogenesis, including

NO, cyclic guanosine monophosphate, SIRT1, PGC-1α, transcription factor A, mitochondrial Tfam, oxidative phosphorylation proteins, porin and mitofilin. These changes were related to a recovery and/or enhancement of DAPC protein levels, most notably dystrophin, β-dystroglycan, utrophin and several sarcoglycans. These changes were accompanied by an improved microstructure according to electron microscopy. The authors propose that these changes are consistent with myofiber regeneration.

Emerging evidence indicates that flavanols, in addition to stimulating eNOS, can also enhance cellular metabolism by activating one of the key kinases, 5′ adenosine monophosphate-activated protein kinase (AMPK),[54] promoting transcription and translocation of the glucose transporter GLUT-4, thereby enhancing glucose uptake and diminishing its serum concentration.[19] AMPK can also stimulate the production of cell energy (as measured by increased ATP levels) by promoting fatty acid oxidation and cholesterol and triglyceride synthesis, regulating insulin synthesis and secretion, modulating satiety and stimulating gene expression and protein synthesis.[32] Downstream, AMPK activates PGC-1α, which is essential to stimulating mitochondrial biogenesis. The latter phenomenon may explain the effects of flavanols on physical exercise, body weight and lipid profiles as observed with CPs or with pure EPI.[19,33]

7.4.6 Stroke

Several lines of evidence suggest an inverse relationship between CP consumption and incidence of stroke. In a meta-analysis including 114 009 participants, Buitrago *et al.*[55] reported a beneficial association between higher levels of CP consumption and cardiometabolic risk. In this report, the highest levels of CP consumption were associated with a 29% reduction in stroke as compared with the lowest levels. Larsson *et al.*[56] (in 37 103 men tracked over 10.2 years of follow-up) reported that high CP consumption was associated with a lower risk of stroke. The multivariable relative risk of stroke comparing the highest quartile of CP consumption (median 62.9 g per week) with the lowest quartile (median 0 g per week) was 0.83 (95% CI: 0.70–0.99).[55,56]

Recently, Walters *et al.*[57] reported results from a randomized double-blind crossover design on the acute effects of 100 g (104 mg of flavanols) of dark or milk chocolate (32 mg of flavanols) on cerebral vasomotor reactivity (CVR) in healthy volunteers. Dark chocolate caused a significant change in CVR at 90 minutes after ingestion, suggesting that the CP-induced vasodilation may contribute to the relationship between chocolate consumption and decreased stroke risk.[57]

Evidence also indicates that the consumption of CPs can beneficially influence cognitive function. Alongside their effects on the vascular system that increase brain blood flow, flavonoids inhibit neuronal apoptosis and increase the expression of neuroprotective and neuromodulatory proteins that increase the number of connections between neurons. Together, these processes may act to maintain the number and quality of synaptic connections. Thus, flavonoids appear to have the potential to prevent the progression of neurodegenerative pathologies and to promote cognition performance.

7.5 Conclusions

Accumulating evidence strongly suggests the salutary effects of flavanol-rich CP consumption on cardiometabolic health parameters including SAH, body weight, lipid profile, insulin resistance and atherosclerosis, amongst others. The mechanisms underlying these effects are currently under extensive investigation and prominently feature the actions of EPI. Candidate signaling pathways that may participate in mediating EPI effects are shown in Figure 7.2. However, conclusive evidence on the salutary effects of CPs require well-designed, controlled, long-term clinical studies encompassing thousands of patients and focused on "hard" endpoints such as cardiovascular or coronary mortality, including relevant surrogate points.

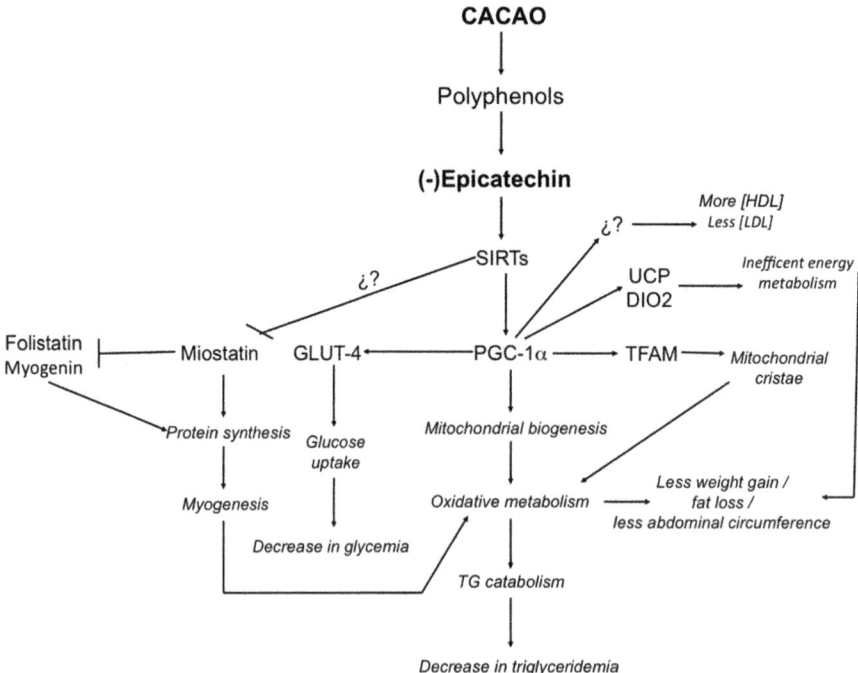

Figure 7.2 Possible pathways mediating the effects of cacao flavanols on metabolism.

References

1. W. Hurst, S. Tarka, T. Powis, F. Valdez and T. Hester, *Nature*, 2002, **418**, 289.
2. T. L. Dillinge, *et al.*, *J. Nutr.*, 2000, **130**(8S Suppl), 2057S.
3. S. K. Agarwal, *Ind. J. Med. Sci.*, 2013, **1**, 13.
4. D. Lippi, *Nutrition*, 2009, **25**, 1100.
5. J. A. Vinson, J. Proch and L. Zubik, *J. Agric. Food Chem.*, 1999, **47**, 4821.
6. U. Bracco, *Am. J. Clin. Nutr.*, 1994, **60**, 1002S.
7. J. F. Hammerstone, S. A. Lazarus and H. H. Schmitz, *J. Nutr.*, 2000, **130**, 2086S.
8. C. J. Kelly, *Am. J. Clin. Nutr.*, 2005, **82**, 486.
9. J. J. Barone and H. R. Roberts, *Food Chem. Toxicol.*, 1996, **34**, 119.
10. E. Lecumberri, R. Mateos, S. Ramos, M. Alia, P. Ruperez, L. Goya, M. Izquierdo-Pulido and L. Bravo, *Nutr. Hosp.*, 2006, **21**, 622.
11. C. Keen, *J. Am. Coll. Nutr.*, 2001, **20**, 436S.
12. C. F. Barnett, *J. Physiol.*, 2011, **589**, 5921.
13. L. Jia, X. Liu, Y. Y. Bai, S. H. Li, K. Sun, C. He, *et al.*, *Am. J. Clin. Nutr.*, 2010, **92**, 218.
14. I. Andújar, M. C. Recio, R. M. Giner and J. L. Ríos, *Oxid. Med. Cell. Longevity*, 2012, 906252.
15. A. Yasuda, M. Natsume, N. Osakabe, K. Kawahata and J. Koga, *J. Agric. Food Chem.*, 2011, **59**, 14706.
16. G. Gutiérrez-Salmeán M. P. Ortiz-Vilchi, C. M. Vacaseydel, I. Rubio-Gayosso, E. Meaney, F. Villarreal, I. Ramírez-Sánchez and G. Ceballos-Reyes, Unpublished data.
17. L. Dauchet, P. Amouyel, S. Hercberg and J. Dallongeville, *J. Nutr.*, 2006, **136**, 2588.
18. M. B. Engler and M. M. Engler, *Nutr. Rev.*, 2006, **64**, 109.
19. M. L. McCullough, K. Chevaux, L. Jackson, M. Preston, G. Martinez, H. H. Schmitz, *et al.*, *J. Cardiovasc. Pharmacol.*, 2006, **47**, S103.
20. N. K. Hollenberg, *Hypertension*, 1997, **29**, 171.
21. K. Ried, T. Sullivan, P. Fakler, O. R. Frank and N. P. Stocks, *BMC Med.*, 2010, **8**, 39.
22. M. G. Shrime, S. R. Bauer, A. C. McDonald, N. H. Chowdhury, C. E. M. Coltart and E. L. Ding, *J. Nutr.*, 2011, **141**, 1982.
23. R. Corti, A. J. Flammer, N. K. Hollenberg and T. F. Lüscher, *Circulation*, 2009, **119**, 1433.
24. B. Buijsse, E. J. M. Feskens, F. J. Kok and D. Kromhout, *Arch. Intern. Med.*, 2006, **166**, 411.
25. S. Baba, M. Natsume, A. Yasuda, Y. Nakamura, T. Tamura, N. Osakabe, M. Kanegae and K. Kondo, *J. Nutr.*, 2007, **137**, 1436.
26. E. Zomer, A. Owen, D. J. Magliano, D. Liew and C. M. Reid, *BMJ*, 2012, **344**, e3657.

27. C. L. Keen, R. R. Holt, P. I. Oteiza, C. G. Fraga and H. H. Schmitz, *Am. J. Clin. Nutr.*, 2005, **81**, 298S.
28. I. Ramirez-Sanchez, H. Aguilar, G. Ceballos and F. Villarreal, *Mol. Cell. Biochem.*, 2012, **370**, 141.
29. D. Taubert, R. Roesen, C. Lehmann, N. Jung and E. Schomig, *JAMA, J. Am. Med. Assoc.*, 2007, **298**, 49.
30. C. G. Fraga, M. C. Litterio, P. D. Prince, V. Calabró, B. Piotrowski and M. Galleano, *J. Clin. Biochem. Nutr.*, 2011, **48**, 63.
31. Y. Gu, W. J. Hurst, D. A. Stuart and J. D. Lambert, *J. Agric. Food Chem.*, 2011, **59**, 5305.
32. D. Grassi, S. Necozione, C. Lippi, G. Croce, L. Valeri, P. Pasqualetti, G. Desideri, J. B. Blumberg and C. Ferri, *Hypertension*, 2005, **46**, 398.
33. S. Oba, C. Nagata, K. Nakamura, K. Fujii, T. Kawachi, N. Takatsuka and H. Shimizu, *Br. J. Nutr.*, 2010, **103**, 453.
34. Y. Gu, W. J. Hurst, D. A. Stuart and J. D. Lambert, *J. Agric. Food Chem.*, 2011, **59**, 5305.
35. K. Hanhineva, *Int. J. Mol. Sci.*, 2010, **11**, 1365.
36. B. K. Chakravarthy, S. Gupta and K. D. Gode, *Life Sci.*, 1982, **31**, 2693.
37. K. Davison, A. M. Coates, J. D. Buckley and P. R. Howe, *Int. J. Obes.*, 2008, **32**, 1289.
38. D. Taubert, R. Roesen, C. Lehmann, N. Jung and E. Schomig, *JAMA, J. Am. Med. Assoc.*, 2007, **298**, 49.
39. I. Andújar, M. C. Recio, R. M. Giner and J. L. Ríos, *Oxid. Med. Cell. Longevity*, 2012, 906252.
40. R. Ventura-Clapier, A. Garnier and V. Veksler, *Cardiovasc. Res.*, 2008, **79**, 208.
41. W. J. Lee, *Biochem. Biophys. Res. Commun.*, 2006, **340**, 291–5.
42. K. Higashida, M. Higuchi and S. Terada, *J. Nutr. Sci. Vitaminol.*, 2009, **55**, 486.
43. G. A. Wolf, *Nutr. Rev.*, 1997, **55**, 178.
44. D. Katz, K. Doughty and A. Ali, *Antioxid. Redox Signaling*, 2011, **15**, 2779.
45. Y. M. K. Farag and M. R. Gaballa, *Nephrol., Dial., Transplant.*, 2011, **26**, 28.
46. P. Pajunen, A. Kotronen, E. Korpi-Hyövälti, S. Keinänen-Kiukaanniemi, H. Oksa, L. Niskanen, *et al.*, *BMC Public Health*, 2011, **11**, 754.
47. B. Isomaa, P. Almgren, T. Tuomi, C. Forsblom, B. Forsén, K. Lahti, *et al.*, *Diabetes Care*, 2001, **24**, 683.
48. D. Grassi, G. Desideri, S. Necozione, C. Lippi, R. Casale, G. Properzi, *et al.*, *J. Nutr.*, 2008, **138**, 1671.
49. R. Muniyappa, G. Hall, T. L. Kolodziej, R. J. Karne, S. K. Crandon and M. Quon, *Am. J. Clin. Nutr.*, 2008, **88**, 1685.
50. N. Osakabe, *J. Clin. Biochem. Nutr.*, 2013, **52**, 186.
51. T. Kurosawa, F. Itoh, A. Nozaki, Y. Nakano, S. Katsuda, N. Osakabe, *et al.*, *J. Atheroscler. Thromb.*, 2005, **12**, 20.
52. K. W. Lee, N. J. Kang, M. H. Oak, M. K. Hwang, J. H. Kim, V. B. Schini-Kerth, *et al.*, *Cardiovasc. Res.*, 2008, **79**, 34.

53. M. Monagas, N. Khan, C. Andres-Lacueva, R. Casas, M. Urpí-Sardà, R. Llorach, *et al.*, *Am. J. Clin. Nutr.*, 2009, **90**, 1144.
54. M. C. Towler and D. Grahame Hardie, *Circ. Res.*, 2007, **100**, 328.
55. A. Buitrago-Lopez, J. Sanderson, L. Johnson, S. Warnakula, A. Wood, E. Di Angelantonio and O. D. Franco, *Br. Med. J.*, 2011, **343**, d4488.
56. S. C. Larsson, J. Virtamo and A. Wolk, *Neurology*, 2012, **79**, 1223.
57. M. R. Walters, C. Williamson, K. Lunn and A. Munteanu, *Neurology*, 2013, **80**, 1173.
58. P. R. Taub, I. Ramirez-Sanchez, T. Ciaraldi, G. Perkins, A. N. Murphy, R. Naviaux, M. Hogan, A. S. Maisel, R. R. Henry, G. Ceballos and F. Villarreal, *Clin. Transl. Sci.*, 2012, **5**, 43.

CHAPTER 8

Chocolate and Exercise Recovery

WILLIAM R. LUNN* AND ALLYSON N. DEROSIER

Southern Connecticut State University, Exercise Science Department, Moore Field House – Human Performance Laboratory, 125 Wintergreen Ave., New Haven, CT 06515, USA
*Email: lunnw1@southernct.edu

8.1 Introduction

At the turn of the 21st century, who would have thought that chocolate milk would be challenging other commercially-available products as the preferred sports beverage, or a chocolate candy bar would be juxtaposed to the popular sports performance bars in the "health food" section of your local supermarket? After all, isn't chocolate milk a "kids' drink" that adults grow out of, and aren't chocolate bars just "junk food"? The surge in popularity of chocolate food products, marketed particularly towards active exercisers and competitive athletes is, refreshingly, not a flash-in-the-pan fad diet with little to no scientific credibility. The fact is, these chocolate products demand little if any modification to fit into the nutritional plan of actively-exercising individuals. Upon reading the previous statement, you may ask yourself, "Chocolate? A health food? How can that be? Did scientists figure out a way to make it good for you?" Well, scientists did not alter chocolate at all, and chocolate really never was "bad" for you in the first place. The proper nutrient composition is there and, as this chapter describes, chocolate food products can provide a viable option for the nutritional goals and demands of the physically active individual, particularly during the postexercise recovery interval.

8.2 Exercise Recovery: Physiology and the Role of Nutrients

Paramount to recovery from an exercise bout is the appropriate provision of nutrients to support repair and regrowth of functional body tissues and energy storage depots. In other words, exercise places tremendous demands on the body's ability to properly contract its muscle tissue and draw upon its sources of energy. Exercise does not negatively act on one's health; rather, the opposite is true. The ability to repeatedly respond to and correct these disturbances of your body's normal state due to exercise is what imparts the health benefits from exercise. As such, postexercise nutrition regimens have focused attention on an easily-accessible source of energy upon which many people depend: stored carbohydrate in skeletal muscle tissue, or glycogen. For individuals who habitually consume at least a moderately high percentage of daily kilocalories from carbohydrates, muscle glycogen depletion occurs in order to support energetic demands for muscle contraction. The carbohydrate is broken down to provide the substrate for the energetic processes of muscle contraction during exercise. The rate of muscle glycogen depletion varies depending on the intensity of exercise: intense exercise markedly accelerates muscle glycogen breakdown. An individual exercising at high intensity for 60 minutes depletes 55% of muscle glycogen stores, whereas continuing to exercise for 2 hours can almost fully deplete muscle and liver glycogen.[1] Therefore, postexercise restoration of stored carbohydrate is sensible if the individual wishes to complete a subsequent exercise bout, either during the same day or on a following day. How does one best restore depleted glycogen? The answer is simple: consume carbohydrates.

It has long been known that replenishing muscle glycogen following depleting exercise can be accomplished by ingesting carbohydrates during the postexercise period,[2] and that varying the content of carbohydrate in the food ingested up to "typical" carbohydrate intake values certainly increases glycogen synthesis rates, but higher intake amounts do not further elevate glycogen storage.[2,3] So when should one start eating carbohydrates following exercise? Simply, the sooner the better. Glycogen can be restored much faster when the carbohydrates are ingested soon after exercise compared to waiting until long after the exercise bout is over, such as 2 hours.[4] Does it matter what amount of carbohydrate is ingested? Indeed; postexercise nutrition consisting of at least 1.2 grams per kilogram body weight per hour ($g\ kg^{-1}\ h^{-1}$) of carbohydrate can optimize the rate of glycogen restoration,[5,6] and this will support glycogen breakdown during a subsequent exercise bout.[6,7] A higher glycemic index carbohydrate (*i.e.*, an indicator of the rate at which glucose can enter the bloodstream) favors an enhanced rate of glycogen resynthesis.[8] However, actual performance in a subsequent exercise bout may not be affected by the glycemic quality of the carbohydrate, as long as the sufficient amount of carbohydrate is ingested.[6,7] Since ingesting carbohydrates is the logical and preferred method of restoring muscle glycogen postexercise, investigators have questioned whether any other nutrient can accelerate the process. Their findings: possibly. Adding protein to the carbohydrate-containing postexercise meal can augment the glycogen storage rate,[9–11] but not if the amount of carbohydrate is already sufficient.[12,13] However, carbohydrate storage is not the only consideration in holistic, nutrition-based recovery from exercise, and postexercise protein consumption is absolutely vital.

The concentric and eccentric muscular contractions involved in exercise result in disrupted metabolic events involved with the contractile components of the skeletal muscle, namely in terms of fewer amino acids being incorporated into the muscle (protein synthesis, or anabolism) than there are amino acids released from the muscle (protein breakdown, or catabolism).[14–16] In other words, exercise results in greater muscle protein breakdown than muscle protein synthesis during the postexercise recovery interval. This condition, when maintained for a prolonged time period, is unfavorable, as it would promote catabolism and atrophy (degeneration) of the muscle protein. A logical nutritional strategy, then, would be to provide the body with amino acids in order to reverse the catabolic condition caused by exercise. It is known that following exercise, protein consumption increases the presence of amino acids in the bloodstream, which results in increased muscle protein synthesis.[17] Like the consideration of carbohydrate restoration mentioned above, popular questions regarding postexercise protein consumption include, "What form or kind of protein is best to increase synthesis and decrease breakdown? How much should be eaten? When should it be eaten?" Each of these questions is addressed below.

What kind of protein is best? Simply providing amino acids in general creates an environment that promotes the uptake of amino acids into the muscle, which will increase synthesis and decrease breakdown. However, ingestion of "high-quality" protein will stimulate the kinetic and cellular signaling events that can reverse the catabolic condition induced by exercise. "High-quality" proteins contain the essential amino acid leucine. Leucine, whether ingested during[18] or following[19,20] an exercise bout, enhances the cellular events promoting muscle protein synthesis. Whether the leucine is provided in the form of free amino acid, protein hydrolysate (an enzymatically predigested form of protein) or complete protein such as soy, animal flesh or whey and casein from dairy, the protein synthesis-stimulating effect is observed.[21,22]

How much protein should be eaten? "More is better" does not necessarily apply to the amount of protein consumed during exercise recovery in the context of muscle protein metabolism. Providing amino acids can reverse the catabolic condition caused by the exercise bout, and there appears to be a "ceiling" amount of 20 g of protein consumed in a single feeding that will maximally stimulate muscle protein synthesis, but no additional synthesis occurs beyond that amount.[23] However, protein breakdown may continue to be suppressed with protein ingestion beyond 20 g.[24]

How frequently should protein be eaten? Now that quantity and quality of postexercise protein consumption has been addressed, the timing of protein intake should be considered. Because physically active individuals should consume 1.2–1.8 g kg^{-1} day^{-1},[25] frequent (four to eight) feedings of 10–20 g of protein favors an anabolic response to promote muscle protein repair.[26]

Given the effect of postexercise carbohydrate and protein consumption on muscle glycogen restoration and muscle protein anabolism respectively, combining both nutrients into one feeding seems plausibly effective. In fact, it is. As noted above, co-ingesting carbohydrate and protein can enhance glycogen resynthesis if the carbohydrate amount is lower than optimal. The combined nutrients consumed postexercise also stimulate events leading to increased protein synthesis.[27–29] However, similar to glycogen synthesis, if the amount of protein consumed is already sufficient, provision of additional carbohydrate will not further stimulate protein synthesis.[30] The bottom line is that one nutrient may not always benefit the postexercise recovery action of the other, but combined ingestion of both will support glycogen and muscle protein recovery independently.

This background information demonstrates the demands that exercise place on the energy storage and mechanics of muscle contraction in the human body, and how nutrition following exercise can respond to those demands. Replenishment of macronutrients such as carbohydrate and protein is a major component to exercise recovery, and chocolate and chocolate-containing foods provide an effective and practical medium and vehicle with which to achieve postexercise recovery goals. Even the disturbances to hydration[31,32] and antioxidant status[33] that result from an acute exercise bout can be relieved with chocolate products. So, enjoy a chocolate bar guilt-free, and discover the benefits of chocolate as a postexercise recovery food.

8.3 Chocolate Milk and Postexercise Recovery

Chocolate milk has become highly popular in the milieu of exercise recovery nutrition options. The drink contains essential amino acids in the form of whey protein, electrolytes and carbohydrate in the lactose sugar, water from the bovine milk, carbohydrate from the added sugar and antioxidant micronutrients from the chocolate. Chocolate milk is clearly a practical and sensible postexercise option even for vegetarian, lactose-intolerant individuals, as soy chocolate milk has a similar nutrient profile to bovine milk. It should be cautioned, however, that in order to ensure the presence of the proper nutrients, the chocolate milk should be soy- or dairy-based. Some vegetarian "milks" are derived from rice or other plant sources and, therefore, have a low quality and/or quantity of protein.

Chocolate milk has all of the previously discussed components of postexercise recovery, including carbohydrate and protein to enhance blood glucose and amino acid levels.[34] Initially, the benefits of chocolate milk were observed in performance measures. Trained cyclists who consumed chocolate milk immediately following an exhausting interval workout experienced a lower time to exhaustion[35,36] and greater total work performed[35] during a subsequent exercise bout following 4 hours of recovery compared to a carbohydrate beverage or water alone. Even if the recovery interval is prolonged to 15–18 hours following the initial exercise, trained athletes who consumed chocolate milk experienced no difference in performance or markers of muscle damage compared to a carbohydrate-only drink.[37] Keep in mind that "no difference" is not bad; if chocolate milk performs just as well at achieving recovery outcomes as the traditional carbohydrate beverage, then that is a success. Chocolate milk's influence is not reserved solely for the endurance athlete; athletes who rely on a combination of power and endurance can similarly benefit. Collegiate soccer players demonstrated similar shuttle run times and lower times to exhaustion in subsequent exercise tests following recovery from practice that included chocolate milk consumption compared to consuming a carbohydrate-only beverage.[38] These observations paved the way for empirical evidence showing why chocolate milk was imparting a beneficial influence on the physiology of exercise recovery.

Exercise that demands high muscular force production can cause notable structural damage to the skeletal muscle tissue, and such damage should be repaired during the postexercise recovery interval. Soccer is an excellent example of such an activity, and evaluating markers of muscle damage in the blood, such as creatine kinase and myoglobin (both indicators of muscle cell disruption), are useful when evaluating the efficacy of postexercise nutrition. Following practice during consecutive days of increased training time, soccer players consumed chocolate milk or a high-carbohydrate beverage and experienced significantly lower creatine kinase levels during recovery when they drank the chocolate milk.[39] Postexercise myoglobin levels and muscle soreness were similar between the two beverages, indicating again that chocolate milk is as effective as traditional beverages regarding recovery nutrition. That chocolate milk rescues disturbances to muscle protein at least as well as a carbohydrate drink is not surprising, since chocolate milk is replete with the protein nutrition necessary for repair. Conversely, protein is absent in the traditional carbohydrate beverage.

So, how does chocolate milk measure up against other popular recovery beverages containing protein? Not surprisingly, chocolate milk measures up very well. As Pritchett *et al.* have shown, over 1 week, trained cyclists consumed either chocolate milk or a commercial recovery beverage containing protein and carbohydrate immediately after the training session.[40] In a performance test at 85% of maximal aerobic capacity (VO_2max) following the training week, no significant difference was noted between beverages in time to exhaustion. Further, there was no difference in pre-test plasma creatine kinase levels between beverages, demonstrating that chocolate milk can afford similar benefits on performance and muscle recovery as a popular, commercially available recovery beverage.

Are there mechanistic explanations for the benefits imparted by postexercise chocolate milk consumption? Indeed; in brief, to stimulate muscle protein synthesis, a series of intracellular signaling events must occur before amino acids are assembled into the intact protein in the skeletal muscle. The activation state of these intracellular molecules determines the success of the signaling cascade that leads to protein synthesis, and quantitative measurement of the activation of the molecules, typically based on the phosphorylation state of the molecule, is a common determination. Ferguson-Stegall et al. conducted a study in which trained cyclists consumed either chocolate milk, a carbohydrate beverage or a placebo immediately following an endurance ride ending with 10 minutes of hard repeats and again consumed the drink 2 hours into recovery.[41] Phosphorylation (activity) of the signaling molecules mammalian target of rapamycin and ribosomal protein S6 was greater during the recovery interval than with the placebo. In addition, glycogen restoration during recovery and 40 km time trial times following recovery were greater than with either the carbohydrate drink or placebo. It is becoming clear exactly how chocolate milk ingestion works regarding the mechanism of recovery that can support glycogen resynthesis, muscle protein metabolism recovery and, therefore, subsequent performance maintenance or enhancement. Provision of carbohydrate during recovery stimulates glycogen restoration, and ingestion of protein provides amino acids that stimulate the molecular pathway of protein synthesis. However, a metabolic component that can comprehensively explain the beneficial nature of postexercise nutrition afforded by chocolate milk is wanting. Measuring favorable signaling activity within the cell is prudent, but the data are fruitless unless they lead to increased kinetics of protein synthesis. In other words, is muscle protein synthesis actually occurring? Furthermore, are catabolic measures being attenuated by chocolate milk consumption? A 2012 study by Lunn et al. answers these questions. In a holistic evaluation of performance, intracellular signaling and kinetic measures of muscle protein synthesis, enzymatic activity of muscle protein breakdown and glycogen replenishment, runners consumed either chocolate milk or a carbohydrate control beverage following a 45 minutes run.[42] During the 3 hours recovery, glycogen restoration following the chocolate milk consumption equaled that from after the carbohydrate drink ingestion (Figure 8.1), and the chocolate milk drinkers ran significantly longer during a subsequent post-recovery time to exhaustion run (Figure 8.2). Moreover, the kinetic measure of muscle protein synthesis (the fractional synthetic rate) was greater during recovery in the milk drinkers (Figure 8.3), which was supported by the activation (phosphorylation) of an important signaling molecule in protein synthesis, eukaryotic initiation factor 4E-binding protein 1 (Figure 8.4). To comprehensively assess the effect of the chocolate milk ingested during recovery, measures of protein breakdown also implicated chocolate milk as a desirable recovery nutrition option. The activity of the enzyme caspase-3, associated with the degradation of contractile proteins, was decreased during chocolate milk recovery, as was the activity of the 26S proteasome, a disposal unit of degraded proteins (Figure 8.4). The activity of FOXO3a, a molecule that can stimulate the generation of other muscle

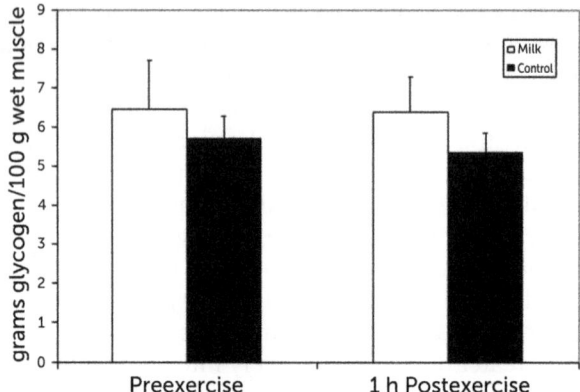

Figure 8.1 Muscle glycogen content preexercise and 1 hour into recovery from a 45 minutes, moderate effort treadmill run. Runners consumed either chocolate milk or a carbohydrate control beverage immediately following the run.

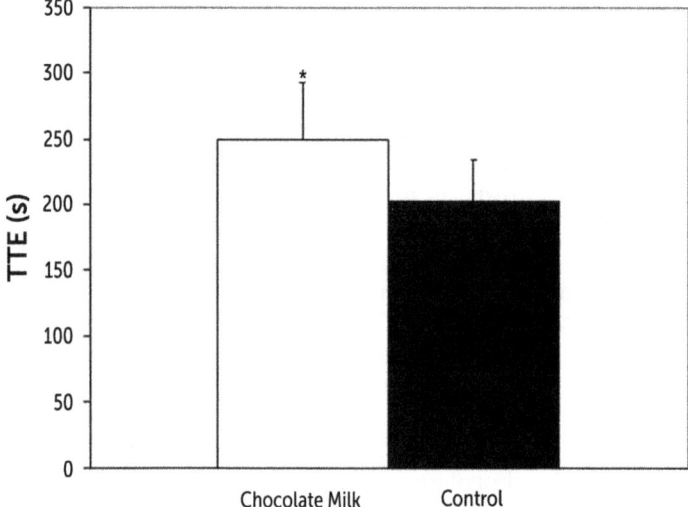

Figure 8.2 Time to exhaustion in a subsequent running test 3 hours following a 45 minutes, moderate effort run. Runners consumed either chocolate milk or a carbohydrate control beverage immediately following the initial run. *Significantly different from Control.

protein-degrading molecules, was also decreased in the milk drinkers, as indicated by greater phosphorylation (Figure 8.4). This study demonstrated that not only did chocolate milk consumption maintain glycogen restoration and improve subsequent exercise performance compared to a carbohydrate beverage, it also created a metabolic environment that promotes anabolism (growth) of the muscle protein.

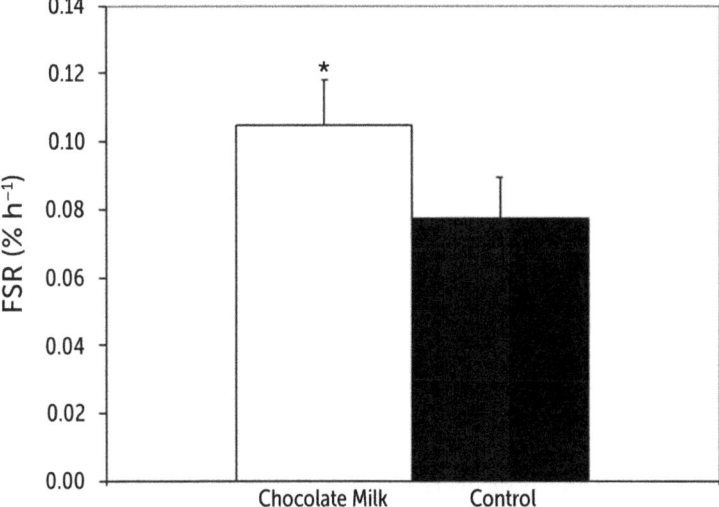

Figure 8.3 Muscle protein synthesis during recovery from a 45 minutes, moderate effort run. Runners consumed either chocolate milk or a carbohydrate control beverage immediately following the run. *Significantly different from Control.

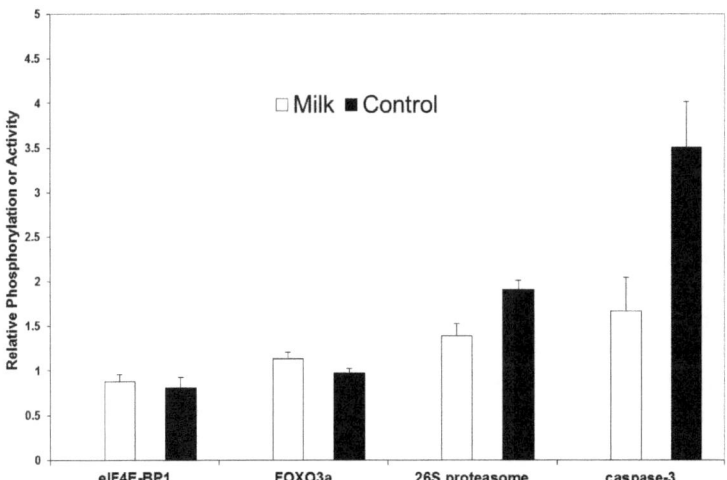

Figure 8.4 Activity of select intracellular signaling proteins and enzymes during recovery from a 45 minutes, moderate effort run. Runners consumed either chocolate milk or a carbohydrate control beverage immediately following the run. *Significantly different from Control.

These unique benefits of chocolate milk ingestion during exercise recovery clearly afford the exercising individual the ability to recover optimally so that he or she can perform a subsequent exercise bout or training session at full metabolic capacity. As such, it is sensible that a training effect can be observed in individuals who habitually consume chocolate milk during their postexercise recovery period. Previously-untrained individuals completed a training regimen of 60 minutes of cycling at 75–80% of VO_2max for 5 days per week for 4.5 weeks, and then consumed either chocolate milk or a carbohydrate beverage following each training session.[27] Individuals who drank the chocolate milk showed reduced fat mass, increased lean mass and more than doubled VO_2max compared to those who consumed carbohydrates only.

To summarize, chocolate milk provides a practical, convenient and palatable package that, when consumed following a single exercise bout, delivers nutrients to the body that are used to reverse damage done to energy stores and muscle protein tissue. As a result, a person may be able to subsequently train or exercise more effectively than when no recovery nutrition was provided or when an alternative beverage, particularly one without protein, was consumed. It is true that chocolate milk is at least equal to its carbohydrate-only or combined protein–carbohydrate peers in some respects of physiologic exercise recovery, but what remains unique about chocolate milk is its appeal to those who prefer a whole-food source of nutrition for their postexercise recovery needs. Chocolate milk is also a less expensive option to pre-formulated milkshakes and beverages that may be impractical to purchase on a regular basis. However, despite chocolate milk's popularity in the exercise recovery market, it is not the only chocolate product that offers benefit, nor is macronutrient replenishment the only recovery consideration when you complete an exercise bout.

8.4 Chocolate Bars and Cocoa

Chocolate milk has indeed cemented itself into its rightful place among postexercise recovery food products on the merit of its nutritional utility, convenience and palatability. But chocolate bars and cocoa have also garnered attention with their effects following consumption during the exercise recovery period. What was once a food item reserved solely for an occasional treat or a reprimand from the dentist may have nutritional benefit in the context of exercise recovery. Furthermore, it appears that ingestion of a chocolate bar pre-exercise affords benefits in the postexercise interval. Chen *et al.* showed in 1996 that young adult males who consumed a chocolate bar 15 minutes before a 60 minutes run at moderate effort experienced greater plasma glucose levels up to 30 minutes into recovery compared to the consumption of a placebo bar.[43] Such a result certainly satisfies the requisite of the elevated plasma glucose needed to support resynthesis of muscle glycogen following an endurance exercise bout. Furthermore, measures of blood lactate concentration, plasma urea nitrogen and rating of perceived exertion were all lower during recovery compared to the placebo, thereby indicating the creation of a positive metabolic environment during recovery that prepares the person for a subsequent exercise bout. Lower lactate concentration implies lower free hydrogen ion levels in the blood and, therefore, a diminished acidic condition, and lower urea nitrogen indicates less protein breakdown occurring during exercise (and, therefore, a smaller catabolic condition during recovery). Similarly, consumption of the chocolate bars during exercise may not elicit metabolic disturbances during the exercise bout that would set up unfavorable postexercise recovery conditions. Following a transient increase of blood glucose concentration shortly after consumption, the levels dropped to pre-exercise levels in trained cyclists performing a 90 minutes ride at moderate effort, and blood lactate, carbohydrate oxidation, fat oxidation and free fatty acid concentration were no different during exercise compared to those who consumed a pre-exercise placebo.[44] In short, any disadvantageous metabolic responses caused by consuming a chocolate bar before exercise were not observed.

Two components of chocolate food products that have attracted attention in the nutritional recovery community is cocoa and cocoa flavanols. While it is known that flavanols can improve vascular dilatation, the use of cocoa to influence exercise-induced changes to blood pressure is a logical curiosity. Older adults who cycled at a moderately-challenging intensity 2 hours after consuming a high-flavanol beverage experienced lower diastolic and mean blood pressures during exercise compared to a low-flavanol drink.[45] What does this mean for the postexercise recovery? A lower exercise blood pressure may elicit a smaller disturbance from pre-exercise values from which the person would need to return. In other words, in the context of blood pressure, the recovery interval may be shorter.

8.5 Chocolate and Postexercise Antioxidation

The flavanol content of cocoa has aroused interest in another popular postexercise recovery endpoint: antioxidation. The exercise paradox stands as the reason for cocoa's popularity in this facet of nutrition research. While myriad health benefits of physical activity are apparent, it is also known that strenuous exercise elicits oxidative stress that produces inflammatory agents such as isoprostanes. These oxidants result from the cellular lipid bilayer peroxidation occurring due to exercise and have been associated with an increased risk of heart attacks. Nutritional strategies to reverse the exercise-induced production of oxidants, therefore, are sensible.

Individuals who consumed a high-flavanol cocoa beverage following a strenuous exercise bout experienced a smaller concentration of plasma isoprostanes up to 4 hours after exercise compared to a low-flavanol cocoa drink.[46] Similarly, healthy adults who consumed a dark chocolate bar (naturally containing flavanols) immediately before a 2.5 hours endurance exercise bout at moderately easy intensity had fewer plasma isoprostanes at 1 hour into their recovery.[47] During recovery, fewer oxidants may mean less work for the body's antioxidation mechanism to return the disturbance to normal, again implying a shorter recovery. So, acute flavanol consumption either before or following exercise has a clear influence on oxidant status during recovery. What if the cocoa flavanols are consumed habitually? Males who regularly consumed dark chocolate for 2 weeks produced significantly less isoprostanes during an exhaustive exercise bout compared to those who ingested a cocoa-free control food, and the attenuated isoprostanes persisted for 1 hour into recovery.[48]

Anecdotally, it follows that such suppression of oxidation during exercise that lingers into the recovery interval may allow an individual to recover quicker, which may allow more rapid, positive adaptations to an endurance exercise stimulus. In other words, can there be an ergogenic benefit derived solely from the flavanol content of chocolate because a person simply recovers faster and can therefore exercise at a higher intensity? For humans, evidence of such a claim remains to be seen. However, animal studies may provide insight into the potential for chocolate flavanol ergogenicity in humans. Mice that were given the chocolate flavanol (−)-epicatechin twice daily over 15 days of treadmill exercise experienced greater mitochondrial and capillary density, greater work performed during exercise, increased resistance to fatigue and more proteins in skeletal muscle cells associated with oxidative phosphorylation compared to exercise alone.[49] These data suggest that chocolate flavanol consumption allows more rapid recovery from exercise, enabling the exerciser to perform more intensely in a subsequent exercise bout and providing the overload stimulus needed to make positive adaptations to metabolic and structural components (*i.e.*, the training effect). While several factors are involved in explaining the holistic effect of nutrition and training on endurance capacity, and it is premature to label flavanol consumption as the pre-eminent cause of the effect, it is interesting to observe that chocolate may play a role in suppressing the oxidation events that may prolong the recovery period.

8.6 Chocolate and Postexercise Mood State

Following the previous discussions of chocolate's impact on the physiology of the human body, it would be easy to overlook chocolate's effect on mood state, but we would be remiss to intentionally omit it. Psychological mood is a crucial component of exercise recovery in order to assess comprehensive preparedness for a subsequent exercise bout. If an individual is not motivated to exercise, the exercise bout may be unsatisfying or may not occur at all. This theory of chocolate's effect on mood was tested on rats, which were provided cocoa flavanols for 2 weeks before performing a forced swimming test. The animals experienced a shorter duration of immobility (recovery) between tests compared to those consuming no flavanol, thereby indicating an antidepressant effect from the flavanol nutrient.[50] In humans, the effect is similar. Adults who consumed chocolate milk following a treadmill run showed reduced mental fatigue and tension when assessed during the postexercise recovery, and although the effect was observed without chocolate consumption, both observations of improved mood were made when pre-exercise breakfast was consumed.[51] In brief, chocolate milk at least supports the benefits of breakfast on postexercise mood and does nothing to detract from it.

A significant concern for many active individuals is a reduction in soreness following an exercise bout. As musculoskeletal soreness persists, so does the diminished motivation to complete subsequent exercise. Can chocolate diminish soreness? Adult males who completed a strenuous, declined-grade treadmill running test drank a cocoa-containing protein and carbohydrate beverage immediately after the bout.[52] The most notable observation was the significant decrease in perceived muscle soreness for up to 48 hours after the exercise compared to a control beverage. While physiological responses drive the mechanics of exercise recovery to support successful completion of a subsequent exercise bout, the psychological benefits apparently afforded by chocolate certainly help to market chocolate food products as viable postexercise recovery nutrition options.

8.7 Conclusion

Gone are the days of the public perception of chocolate bars and chocolate milk as an occasional, guilt-inducing treat. Science has recognized the nutrient composition of chocolate foods. Furthermore, it has shown this nutritional profile to be the same as that which is mandatory for the comprehensive recovery of human physiology following an exercise bout. Whether it is the high-quality protein and carbohydrate in chocolate milk that support positive changes in muscle protein metabolism and carbohydrate storage, respectively, or the antioxidant micronutrients that can elevate mood state, contemporary research in the use of chocolate in exercise recovery nutrition has resulted in convincing and encouraging evidence. At the very least, science has demonstrated that chocolate deserves its rightful place among other traditional postexercise recovery foods.

References

1. P. Felig and J. Wahren, *N. Engl. J. Med.*, 1975, **293**, 1078.
2. J. D. MacDougall, G. R. Ward and J. R. Sutton, *J. Appl. Physiol.*, 1977, **42**, 129.
3. J. L. Ivy, M. C. Lee, J. T. Brozinick Jr. and M. J. Reed, *J. Appl. Physiol.*, 1988, **65**, 2018.
4. J. L. Ivy, A. L. Katz, C. L. Cutler, W. M. Sherman and E. F. Coyle, *J. Appl. Physiol.*, 1988, **64**, 1480.
5. R. Jentjens and A. Jeukendrup, *Sports Med.*, 2003, **33**, 117.
6. K. Tsintzas, C. Williams, L. Boobis, S. Symington, J. Moorehouse, P. Garcia-Roves and C. Nicholas, *Int. J. Sports Med.*, 2003, **24**, 452.
7. L. J. Brown, A. W. Midgley, R. V. Vince, L. A. Madden and L. R. McNaughton, *J. Sci. Med. Sport*, 2013, **16**, 450.
8. L. M. Burke, G. R. Collier and M. Hargreaves, *J. Appl. Physiol.*, 1993, **75**, 1019.
9. K. M. Zawadzki, B. B. Yaspelkis III and J. L. Ivy, *J. Appl. Physiol.*, 1992, **72**, 1854.
10. M. Williams, P. B. Raven, D. L. Fogt and J. L. Ivy, *J. Strength Cond. Res.*, 2003, **17**, 12.

11. L. J. van Loon, W. H. Saris, M. Kruijshoop and A. J. Wagenmakers, *Am. J. Clin. Nutr.*, 2000, **72**, 106.
12. T. P. Gunnarsson, M. Bendiksen, R. Bischoff, P. M. Christensen, B. Lesivig, K. Madsen, F. Stephens, P. Greenhaff, P. Krustup and J. Bangsbo, *Scand. J. Med. Sci. Sports*, 2013, **23**, 508.
13. R. L. Jentjens, L. J. van Loon, C. H. Mann, A. J. Wagenmakers and A. E. Jeukendrup, *J. Appl. Physiol.*, 2001, **91**, 839.
14. S. M. Phillips, K. D. Tipton, A. Aarsland, S. E. Wolf and R. R. Wolfe, *Am. J. Physiol.*, 1997, **273**, E99.
15. H. C. Dreyer, S. Fuijita, J. G. Cadenas, D. L. Chinkes, E. Volpi and B. B. Rasmussen, *J. Physiol.*, 2006, **576**, 613.
16. J. R. Poortmans, A. Carpenter, L. O. Pereira-Lancha and A. Lancha Jr., *Braz. J. Med. Biol. Res.*, 2012, **45**, 875.
17. D. W. West, N. A. Burd, V. G. Coffey, S. K. Baker, L. M. Burke, J. A. Hawley, D. R. Moore, T. Stellingwerff and S. M. Phillips, *Am. J. Clin. Nutr.*, 2011, **94**, 795.
18. S. M. Pasiakos, H. L. McClung, J. P. McClung, L. M. Margolis, N. E. Andersen, G. J. Cloutier, M. A. Pikosky, J. C. Rood, R. A. Fielding and A. J. Young, *Am. J. Clin. Nutr.*, 2011, **94**, 809.
19. H. C. Dreyer, M. J. Drummond, B. Pennings, S. Fujita, E. L. Glynn, D. L. Chinkes, S. Dhanani, E. Volpi and B. B. Rasmussen, *Am. J. Physiol.: Endocrinol. Metab.*, 2008, **294**, E392.
20. R. Koopman, L. Verdijk, R. J. Manders, A. P. Gijsen, M. Gorselink, E. Pijpers, A. J. Wagenmakers and L. J. van Loon, *Am. J. Clin. Nutr.*, 2006, **84**, 623.
21. L. J. van Loon, *Int. J. Sport Nutr. Exercise Metab.*, 2007, **17**, S104.
22. P. T. Reidy, D. K. Walker, J. M. Dickinson, D. M. Gundermann, M. J. Drummond, K. L. Timmerman, C. S. Fry, M. S. Borack, M. B. Cope, R. Mukherjea, K. Jennings, E. Volpi and B. B. Rasmussen, *J. Nutr.*, 2013, **143**, 410.
23. D. R. Moore, M. J. Robinson, J. L. Fry, J. E. Tang, E. I. Glover, S. B. Wilkinson, T. Prior, M. A. Tarnopolsky and S. M. Phillips, *Am. J. Clin. Nutr.*, 2009, **89**, 161.
24. N. E. Deutz and R. R. Wolfe, *Clin. Nutr.*, 2013, **32**, 309.
25. M. A. Tarnopolsky, J. D. MacDougall and S. A. Atkinson, *J. Appl. Physiol.*, 1988, **64**, 187.
26. J. L. Areta, L. M. Burke, M. L. Ross, D. M. Camera, D. W. West, E. M. Broad, N. A. Jeacocke, D. R. Moore, T. Stellingwerff, S. M. Phillips, J. A. Hawley and V. G. Coffey, *J. Physiol.*, 2013, **591**, 2319.
27. L. Ferguson-Stegall, E. McCleave, Z. Ding, P. G. Doerner III, Y. Liu, B. Wang, M. Healy, M. Kleinert, B. Dessard, D. G. Lassiter, L. Kammer and J. L. Ivy, *J. Nutr. Metab.*, 2011, **2011**, 1.
28. R. Koopman, A. J. Wagenmakers, R. J. Manders, A. H. Zorenc, J. M. Senden, M. Gorselink, H. A. Keizerand and L. J. van Loon, *Am. J. Physiol.: Endocrinol. Metab.*, 2005, **288**, E645.
29. K. R. Howarth, N. A. Moreau, S. M. Phillips and M. J. Gibala, *J. Appl. Physiol.*, 2009, **106**, 1394.

30. R. Koopman, M. Beelen, T. Stellingwerff, B. Pennings, W. H. Saris, A. K. Kies, H. Kuipers and L. J. van Loon, *Am. J. Physiol.: Endocrinol. Metab.*, 2007, **293**, E833.
31. J. Peter and C. H. Wyndham, *J. Physiol.*, 1966, **187**, 583.
32. E. R. Nadel, S. M. Fortney and C. B. Wenger, *J. Appl. Physiol.*, 1980, **49**, 715.
33. A. Hernandez, A. Cheng and H. Westerblad, *Front. Physiol.*, 2012, **3**, 46.
34. T. J. Smith, S. J. Montain, D. Anderson and A. J. Young, *Int. J. Sport Nutr. Exercise Metab.*, 2009, **19**, 1.
35. J. R. Karp, J. D. Johnston, S. Tecklenburg, T. D. Mickleborough, A. D. Fly and J. M. Stager, *Int. J. Sport Nutr. Exercise Metab.*, 2006, **16**, 78.
36. K. Thomas, P. Morris and E. Stevenson, *Appl. Physiol. Nutr. Metab.*, 2009, **34**, 78.
37. K. L. Pritchett, P. Bishop, R. C. Pritchett, J. M. Green and C. Katica, *Appl. Physiol., Nutr., Metab.*, 2009, **34**, 1017.
38. K. J. Spaccarotella and W. D. Andzel, *J. Strength Cond. Res.*, 2011, **25**, 3456.
39. S. F. Gilson, M. J. Saunders, C. W. Moran, R. W. Moore, C. J. Womack and M. K. Todd, *J. Int. Soc. Sports Nutr.*, 2010, **7**, 19.
40. K. L. Pritchett, R. C. Pritchett, J. M. Green, C. Katica, B. Combs, M. Eldridge and P. Bishop, *J. Exerc. Physiol.*, 2011, **14**, 29.
41. L. Ferguson-Stegall, E. L. McCleave, Z. Ding, P. G. Doerner III, B. Wang, Y. H. Liao, L. Kammer, Y. Liu, J. Hwang, B. M. Dessard and J. L. Ivy, *J. Strength Cond. Res.*, 2011, **25**, 1210.
42. W. R. Lunn, S. M. Pasiakos, M. R. Colletto, K. E. Karfonta, J. W. Carbone, J. M. Anderson and N. R. Rodriguez, *Med. Sci. Sports Exercise*, 2012, **44**, 682.
43. J. D. Chen, H. Ai, J. D. Shi, Y. Z. Wu and Z. M. Chen, *Biomed. Environ. Sci.*, 1996, **9**, 247.
44. J. C. Alberici, P. A. Farrell, P. M. Kris-Etherton and C. A. Shively, *Int. J. Sport Nutr.*, 1993, **3**, 323.
45. N. M. Berry, K. Davison, A. M. Coates, J. D. Buckley and P. R. Howe, *Br. J. Nutr.*, 2010, **103**, 1480.
46. I. Wiswedel, D. Hirsch, S. Kropf, M. Gruening, E. Pfister, T. Schewe and H. Sies, *Free Radical Biol. Med.*, 2004, **37**, 411.
47. G. Davison, R. Callister, G. Williamson, K. A. Cooper and M. Gleeson, *Eur. J. Nutr.*, 2012, **51**, 69.
48. J. Allgrove, E. Farrell, M. Gleeson, G. Williamson and K. Cooper, *Int. J. Sport Nutr. Exercise Metab.*, 2011, **21**, 113.
49. L. Nogueira, I. Ramirez-Sanchez, G. A. Perkins, A. Murphy, P. R. Taub, G. Ceballos, F. J. Villarreal, M. C. Hogan and M. H. Malek, *J. Physiol.*, 2011, **589**, 4615.
50. M. Messaoudi, J. F. Bisson, A. Nejdi, P. Rozan and H. Javelot, *Nutr. Neurosci.*, 2008, **11**, 269.
51. R. C. Veasey, J. T. Gonzalez, D. O. Kennedy, C. F. Haskell and E. J. Stevenson, *Appetite*, 2013, **68**, 38.
52. N. M. McBrier, G. L. Vairo, D. Bagshaw, J. M. Lekan, P. L. Bordi and P. M. Kris-Etherton, *J. Strength Cond. Res.*, 2010, **24**, 2203.

CHAPTER 9

Chocolate, Cocoa and Women's Health

ELEONORA BRILLO AND GIAN CARLO DI RENZO*

University of Perugia, Italy, Dept. of Obstetrics and Gynecology and Centre for Perinatal and Reproductive Medicine, S. Maria della Misericordia University Hospital, 06132 Perugia, Italy
*Email: direnzo@unipg.it

9.1 Introduction

Cocoa is obtained from a long and complex work process to which cacao beans are subjected (Figure 9.1). The Swedish botanist Carl Linnaeus named the plant *Theobroma cacao*, a combined term in which the generic name (*i.e.*, the genus) is derived from the Greek for "food of the gods" (from *theos*, meaning "god", and *broma*, meaning "food"), whereas the specific name (*i.e.*, the species) most likely derives from the words *kakaw*, *cacahuatl* or *kagaw*, which were used to identify the cacao plant in indigenous Mesoamerican languages. According to the Directive 2000/36/EC, chocolate is defined as a food obtained from cocoa and sugars, containing at least 35% total dry cocoa, comprised of not less than 18% cocoa butter and not less than 14% degreased dry cocoa. In addition to the minimum percentages of cocoa butter and total dry cocoa, a maximum of 5% fat cocoa butter equivalents by weight of the finished product is allowed.

In the course of its 3000 year history, cocoa has met admirers from all cultural environments and walks of life. Long before being so named, *Theobroma cacao* fruits were considered "food of the gods" among pre-Columbian populations. The Aztecs believed that the cacao tree pods symbolized life and fertility and that, by feeding on cocoa, they could gain

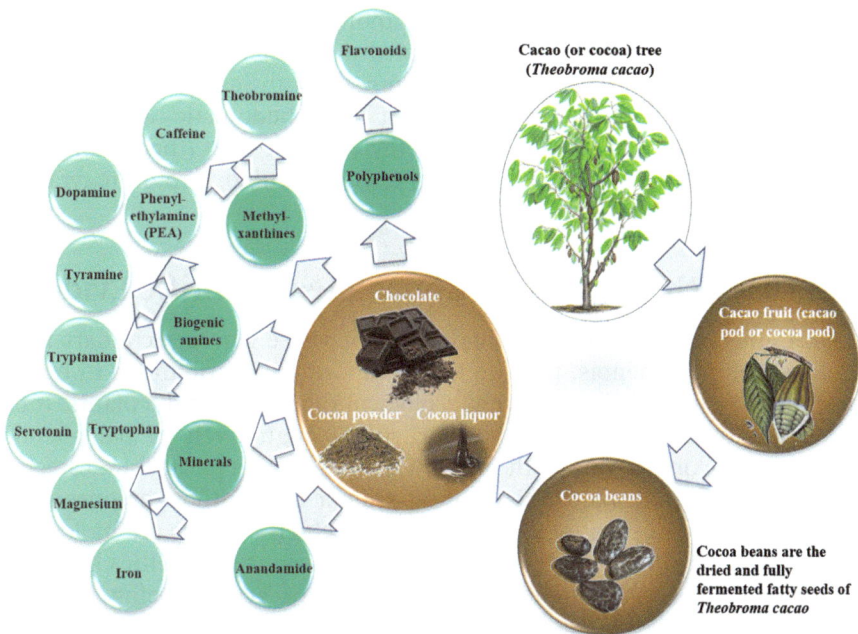

Figure 9.1 Representation of the transformation of cocoa into chocolate and of the biologically active substances in cocoa-derived products.

wisdom and power. An infusion of cacao beans was flavored with spices and enriched with other foods to create a bitter drink, which was considered to be an elixir of health.

Over the millennia, a multi-cultural following has developed around this product that appreciates its hedonic value as well as its health and therapeutic properties. Among these healing properties are its recognition as a tonic, energetic, stimulant, aphrodisiac, analeptic, anti-inflammatory agent, promoter of fertility and digestion and anti-depressant, among many other.[1–3] However, the appreciation and the popularity of these healing properties have fluctuated widely over the centuries.[3] The recent resurgence of interest in cocoa's real health effects was derived from epidemiological studies that have shown how dietary patterns in populations of San Blas Kuna islanders, individuals who consumed considerable quantities of cocoa-derived food, were correlated with a low prevalence of hypertension, atherosclerosis, dyslipidemia and diabetes,[4–6] as well as cancer-related mortality and morbidity in general.[7]

A multitude of studies distinct from the epidemiological studies have shown the relevance of the phytocomplex *T. cacao* for general well-being and, in particular, for the cardiovascular system.[8–11] These positive effects have overturned the negative aura surrounding foods containing cocoa that had fueled widespread opinion for years that cocoa was responsible for uncontrolled weight gain, acne and dental caries. Further, biologically active components capable of interacting through multiple mechanisms were identified in cocoa and chocolate. Key among the healthful components of cocoa are polyphenols that, when coupled with other bioactive substances, are able to intervene in neurotransmitter pathways with consequent psychostimulant, tonic, anti-depressant and hedonistic effects. These substances chemically belong to the methylxanthine class (theobromine and caffeine)[12,13] of the biogenic amines (phenylethylamine, serotonin, tyramine and tryptamine).

9.2 Health, Cocoa and Chocolate: Scientific Evidence

Cocoa and its derivatives are obtained from a tree whose secondary metabolites, the polyphenols, provide many important contributions to human health, as well as performing vital ecological functions for the plants. It has been found that cocoa contains particularly high levels of polyphenols, especially flavonoids[14,15] (with prevailing amounts of epicatechin, catechin, procyanidin B2 and B5, trimer C and tetramer D[16]). Although concerns regarding the bioavailability and the extent to which phenols are biologically active have not been completely clarified,[17–23] various capabilities have been assigned to cocoa flavonoids, with anti-oxidant activity being the most predominant. Several studies analyzing the biological properties of these molecules have provided considerable supportive evidence of flavonoid's role in improving vascular functions, preventing related endothelial dysfunctions and diseases,[24–27] reducing insulin-resistance indices[26,28] (probably as a result of a decrease in insulin secretion[28]), increasing insulin sensibility,[26,29] inhibiting platelet aggregation and activation[30–34] and decreasing blood pressure.[9,23,25,29,34–38] However, a meta-analysis of randomized trials published in 2012 did not confirm the flavonoid effect on systolic blood pressure.[28] In addition to the extensive list of proposed health properties, one also finds cocoa contributing to anti-inflammatory reactions[39–41] and improving total plasma anti-oxidation,[42,43] cognitive capacity[44] and depressive states.[45] Multiple trials have emphasized that cocoa has the ability to disrupt the lipid fraction not only by limiting oxidation, but also by directly altering the lipid profile of plasma by increasing high-density lipoproteins[23,46–50] and reducing low-density lipoprotein (LDL)[23,41,48–51] and total cholesterol.[23,50,52,53] The qualitative and quantitative characteristics of the flavonoid molecules are influenced by *T. cacao*'s genetic, climatic and agronomic factors.[54] They are also influenced by the processing of cocoa beans and by the bean's ultimate transformation into food derivatives.[16,55–59] The different types of chocolate on the market have a variable amount of phenols, which is generally proportional to the cocoa content; therefore, the amount of phenols is usually higher in dark chocolate than in milk chocolate, and it is usually irrelevant in white chocolate.[60] These phytochemicals represent only a fraction of chocolate's complex chemical composition, as it has more than 600 identified compounds including minerals, vitamins, methylxanthines and biogenic amines, as well as some anti-nutritional components including phytic acid, oxalic acid, chlorogenic acid, caffeic acid and select polyphenols.

The protection of cardiovascular and endothelial tissue as a consequence of chocolate consumption may arise due to the synergetic action of flavonoids, theobromine and magnesium;[61] indeed, chocolate contains a high concentration of these molecules[12,14,62] and, in practice, it is the only possessor of 3,7-dimetilxantina.[27,61] In any case, cocoa flavonoids have demonstrated an intense activity at the endothelial level,[26,29] expressed through different mechanisms, including anti-inflammation and the modulation of the synthesis and expression of the proteins and secondary messengers involved in many cellular reactions.

Flavonoids are efficient activators of nitric oxide (NO) enzyme synthase, thereby showing a capacity to improve vascular function and to prevent endothelial dysfunction and related diseases.[2–26,37] Cocoa flavonoid activity, as expressed through alternative or additional control pressure mechanisms, may be involved in the phenolic effects of angiotensin-converting enzyme inhibition.[63]

Cocoa procyanidins have the ability to influence levels of leukotrienes (potent vasoconstrictors, proinflammatory agents and stimulants of platelet aggregation) and prostacyclins (vasodilators and inhibitors of platelet aggregation).[39] *In vitro* studies have shown that chocolate's components are able to inhibit the lipoxygenase pathway, which leads to an increase of proinflammatory leukotrienes.[40,64] Cocoa flavonoids appear to influence the inflammatory process through mechanisms ranging from the reduction of the interleukin-2 secretion to the control of other cytokines (*e.g.*, IL-5, TNF-α and TGF-β).[42,65] Enhancements in psychological functioning induced by consuming chocolate with a high cocoa content are due, at least in part, to the combination of methylxanthines,[12] biogenic amines,[66] anandamide (*N*-arachidonoyl-ethanolamine) and *N*-acylethanolamines actions.

The main monoamines contained in cocoa are phenylethylamine, tyramine and tryptamine.

Tyramine and phenylethylamine have the ability to act on different areas of the brain that are responsible for mood control and the waking state. They are able to delay fatigue and induce the release of catecholamines (norepinephrine and dopamine) at the synaptic level, which consequently produces stimulation similar to that induced by amphetamine, including an attenuation of the sensation of hunger. Phenylethylamine, assisted by magnesium (found in abundance in chocolate) and small amounts of serotonin, acts on mood. The mechanism of action of these amines on improving depressive states consists of an increase of catecholaminergic transmission.

The modulation of catecholaminergic transmission is also the aim of the therapeutic approaches based on current neurobiological theories that are based on the causal relationship between deficient monoaminergic transmission in the central nervous system and endogenous depression. Anandamide, an endogenous lipid belonging to the class of endocannabinoids, acts as a cannabinoid receptor agonist, mimicking the central and peripheral action of cannabinoids such as Δ9-tetrahydrocannabinol (the active ingredient of cannabis extracts). Although an interesting double-blind trial has recognized a marginal role of all of the psychoactive substances apart from xanthine,[12] the biological activity of the amines and anandamide may contribute to the feeling of reward derived from the act of consuming chocolate and the pleasant post-consumption euphoria. Studies carried out on the abuse of stimulants suggest a causal relationship with self-therapy for mood control resulting from cocoa intake.[67] An apparent desire for chocolate intake can become uncontrollable, presenting as a condition known as "chocolate craving". The pleasant sensations derived from the fulfillment of this intense need can lead to a loss of intake control, a condition termed "chocolate binge",[68] which can flow into "chocoholism", which is the dependence on chocolate consumption, a condition that is also termed "chocolate addiction".

Similar to questioning whether and how the pharmacological effects of chocolate on mood control are directly related to the admittedly low levels of psychoactive substances it contains, discerning whether the phenomenon of addiction can derive exclusively from a chemical–pharmacological mechanism also remains uncertain. More likely, human dependency upon chocolate derives from an association of the effects due to the chemical substances and those due to psychological mechanisms. Nevertheless, some evidence suggests that inducing mechanisms of craving[12,69] remains the sole responsibility of taste and orosensory satisfaction regarding food itself. Macronutrients play an important role in chocolate beyond mere caloric intake. Regarding the psychological aspect, it is the carbohydrate component of chocolate that produces a sizeable effect upon the mechanism of addiction. Firstly, the taste of sugar leads the body to experience sensory pleasure, then the carbohydrates and lipids induce the release of endorphins, the neurotransmitters that produce gratification, satisfaction and euphoria. The desire to prolong the pleasure and the search for new reward cues are inherent in the brain's biological mechanisms that lead to the consumption of substances that are capable of satisfying hedonic needs. Although not all chocolate contains the same percentage of fat, the lipid portion is consistent in each case, and this is what is responsible for the high energy value of the food. However, this value is often influenced by an interaction of food components that may be different from that based solely on the proportional valuation of the cocoa alkaloid molecules themselves.[70]

The quantity and quality of the lipid components of chocolate are protected by European Community standards designed to ensure the maintenance of chocolate's natural fats (*i.e.*, cocoa butter). This is made up of 97–98% triglycerides; about two-thirds of the triglyceride fatty acid content is saturated (palmitic and stearic) and the rest is monounsaturated (oleic), with only a tiny percentage being polyunsaturated (linoleic and arachidonic). Although the proportion of saturated fats is conspicuous, no direct correlations can be made between the amount and the extent of the known cholesterol-raising and atherogenic effects of these fats. The atypia of the saturated fat behavior is due to the activity of stearic acid, activity that is

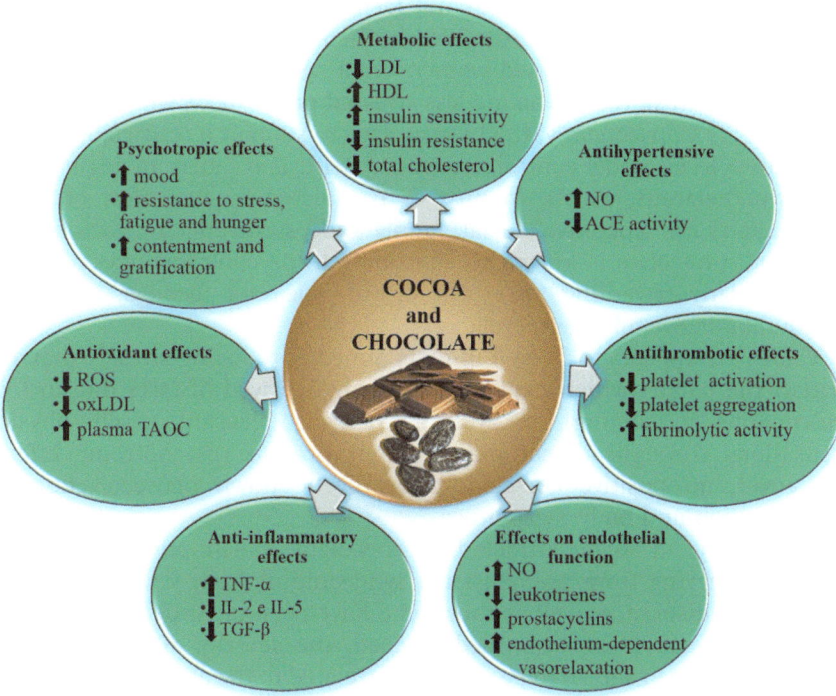

Figure 9.2 Positive effects of cocoa and chocolate. (LDL: low-density lipoprotein; HDL: high-density lipoprotein; NO: nitric oxide; ACE: angiotensin-converting enzyme; TNF-α: tumor necrosis factor-α; IL: interleukin; TGF-β: transforming growth factor-β; ROS: reactive oxygen species; TAOC: total anti-oxidant capacity).

more attributable to the unsaturated than monounsaturated moiety[71–76] due to the desaturation experienced in the body and its consequent transformation into oleic acid. Oleic acid is considered to be a cardiovascular disease prevention element[77,78] that reduces total cholesterol[79,80] and LDL levels. It follows that most of the fatty acids of cocoa butter assume a neutral, if not positive, role on health.[81] At any rate, many types of chocolate on the market have a reduced calorie content and offer low-carbohydrate or low-fat contents. The overall favorable effects of cocoa and chocolate are schematically represented in Figure 9.2.

9.3 Female Mental Well-being

In some cultures, the relationship between food and female gender is conflicted and, at times, represents a major source of stress in females. In the Western world, a great number of women are dissatisfied with their body image, and they spend considerable time counting calories and restricting dietary intake in order to reduce body weight.[82] Constantly watching over nutrition and body balance does not facilitate the diet itself; rather, it increases "stress hormone" (cortisol) levels, thereby triggering a higher demand for caloric foods.[83–85] Stressed women of all ages prefer sweet and high-fat foods; by contrast, those who are less stressed prefer foods with a lower lipid content.[86–88] Consequently, the best strategy to support a restricted diet is to keep stress levels under control. Hunger, appetite and satiety are regulated *via* the hypothalamus, which, in turn, is modulated by the effect of neurotransmitters and cognitive, social, family and psycho-affective factors, such as mood. Chocolate is known to contribute to psychotropic effects, deliver a particularly rewarding taste and provide pleasant feelings, as well as re-equilibrate overall stress levels.

Chocolate appears to fit well in diets aimed at reducing stress[89] that also satisfy the palate (which may be commonly deprived of taste due to food restrictions in dieting), as well as offering pleasant psycho-physical sensations through its components that produce psychotropic effects. Moreover, no positive correlation has been shown between chocolate consumption and body mass index.[90]

Females seem to be attracted to chocolate consumption,[90] especially women suffering from premenstrual syndrome or from seasonal affective disorder or disease, as well as those who have been identified as dysphoric, "emotional eaters" or depressed. Indeed, the severity of the depressive disorder has been shown to be directly correlated to the amount of chocolate consumed in order to counter such conditions.[91] Genetic and biological factors, especially hormonal fluctuations, affect female mood swings.[92] Women are more prevalent than men for unipolar depression,[92] thus presenting a greater opportunity for exploring the feminine sphere of chocolate consumption. The affinity between women and chocolate is probably due in part to mood. The female psyche is particularly susceptible to fascination regarding this type of food, and its consequent consumption is often noted as a self-treatment type of psychological approach among women.

Gender differences also exist in the phenomenon of craving chocolate, which has a higher incidence in women than men,[93–96] with women experiencing their highest cravings during the perimenstrual period,[96] though hormone variation has not been proven to be the root cause.[95] An observational study also demonstrated a high frequency of this phenomenon among the post-menopausal population.[95] In addition to the positive effects that cocoa and chocolate provide to mental and physical health, particularly in relation to improvements in cardiovascular health,[97] cocoa's virtues also demonstrate gender-specific effects.

9.4 Cocoa and Reproduction

Cocoa and chocolate are at least indirectly associated with human reproduction due to their possible influence upon sexual desire and upon possible factors that stimulate fertility. An aphrodisiac value has been attributed to these substances since ancient times, when they were consumed by pre-

Columbian peoples.[1,3] The use of cocoa by the Tlatoani Aztec, Montezuma II, has become legendary primarily because he employed it as a tonic to support ongoing sexual relations with his many wives. Famous lustful lovers in later eras, including the incomparable Giacomo Girolamo Casanova, have also praised chocolate as an erotic stimulant, as have the gastronome Jean Anthelme Brillat-Savarin and the Italian nationalist poet Gabriele d'Annunzio. Although scientific evidence has yet to support this reputed erotic quality of chocolate, its vasoprotective, anti-depressant and tonic-stimulating qualities likely function to maintain sexual health. Male and female sexual health are complex phenomena in which physical, mental, relational and cultural characteristics interact, and chocolate seems to positively contribute to this complexity.

The sensory characteristics of cocoa create an intimate connection between chocolate and sex; thus, we can consider chocolate as a sexual food because it carries sensorial stimulation. On the other hand, the multi-sensory experience that chocolate provides is a strong stimulus for the most powerful erogenous organ: the brain. Today, an increasing number of women claim to receive more gratification from chocolate than sex such that, between the two, chocolate was preferred by 60% of Italian women who participated in an online survey supervised by the Italian Society of Obstetrics and Gynecology in 2009.[98] London females expressed only a slightly different numerical finding. The Agence France-Presse reported results in 2007 of a survey backed by the Cadbury Chocolate Company that showed a preference of chocolate over sex in 52% of the female subjects in the sample.

Chocolate as an *entrée* to coital rapport marks the meeting of gluttony and lust, providing consummatory satisfaction not only of the palate, but also in terms of igniting sexual desire. The organoleptic characteristics of chocolate can provide intense and varied types of sensory stimulation, revealing an intrinsic voluptuousness correlated with the female and male libido. Although scientifically demonstrating a direct association between female sexual ardor and chocolate consumption has not always led to significant results,[99] this does not discount the contribution that chocolate may well provide to love biochemistry. Chocolate, the "food of the gods", is a pleasure food with sex appeal that also makes those who consume it feel better.

Cocoa consumption affects reproduction not only due to its effect upon sexual appetite, but also in possibly interfering with the etiology of subfertility. Cocoa polyphenols have been shown to be potent anti-oxidants, and oxidative stress (an imbalance between pro-oxidants and anti-oxidants due to decreased anti-oxidant defense mechanisms or an increase in reactive oxygen and/or reactive nitrogen species) is known to be involved in the causes of male and female infertility, reproductive diseases (including endometriosis, polycystic ovary syndrome and unexplained infertility) and pregnancy complications (namely pre-eclampsia and miscarriages).[100]

Although attempts have been made to prevent reproductive disorders through anti-oxidant supplementation, the effectiveness of this strategy has not been clearly demonstrated. For this reason, positive contributions of the phytochemical cocoa in preventing reproductive problems can only be speculated about, but the possible correlation continues to fuel its exploration through high-quality clinical trials. Moreover, the bioactive constituents of cocoa may contribute to reducing reproductive difficulties through actions directly exerted on the vascular endothelium and circulation. In further regards to its effects on reproductive capacity, it must also be noted that even in low doses, cocoa contains caffeine, a molecule that is known to cross the placental barrier freely[101] and is slowly metabolized during pregnancy. The fetus does not have sufficient enzymes to inactivate caffeine,[102,103] thus its metabolites accumulate in the fetal brain.[101,104] Caffeine, if consumed in large amounts, may seemingly result in reduced fecundity,[105,106] though other prospective studies have demonstrated either little or no effect[107,108] or even increased fecundity.[109] Caffeine has also been positively correlated with spontaneous abortion, congenital malformation, fetal death, fetal growth restriction, preterm delivery and decreased birth weight. Again, some studies have yielded conflicting results,[110–113] though caffeine has been conclusively demonstrated to decrease fetal weight and increase the risk of small-for-gestational-age fetal development.[114] In any case, caffeine seems to compromise normal reproductive function and to increase embryo–fetal risks. Accordingly, the World Health Organization's recommended threshold of caffeine consumption[114] (200 mg day^{-1} in the Nordic countries and USA,[115,116] with 300 mg day^{-1} equivalent to 4.6 mg kg^{-1} of body weight per day in a 65 kg person) should be adhered to.[117] Chocolate's added contribution to caffeine intake should be noted, particularly when other drinks that are rich in caffeine are already included in a diet.

9.5 Cocoa and Pregnancy

During pregnancy, so many hemodynamic and metabolic changes occur that considerable attention to maternal and fetal nutrition is warranted. From Prochownich's early 20th-century publication regarding the relationship of diet to pregnancy, attention to maternal–fetal nutrition has grown in terms of improving the endpoint of fetal health globally. During gestation, biomolecular metabolism and cellular redox activity undergo changes that shift the balance in favor of oxidizing agents and pro-oxidants,[118–120] thereby reducing the total plasma anti-oxidant capacity in advanced gestation.[121] The oxidation–reduction imbalance leads to oxidative stress-linked pathological conditions concerning both the mother and the fetus and, relatedly, the susceptibility of embryonic and syncytiotrophoblastic cells to oxidative damage. The alteration of the redox status is a constant of gestation, but its extent is significantly higher in cases leading to spontaneous abortion and in pregnancies with pre-eclampsia or with hypertension alone.[122] The gravidarum oxidative imbalance and the need for additional caloric intake between the 10th and 13th week of gestation[123] led to scrupulous food choices that combine a modest energetic intake (between 100 and 400 kcal, as a function of the index of pre-pregnancy body mass) with a high content of anti-oxidant molecules. The availability of foods with these characteristics and the inclusion of these foods in the diet of pregnant women are considered strategies for rationalizing economic and nutritional profiles. The available scientific evidence suggests that chocolate with a high cocoa content consumed daily in small quantities may properly fit into this nutritional strategy without entailing negative consequences in terms of weight during various trimesters.[124] Chocolate supplementation during pregnancy appears to reduce systolic and diastolic blood pressure during gestation,[124] (Figure 9.3) though this association was not found in all studies.[125] Considering the characteristics and risk factors of pre-eclampsia, including maternal hypertension, placental disease, endothelial

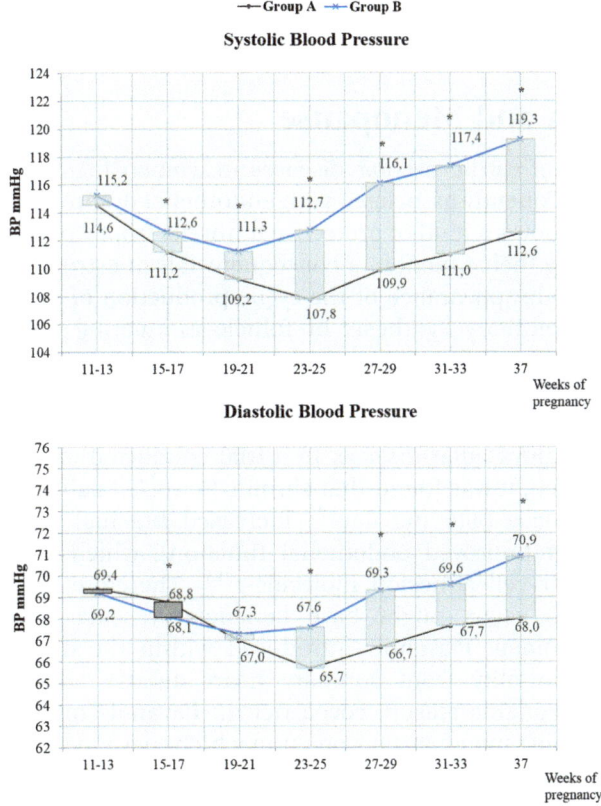

Figure 9.3 Blood pressure (systolic and diastolic) recorded from a sample of 90 pregnant women from a randomized controlled trial (Di Renzo et al., 2012).[124] Group A comprised 46 pregnant women with daily supplementation of 30 g of dark chocolate (70% cocoa). Group B comprised 44 pregnant women without chocolate supplementation. BP: blood pressure. *Statistically significant differences between the two groups ($p<0.05$).

dysfunction, oxidative stress, lack of NO,[126-128] the theoretical possibility of preventing pre-eclampsia by up-regulating NO availability due to antioxidant activity and the induction of NO-dependent vasodilatation by cocoa, it has been suggested that chocolate consumption could be a reasonable strategy for preventing this disease. In this regard, two recent studies—a cohort study and a case–control study—have noted chocolate consumption's contribution to reducing this disease risk,[61,129] though this assumption has not been confirmed by a further case–control study.[130] Although not all of cocoa's potential health benefits for pregnant women have been clearly confirmed, chocolate consumption during pregnancy has shown to be positive in some accounts and harmless in others.[124] Overall, there is a natural female preference for chocolate that becomes markedly more apparent during pregnancy, with an increasing trend that follows the progress of gestation.[131] Chocolate in a balanced diet of a pregnant woman can instill psychological well-being to both the pregnant woman (typically during a time of high emotional lability) and the future child.[132]

9.6 Cocoa and Menopause

The incidence of cardiovascular diseases in women increases after menopause and is frequently associated with endothelial dysfunction.[133] Entering menopause removes a kind of protective umbrella from woman in that when the ovaries stop their function, a consequent fall in estrogen levels occurs, along with the disappearance of the cyclic production of progesterone and an increased pituitary synthesis of follicle-stimulating hormone.[134] The disappearance of sex hormones, which typically offer a cardio-protective effect, correlates to an increase in LDL and an increase in obesity and carbohydrate disorder probabilities,[134] both of which are risk factors for developing cardiovascular disease. In a trial conducted on 32 hypercholesterolemic women, it was found that chronic (6 weeks) daily consumption of flavanol-rich cocoa (446 mg day^{-1}) increased brachial artery hyperemic blood flow (via improved endothelial function), which reduced plasma-soluble vascular cell adhesion molecule-1.[135] Therefore, a regular intake of flavanol-rich cocoa might have beneficial vascular effects in post-menopausal women. Flavanols introduced within the diet may decrease the risk of mortality from coronary heart disease and cardiovascular disease.[134,135] For this reason, cocoa rich in flavanols may be particularly important for post-menopausal women in which these metabolic-associated diseases have a higher incidence.

9.7 Cocoa and Beauty

Skin care has haunted women for centuries. In dermatologic research exploring products for the treatment of skin imperfections, cocoa components are frequently tested. Once chocolate was acquitted of causing skin blemishes, increasing attention was been directed to recognizing its positive properties for skin beauty. Knowledge of cocoa's ability to enhance the integumentary system has been handed down over time. The emollient and restorative properties of cocoa butter in particular were celebrated in the past centuries, though other properties of cocoa that are useful to dermatological application were only discovered in recent decades. Such experiments have shown that chocolate's biomolecules express cosmetic virtues, both *via* external and internal administration. A diet rich in anti-oxidants has a great value for the maintenance of skin integrity,[136,137] as well as for making important contributions to improving endogenous photoprotection systems.[138,139] This effect is attributed to two mechanisms: ultraviolet (UV) radiation absorption and free radical scavenging activity towards the reactive oxygen species formed during fat oxidation.[138] Women who consume these flavonoids daily may experience an increased density of epidermal and dermal hydration,[140] increased blood flow in the skin and subcutaneous tissues,[138,141] and decreased sensitivity to UV radiation.[138,139] Cocoa molecules thereby combat skin aging, reduce the processes of photoaging and oppose chronoaging. Systemic and/or topical application of cocoa flavanols help improve skin texture and functionality.

9.8 Conclusions

The bond linking psychological and physical wellness to diet is well known. Food is not only a source of exogenous anabolic reagents, but also a great supplier of biologically active compounds. Scientific research supports the hypothesis that chocolate constitutes one of the most rewarding and fulfilling vehicles for both genders, through which we surmise that molecules with biological properties are useful for the maintenance and rehabilitation of mental and physical well-being. Chocolate with a high cacao content has a great number of bioactive phytochemical anti-oxidants and considerable psychotropic substances that stimulate and regulate mood. Its high nutritional value is the limiting factor to chocolate consumption; however, it has been shown that products with a high cacao concentrations can provide an anti-oxidant contribution with only little concomitant energetic intake. In analyzing the positive and negative aspects of chocolate, it is concluded that consuming chocolate in moderation is good for human health! Chocolate's rehabilitative effects are due to cholesterol and blood pressure reductions, improved endothelial function, increased anti-oxidant power and mood improvement. Chocolate can also be used in the diet of pregnant woman to maximize potential nutrition, reduce the risk of certain pathological changes and optimize psychophysical wellness in the maternal–fetal stages and in the future child. Chocolate has sensorial and nutritional value, but it also carries great traditional and symbolic meaning that is closely correlated with love and sexual pleasure. Chocolate's chemical components produce intense multisensory stimulation and serve as a remedy to anhedonia. Though presenting cocoa as a panacea is scientifically unacceptable, at the same time, cocoa's multi-functionality can provide a wide range of potential health benefits.

References

1. T. L. Dillinger, P. Barriga, S. Escárcega, M. Jimenez, D. Salazar Lowe and L. E. Grivetti, *J. Nutr.*, 2000, **130**(Suppl 8), 2057S–2072S.
2. D. L. Pucciarelli and L. E. Grivetti, *Mol. Nutr. Food Res.*, 2008, **52**(10), 1215–1227.
3. D. Lippi, *Nutrition*, 2009, **25**(11–12), 1100–1103.
4. N. K. Hollenberg, *J. Cardiovasc. Pharmacol.*, 2006, **47**(Suppl 2), S99–S102, discussion S119–S121.
5. C. Ferri and G. Grassi, *J. Hypertens.*, 2003, **21**, 2231–2234.
6. M. L. McCullough, K. Chevaux, L. Jackson, M. Preston, G. Martinez, H. H. Schmitz, et al., *J. Cardiovasc. Pharmacol.*, 2006, **47**(Suppl 2), 103S–109S.
7. N. K. Hollenberg, N. D. Fisher and M. L. McCullough, *J. Am. Soc. Hypertens.*, 2009, **3**(2), 105–112.
8. P. M. Kris-Etherton and C. L. Keen, *Curr. Opin. Lipidol.*, 2002, **13**, 14–49.
9. B. Buijsse, E. J. Feskens, F. J. Kok, et al., *Arch. Intern. Med.*, 2006, **166**(4), 411–417.
10. V. Bayard, F. Chamorro, J. Motta, et al., *Int. J. Med. Sci.*, 2007, **4**, 53–58.
11. I. Janszky, K. J. Mukamal, R. Ljung, et al., *J. Intern. Med.*, 2009, **266**(3), 248–257.
12. H. J. Smit, E. A. Gaffan and P. J. Rogers, *Psychopharmacology*, 2004, **176**(3–4), 412–419.
13. H. J. Smit and R. J. Blackburn, *Psychopharmacology*, 2005, **181**(1), 101–106.
14. K. W. Lee, Y. J. Kim, H. J. Lee and C. Y. Lee, *J. Agric. Food Chem.*, 2003, **51**(25), 7292–7295.
15. R. M. Lamuela-Raventós, A. I. Romero-Pérez, C. Andrés-Lacueva and A. Tornero, *Food Sci. Technol. Int.*, 2005, **11**(3), 159–176.
16. K. A. Cooper, E. Campos-Gimenez, D. Jimenez Alvarez, K. Nagy, J. L. Donovan and G. Williamson, *J. Agric. Food Chem.*, 2007, **55**, 2841–2847.
17. M. Richelle, I. Tavazzi, M. Enslen and E. A. Offord, *Eur. J. Clin. Nutr.*, 1999, **53**, 22–26.
18. P. C. Hollman and M. B. Katan, *Food Chem. Toxicol.*, 1999, **37**, 937–942.
19. J. P. E. Spencer, F. Chaudry, A. S. Pannala, S. K. Srai, E. Debnam and C. Rice-Evans, *Biochem. Biophys. Res. Commun.*, 2000, **272**, 236–241.
20. J. F. Wang, D. D. Schramm, R. R. Holt, J. L. Ensunsa, C. G. Fraga, H. H. Schmitz, et al., *J. Nutr.*, 2000, **130**(Suppl 8), 2115S–2119S.
21. S. Déprez, I. Mila, J. F. Huneau, D. Tomé and A. Scalbert, *Antioxid. Redox Signaling*, 2001, **3**(6), 957–967.
22. L. Y. Rios, R. N. Bennett, S. A. Lazarus, C. Rémésy, A. Scalbert and G. Williamson, *Am. J. Clin. Nutr.*, 2002, **76**(5), 1106–1110.

23. C. G. Fraga, L. Actis-Goretta, J. I. Ottaviani, F. Carrasquedo, S. B. Lotito, S. Lazarus, et al., Clin. Dev. Immunol., 2005, **12**(1), 11–17.
24. N. D. Fisher, M. Hughes, M. Gerhard-Herman and N. K. Hollenberg, J. Hypertens., 2003, **21**(12), 2281–2286.
25. C. Heiss, A. Dejam, P. Kleinbongard, T. Schewe, H. Sies and M. Kelm, JAMA, J. Am. Med. Assoc., 2003, **290**, 1030–1031.
26. D. Grassi, S. Necozione, C. Lippi, G. Croce, L. Valeri and P. Pasqualetti, Hypertension, 2005, **46**, 398–405.
27. M. Karim, K. McCormick and C. T. Kappagoda, J. Nutr., 2000, **130**(Suppl 8), 2105S–2108S.
28. L. Hooper, C. Kay, A. Abdelhamid, P. A. Kroon, J. S. Cohn, E. B. Rimm and A. Cassidy, Am. J. Clin. Nutr., 2012, **95**(3), 740–751, DOI: 10.3945/ajcn.111.023457.
29. D. Grassi, C. Lippi, S. Necozione, G. Desideri and C. Ferri, Am. J. Clin. Nutr., 2005, **81**(3), 611–614.
30. D. Rein, T. G. Paglieroni, T. Wun, D. A. Pearson, H. H. Schmitz, R. Gosselin, et al., Am. J. Clin. Nutr., 2000, **72**(1), 30–35.
31. D. Rein, T. G. Paglieroni, D. A. Pearson, T. Wun, H. H. Schmitz, R. Gosselin, et al., J. Nutr., 2000, **130**(Suppl 8), 2120S–2126S.
32. D. A. Pearson, T. G. Paglieroni, D. Rein, T. Wun, D. D. Schramm, J. F. Wang, et al., Thromb. Res., 2002, **106**(4–5), 191–197.
33. R. R. Holt, D. D. Schramm, C. L. Keen, S. A. Lazarus and H. H. Schmitz, JAMA, J. Am. Med. Assoc., 2002, **287**(17), 2212–2213.
34. K. J. Murphy, A. K. Chronopoulos, I. Singh, M. A. Francis, H. Moriarty, M. J. Pike, et al., Am. J. Clin. Nutr., 2003, **77**(6), 1466–1473.
35. D. Taubert, R. Berkels, R. Roesen and W. Klaus, JAMA, J. Am. Med. Assoc., 2003, **290**(8), 1029–1030.
36. L. Actis-Goretta, J. I. Ottaviani, C. L. Keen and C. G. Fraga, FEBS Lett., 2003, **555**(3), 597–600.
37. D. Taubert, R. Roesen, C. Lehmann, N. Jung and E. Schömig, JAMA, J. Am. Med. Assoc., 2007, **298**(1), 49–60.
38. D. Grassi, G. Desideri, S. Necozione, C. Lippi, R. Casale and G. Properzi, J. Nutr., 2008, **138**(9), 1671–1676.
39. D. D. Schramm, J. F. Wang, R. R. Holt, J. L. Ensunsa, J. L. Gonsalves, S. A. Lazarus, et al., Am. J. Clin. Nutr., 2001, **73**(1), 36–40.
40. R. H. Liu, J. Nutr., 2004, **134**(Suppl 12), 3479S–3485S.
41. T. P. Kenny, C. L. Keen, H. H. Schmitz and M. E. Gershwin, Exp. Biol. Med., 2007, **232**(2), 293–300.
42. M. Serafini, R. Bugianesi, G. Maiani, S. Valtuena, S. De Santis and A. Crozier, Nature, 2003, **424**(6952), 1013.
43. D. Rein, S. Lotito, C. G. Fraga, H. H. Schmitz and C. L. Keen, J. Nutr., 2000, **130**, 2109S–2114S.
44. E. Nurk, H. Refsum, C. A. Drevon, G. S. Tell, H. A. Nygaard, K. Engedal, et al., J. Nutr., 2009, **139**(1), 120–127.
45. M. Messaoudi, J. F. Bisson, A. Nejdi, P. Rozan and H. Javelot, Nutr. Neurosci., 2008, **11**(6), 269–276.

46. J. F. Hammerstone, S. A. Lazarus and H. H. Schmitz, *J. Nutr.*, 2000, **130**(Suppl 8), 2086S–2092S.
47. S. Baba, N. Osakabe, Y. Kato, M. Natsume, A. Yasuda and T. Kido, *Am. J. Clin. Nutr.*, 2007, **85**(3), 709–717.
48. S. Baba, M. Natsume, A. Yasuda, Y. Nakamura, T. Tamura, N. Osakabe, *et al.*, *J. Nutr.*, 2007, **137**(6), 1436–1441.
49. Y. Wan, J. A. Vinson, T. D. Etherton, J. Proch, S. A. Lazarus and P. M. Kris-Etherton, *Am. J. Clin. Nutr.*, 2001, **74**(5), 596–602.
50. A. Ruzaidi, I. Amin, A. G. Nawalyah, M. Hamid and H. A. Faizul, *J. Ethnopharmacol.*, 2005, **98**(1–2), 55–60.
51. S. Mathur, S. Devaraj, S. M. Grundy and I. Jialal, *J. Nutr.*, 2002, **132**(12), 3663–3667.
52. N. Osakabe and M. Yamagishi, *J. Clin. Biochem. Nutr.*, 2009, **45**(2), 131–136.
53. A. Yasuda, M. Natsume, K. Sasaki, S. Baba, Y. Nakamura, M. Kanegae, *et al.*, *BioFactors*, 2008, **33**(3), 211–223.
54. F. A. Tomás-Barberán and J. C. Espín, *J. Sci. Food Agric.*, 2001, **81**(9), 853–876.
55. C. Summa, F. C. Raposo, J. McCourt, R. Lo Scalzo, K.-H. Wagner, I. Elmadfa and E. Anklam, *Eur. Food Res. Technol.*, 2006, **222**, 368–375.
56. N. Ortega, M. P. Romero, A. Macià, J. Reguant, N. Anglès, J. R. Morelló and M. J. Motilva, *J. Agric. Food Chem.*, 2008, **56**(20), 9621–9627.
57. L. Gu, S. E. House, X. Wu, B. Ou and R. L. Prior, *J. Agric. Food Chem.*, 2006, **54**(11), 4057–4061.
58. C. Counet, C. Ouwerx, D. Rosoux and S. Collin, *J. Agric. Food Chem.*, 2004, **52**(20), 6243–6249.
59. F. Luna, D. Crouzillat, L. Cirou and P. Bucheli, *J. Agric. Food Chem.*, 2002, **50**(12), 3527–3532.
60. J. A. Vinson, J. Proch and L. Zubik, *J. Agric. Food Chem.*, 1999, **47**(12), 4821–4824.
61. E. W. Triche, L. M. Grosso, K. Belanger, A. S. Darefsky, N. L. Benowitz and M. B. Bracken, *Epidemiology*, 2008, **19**(3), 459–464.
62. United States Department of Agriculture [Internet]. USDA National Nutrient Database for Standard Reference Release 25. Available from: http://ndb.nal.usda.gov/ndb/search/list.
63. L. Actis-Goretta, J. I. Ottaviani and C. G. Fraga, *J. Agric. Food Chem.*, 2006, **54**(1), 229–234.
64. T. Schewe, C. Sadik, L. O. Klotz, *et al.*, *Biol. Chem.*, 2001, **382**(12), 1687–1696.
65. T. K. Mao, J. Van De Water, C. L. Keen, *et al.*, *Exp. Biol. Med.*, 2003, **228**(1), 93–99.
66. P. Pastore, G. Favaro, D. Badocco, A. Tapparo, S. Cavalli and G. Saccani, *J. Chromatogr. A*, 2005, **1098**(1–2), 111–115.
67. F. Schifano and G. Magni, *Biol. Psychiatry*, 1994, **36**(11), 763–767.

68. P. Maccioni, D. Pes, M. A. Carai, G. L. Gessa and G. Colombo, *Behav. Pharmacol.*, 2008, **19**(3), 197–209.
69. W. Michener and P. Rozin, *Physiol. Behav.*, 1994, **56**(3), 419–422.
70. N. Ramli, A. M. Yatim, M. Said and H. C. Hok, *Malays. J. Anal. Sci.*, 2001, **7**(2), 377–386.
71. R. P. Mensink, P. L. Zock, A. D. Kester and M. B. Katan, *Am. J. Clin. Nutr.*, 2003, **77**(5), 1146–1155.
72. M. A. Thijssen and R. P. Mensink, *Am. J. Clin. Nutr.*, 2005, **82**(3), 510–516.
73. K. A. Hunter, L. C. Crosbie, A. Weir, G. J. Miller and A. K. Dutta-Roy, *J. Nutr. Biochem.*, 2000, **11**(7–8), 408–416.
74. K. A. Hunter, L. C. Crosbie, G. W. Horgan, G. J. Miller and A. K. Dutta-Roy, *Br. J. Nutr.*, 2001, **86**(2), 207–215.
75. F. D. Kelly, A. J. Sinclair, N. J. Mann, A. H. Turner, L. Abedin and D. Li, *Eur. J. Clin. Nutr.*, 2001, **55**(2), 88–96.
76. F. D. Kelly, A. J. Sinclair, N. J. Mann, A. H. Turner, F. L. Raffin, M. V. Blandford and M. J. Pike, *Eur. J. Clin. Nutr.*, 2002, **56**(6), 490–499.
77. F. B. Hu and W. C. Willett, *JAMA, J. Am. Med. Assoc.*, 2002, **288**(20), 2569–2578.
78. P. M. Kris-Etherton, S. R. Daniels, R. H. Eckel, M. Engler, B. V. Howard, R. M. Krauss, *et al.*, *Circulation*, 2001, **103**(7), 1034–1039.
79. P. M. Kris-Etherton, T. A. Pearson, Y. Wan, R. L. Hargrove, K. Moriarty, V. Fishell, *et al.*, *Am. J. Clin. Nutr.*, 1999, **70**(6), 1009–1015.
80. P. M. Kris-Etherton, A. E. Binkoski, G. Zhao, S. M. Coval, K. F. Clemmer, K. D. Hecker, *et al.*, *Proc. Nutr. Soc.*, 2002, **61**(2), 287–298.
81. E. L. Ding, S. M. Hutfless, X. Ding and S. Girotra, *Nutr. Metab.*, 2006, **3**, 2.
82. D. T. Barry, C. M. Grilo and R. M. Masheb, *Int. J. Eat. Disord.*, 2002, **31**(1), 63–70.
83. C. H. Gilhooly, S. K. Das, J. K. Golden, M. A. McCrory, G. E. Dallal, E. Saltzman, F. M. Kramer and S. B. Roberts, *Int. J. Obes.*, 2007, **31**(12), 1849–1858.
84. M. E. Gluck, A. Geliebter and M. Lorence, *Ann. N. Y. Acad. Sci.*, 2004, **1032**, 202–207.
85. M. E. Gluck, A. Geliebter, J. Hung and E. Yahav, *Psychosom. Med.*, 2004, **66**(6), 876–881.
86. R. T. Mikolajczyk, W. El Ansari and A. E. Maxwell, *Nutr. J.*, 2009, **15**(8), 31.
87. N. Sudo, D. Degeneffe, H. Vue, K. Ghosh and M. Reicks, *Appetite*, 2009, **52**(1), 137–146.
88. E. Epel, R. Lapidus, B. McEwen and K. Brownell, *Psychoneuroendocrinology*, 2001, **26**(1), 37–49.
89. F. P. Martin, S. Rezzi, E. Peré-Trepat, B. Kamlage, S. Collino, E. Leibold, *et al.*, *J. Proteome Res.*, 2009, **8**(12), 5568–5579.
90. C. Donfrancesco, C. Lo Noce, O. Brignoli, G. Riccardi, P. Ciccarelli, F. Dima, *et al.*, *BMC Fam. Pract.*, 2008, **9**, 53.

91. N. Rose, S. Koperski and B. A. Golomb, *Arch. Intern. Med.*, 2010, **170**(8), 699–703.
92. World Health Organization. *Mental Health Aspects of Women's Reproductive Health: A Global Review of the Literature*, World Health Organization Press, Geneva, Switzerland, 2009.
93. J. L. Osman and J. Sobal, *Appetite*, 2006, **47**(3), 290–301.
94. S. Rodríguez, C. S. Warren, S. Moreno, A. Cepeda-Benito, D. H. Gleaves, M. del C. Fernández, *et al.*, *Appetite*, 2007, **49**(1), 245–250.
95. J. M. Hormes and P. Rozin, *Appetite*, 2009, **53**(2), 256–259.
96. D. A. Zellner, A. Garriga-Trillo, S. Centeno and E. Wadsworth, *Appetite*, 2004, **42**(1), 119–121.
97. L. I. Mennen, D. Sapinho, A. de Bree, N. Arnault, S. Bertrais, P. Galan, *et al.*, *J. Nutr.*, 2004, **134**(4), 923–926.
98. Società Italiana di Ginecologia e Ostetricia (SIGO) [Internet]. Available from: http://www.sigo.it.
99. A. Salonia, F. Fabbri, G. Zanni, M. Scavini, G. V. Fantini, A. Briganti, *et al.*, *J. Sex. Med.*, 2006, **3**(3), 476–482.
100. A. Agarwal, A. Aponte-Mellado, B. J. Premkumar, A. Shaman and S. Gupta, *Reprod. Biol. Endocrinol.*, 2012, **10**, 49, DOI: 10.1186/1477-7827-10-49.
101. A. Aldridge, J. Bailey and A. H. Neims, *Semin. Perinatol.*, 1981, **5**, 310–314.
102. J. R. Oesterheld, *J. Child Adolesc. Psychopharmacol.*, 1998, **8**, 161–174, DOI: 10.1089/cap.1998.8.161.
103. H. Andersson, H. Hallström and B. A. Kihlman, *Intake of Caffeine and Other Methylxanthines during Pregnancy and Risk for Adverse Effects in Pregnant Women and their Foetuses*, Copenhagen, Denmark, Nordic Council of Ministers; 2004.
104. J. M. Wilkinson and I. Pollard, *Brain Res. Dev. Brain Res.*, 1993, **75**, 193–199, DOI: 10.1016/0165-3806(93)90023-4.
105. A. Wilcox, C. Weinberg and D. Baird, *Lancet*, 1988, **2**(8626–8627), 1453–1456.
106. T. K. Jensen, T. B. Henriksen, N. H. Hjollund, T. Scheike, H. Kolstad, A. Giwercman, *et al.*, *Reprod. Toxicol.*, 1998, **12**(3), 289–295.
107. R. B. Hakim, R. H. Gray and H. Zacur, *Fertil. Steril.*, 1998, **70**(4), 632–637.
108. K. C. Taylor, C. M. Small, C. E. Dominguez, L. E. Murray, W. Tang, M. M. Wilson, *et al.*, *Ann. Epidemiol.*, 2011, **21**(11), 864–872, DOI: 10.1016/j.annepidem.2011.04.011.
109. E. I. Florack, G. A. Zielhuis and R. Rolland, *Prev. Med.*, 1994, **23**(2), 175–180.
110. S. Cnattingius, L. B. Signorello, G. Annerén, B. Clausson, A. Ekbom, E. Ljunger, *et al.*, *N. Engl. J. Med.*, 2000, **343**(25), 1839–1845.
111. W. Wen, X. O. Shu, D. R. Jacobs Jr and J. E. Brown, *Epidemiology*, 2001, **12**(1), 38–42.
112. B. H. Bech, E. A. Nohr, M. Vaeth, T. B. Henriksen and J. Olsen, *Am. J. Epidemiol.*, 2005, **162**(10), 983–990.

113. J. D. Peck, A. Leviton and L. D. Cowan, *Food Chem. Toxicol.*, 2010, **48**(10), 2549–2576, DOI: 10.1016/j.fct.2010.06.019.
114. V. Sengpiel, E. Elind, J. Bacelis, S. Nilsson, J. Grove, R. Myhre, *et al.*, *BMC Med.*, 2013, **11**, 42, DOI: 10.1186/1741-7015-11-42.
115. NNR Project group, *Nordic Nutrition Recommendations, Integrating Nutrition and Physical Activity*, Nordic Council of Ministers, Copenhagen, Denmark, 2004.
116. American College of Obstetricians and Gynecologists, ACOG Committee Opinion No. 462, *Obstet. Gynecol.*, 2010, **16**(2 Pt 1), 467–468, DOI: 10.1097/AOG.0b013e3181eeb2a1.
117. World Health Organization, *The World Health Report, Reducing Risks, Promoting Healthy Life*, World Health Organization, Geneva, Switzerland, 2002.
118. O. Ishihara, M. Hayashi, H. Osawa, K. Kobayashi, S. Takeda, B. Vessby, *et al.*, *Free Radical Res.*, 2004, **38**(9), 913–918.
119. M. Palm, O. Axelsson, L. Wernroth and S. Basu, *Free Radical Res.*, 2009, **43**(6), 546–552.
120. V. Toescu, S. L. Nuttall, U. Martin, M. J. Kendall and F. Dunne, *Clin. Endocrinol.*, 2002, **57**(5), 609–613.
121. A. Alberti–Fidanza, G. C. Di Renzo, G. Burini, G. Antonelli and G. Perriello, *J. Matern.-Fetal Neonat. Med.*, 2002, **12**(1), 59–63.
122. E. Jauniaux, L. Poston and G. J. Burton, *Hum. Reprod. Update*, 2006, **12**(6), 747–755.
123. Institute of Medicine, *DRI Dietary Reference Intakes for Energy, Carbohydrate, Fiber, Fat, Fatty Acids, Cholesterol, Protein, and Amino Acids*, National Academy Press, Washington, DC, 2002.
124. G. C. Di Renzo, E. Brillo, M. Romanelli, G. Porcaro, F. Capanna, T. T. Kanninen, *et al.*, *J. Matern.-Fetal Neonat. Med.*, 2012, **25**(10), 1860–1867, DOI: 10.3109/14767058.2012.683085.
125. J. A. Mogollon, E. Bujold, S. Lemieux, M. Bourdages, C. Blanchet, L. Bazinet, *et al.*, *Nutr. J.*, 2013, **12**(1), 41.
126. K. P. Conrad, G. M. Joffe, H. Kruszyna, R. Kruszyna, L. G. Rochelle, R. P. Smith, J. E. Chavez and M. D. Mosher, *FASEB J.*, 1993, **12**(6), 566–571.
127. R. R. Magness, C. E. Shaw, T. M. Phernetton, J. Zheng and I. M. Bird, *Am. J. Physiol.*, 1997, **12**(4 Pt 2), H1730–H1740.
128. M. E. Widlansky, N. Gokce, J. F. Keaney Jr and J. A. Vita, *J. Am. Coll. Cardiol.*, 2003, **12**(7), 1149–1160, DOI: 10.1016/S0735-1097(03)00994-X.
129. A. F. Saftlas, E. W. Triche, H. Beydoun and M. B. Bracken, *Ann. Epidemiol.*, 2010, **20**(8), 584–591, DOI: 10.1016/j.annepidem.2010.05.010.
130. M. A. Klebanoff, J. Zhang, C. Zhang and R. J. Levine, *Epidemiology*, 2009, **20**(5), 727–732, DOI: 10.1097/EDE.0b013e3181aba664.
131. S. R. Crozier, S. M. Robinson, K. M. Godfrey, C. Cooper and H. M. Inskip, *J. Nutr.*, 2009, **139**(10), 1956–1963.

132. K. Räikkönen, A. K. Pesonen, A. L. Järvenpää and T. E. Strandberg, *Early Hum. Dev.*, 2004, **76**(2), 139–145.
133. J. F. Wang-Polagruto, A. C. Villablanca, J. A. Polagruto, L. Lee, R. R. Holt, H. R. Schrader, *et al.*, *J. Cardiovasc. Pharmacol.*, 2006, **47**(Suppl 2), S177–S186, discussion S206–S209.
134. J. C. Stevenson, *Maturitas*, 2011, **70**(2), 197–205, DOI: 10.1016/j.maturitas.2011.05.017.
135. P. J. Mink, C. G. Scrafford, L. M. Barraj, *et al.*, *Am. J. Clin. Nutr.*, 2007, **85**(3), 895–909.
136. E. Boelsma, L. P. Van de Vijver, R. A. Goldbohm, I. A. Klöpping-Ketelaars, H. F. Hendriks and L. Roza, *Am. J. Clin. Nutr.*, 2003, **77**(2), 348–355.
137. H. Sies and W. Stahl, *Annu. Rev. Nutr.*, 2004, **24**, 173–200.
138. U. Heinrich, K. Neukam, H. Tronnier, H. Sies and W. Stahl, *J. Nutr.*, 2006, **136**(6), 1565–1569.
139. S. Williams, S. Tamburic and C. Lally, *J. Cosmet. Dermatol.*, 2009, **8**(3), 169–173.
140. P. Gasser, E. Lati, L. Peno-Mazzarino, D. Bouzoud, L. Allegaert and H. Bernaert, *Int. J. Cosmet. Sci.*, 2008, **30**(5), 339–345.
141. K. Neukam, W. Stahl, H. Tronnier, H. Sies and U. Heinrich, *Eur. J. Nutr.*, 2007, **46**(1), 53–56.

CHAPTER 10

Chocolate and Skin Health: Effects of Dietary Cocoa Polyphenols

ULRIKE HEINRICH*[a] AND WILHELM STAHL[b]

[a] Institut für Experimentelle Dermatologie, Universität Witten-Herdecke, Alfred-Herrhausen-Str. 44, D-58455, Witten, Germany; [b] Institut für Biochemie und Molekularbiologie I, Medizinische Fakultät, Heinrich-Heine-Universität Düsseldorf, P.O.Box 101007, D-40001, Düsseldorf, Germany
*Email: ulrike.heinrich@uni-wh.de

10.1 Introduction

In addition to other factors, skin health depends on an appropriate supply of different tissue layers and cells with energy from dietary sources, lipids to maintain specific dermal structures and vitamins providing enzymatic cofactors or acting as signaling molecules.[1] Human intervention studies provide evidence that secondary plant constituents such as carotenoids and polyphenols (*e.g.* flavonoids) modulate the skin's properties, thus improving its physiological function and appearance.[2] In this context, the impact of our diet or selected dietary factors on dermal health, aging and appearance has prompted considerable research in the fields of medicine, nutrition and cosmetics.

Topical treatment with oils and creams enriched with herbal extracts has a long cross-cultural tradition, and such products still constitute a major part of the cosmetics that are sold worldwide. Many studies have been performed to prove the effects of topically applicable items and their active constituents. However, more recently, a strategy was developed to supply the skin with active nutrients *via* systemic circulation. It has been demonstrated that biologically active micronutrients are very effective when they are provided as dietary supplements or as enrichments of dietary products. New insights into the effects of orally-administered, biologically active molecules on skin functions have stimulated growing interest in the development of nutritional supplements and functional food products to benefit human skin.

Cosmeceuticals represent products combining pharmaceutical and cosmetic properties that are mainly related to ingredients that modulate the biological function of the skin.[3] They are designed to improve skin tone, texture and radiance while reducing wrinkling.

Various products claiming to improve skin health and function are on the market, but evidence for their efficacy is not always convincing. Reliable data are still required to prove their effects and substantiate their claims.[4]

This chapter provides a brief overview of the use of dietary constituents (vitamins and micronutrients) in skin care and will focus on selected human studies that demonstrate that a systemic application of cocoa products rich in cocoa-specific polyphenols acutely and persistently modify skin properties and function. Independent and objective methods were applied to determine dermal changes in all of these studies, all of which are described and evaluated in this chapter.

10.2 Skin Structure and Function

The skin is the largest organ of the human body with a surface of about 1.5–2.0 m^2. It is composed of three major layers: the epidermis, dermis and hypodermis (subcutaneous fat tissue).[5] The outer layer of skin is the epidermis, which functions as a protective shield for the body, and it completely renews itself approximately every 28 days. Keratinocytes, the most frequent cell type of the epidermis, are generated in the basal structure (*stratum basale*) and move upwards to the skin surface. During this process, the cells undergo sequential changes in chemical composition and form further substructures of the epidermis (*e.g.* the *stratum corneum* [horny layer]). Cellular water content decreases from about 80% in the lower layers to 10–15% in the *stratum corneum*. The epidermis is a complex system of cells and specialized extracellular components that prevent water loss by evaporation, act as a receptor organ and provide a protective barrier for the underlying tissues.

Fibroblasts are the major cell type in the dermis, which is located between the epidermis and hypodermis, and it acts as a fibrous network, composed of structural proteins (collagen and elastin) that are important for maintaining skin structure and resilience. Collagen fibers contribute to skin strength and flexibility, as does elastin to elasticity. Both collagen and elastin fibers are produced by the fibroblasts and secreted to form the extracellular matrix. Blood vessels, lymph vessels and mast cells are embedded in the dermis. Blood vessels in the dermis provide oxygen and nutrients to the lower layers of the epidermis; however, they do not extend into that layer. Mast cells and the lymphatic network are involved in immune function, thereby constituting part of the skin's defense system.

The junction between the dermis and epidermis is a wave-like border that provides an increased surface for the exchange of oxygen and nutrients between both major layers. Nutrients and oxygen diffuse into the lower regions of the epidermis. The hypodermis, the deepest section of the skin, comprises a lipid-rich connective tissue with major blood vessels and nerves. Its main tasks are shock absorption and thermal insulation. Fat cells therein provide important lipid energy storage.

Skin aging is a slow process that affects all skin layers. As skin ages, the hypodermis begins to atrophy, thus contributing to the thinning process. Wrinkles are formed as a consequence of collagen degradation in the dermal extracellular matrix. Modification of water homeostasis leading to decreased skin hydration is attributed to changes in the epidermal structure.

Other important skin cells are the melanocytes, located in the *stratum basale*, which synthesize the skin pigment melanin. Melanin is responsible for the light–dark component of the skin color, and it acts as a photoprotective agent against ultraviolet (UV) radiation (*i.e.* sunlight). The presence of an adequate amount of water in the *stratum corneum* is important for the general appearance of the skin, which contributes to its soft, smooth and healthy qualities. An intact skin barrier is associated with a low rate of transepidermal water loss (TEWL). A decrease of TEWL may be observed by cosmetic treatments with suitable emulsions. This decrease is generally accompanied by a slow increase in skin hydration. Improvement of skin hydration and barrier function was observed in intervention studies with "nutricosmeticals" (*i.e.* dietary supplements like carotenoids, polyphenols or vitamins, which modulate skin properties).

10.3   Skin and Nutrition

Based on their chemical and biological properties, micronutrients like carotenoids, flavonoids and vitamins E and C are suitable compounds for photoprotection in humans.[6] Human intervention studies with carotenoid supplements or diets rich in carotenoids reveal that this class of compounds contributes to systemic photoprotection ameliorating UV-induced erythema. After ingestion of lycopene or tomato-derived products rich in lycopene, photoprotective effects have been demonstrated. Within 10–12 weeks of intervention, a decrease in the sensitivity towards UV-induced erythema was observed. However, it should be noted that photoprotection through individual dietary components in terms of sun protection factor is considerably lower than that achieved by using topical sunscreens.

Supplementation with micronutrients modulates skin structure and texture, thereby contributing to the skin's resistance to environmental stress and to improving general parameters that are indicative of skin health. After ingestion of an antioxidant mixture composed of carotenoids, vitamin E and selenium, a significant increase of skin density and thickness was observed.[7] Also, skin surface parameters regarding roughness and scaling were improved.

In addition to the supply via epidermal blood vessels, alpha- and gamma-tocopherol are continuously secreted with the human sebum.[8] Thus, vitamin E is also a constituent of the antioxidant network of the *stratum corneum*, the first line of defense against exogenous oxidants such as ozone.[9] Various endpoints indicating phototoxic damage, like UV-dependent erythema, formation of sunburn cells, skin wrinkling, lipid oxidation and DNA damage, can be modulated.[10] Preparations of topically applicable sunscreen have been shown to be more effective by combining key vitamins rather than using single ingredients.[11]

Retinoids, mainly retinoic acid and several natural or synthetic derivatives, are widely applied in treating skin disorders such as acne, psoriasis, ichthyosis and keratodermatosis.[12] Clinical trials using retinoic acid and related compounds in treating skin cancers have led to conflicting data. Although retinoic acid triggers pathways of cell differentiation and growth, its use as a preventive agent is limited due to toxicological concerns. The major retinoid present in human blood and tissues is retinol (vitamin A) and its fatty acid esters. In a small case–control study, the level of serum retinol was found to be inversely correlated with the occurrence of non-melanoma skin cancer.[13]

Some polyunsaturated fatty acids are essential for the human organism and serve several biological functions. They are important components of cellular membranes, contributing to their fluidity, rigidity, permeability and function. Supplementation with certain fatty acids can influence the fatty acid pattern of the skin and affect its sensitivity towards photo-oxidation[14] and may also contribute to preventing non-melanoma skin cancers.[15] Epidermal lipids are known to play an important role in mediating normal desquamation, and a deficiency of essential fatty acids has been reported in cutaneous scaling disorders such as senile xerosis, psoriasis and atopic dermatitis.[16] Supplementation with linoleic acid (LA) and γ-linolenic acid (GLA) leads to decreased TEWL and to less itchy, dry-appearing skin.[17] In an intervention study with women ingesting flaxseed and borage oil supplements, skin conditions were improved.[18] Skin hydration was significantly increased after treatment and TEWL was also decreased. Surface evaluation of living skin (SELS) revealed that its roughness and scaling were significantly diminished following ingestion of flaxseed and borage oil.

10.4 Cocoa Constituents with Dermal Activity

It is beyond the scope of this chapter to present data on the different constituents of cocoa and their patterns in various products. With respect to skin effects, it is the cocoa polyphenols and cocoa lipids that remain of special interest.

Polyphenols is the collective term for aromatic compounds that have at least two hydroxyl groups attached to a benzene ring.[19] They are further assigned to a series of subgroups: the flavonols, flavanones, flavones, isoflavones, flavanols and anthocyanins. As secondary plant constituents, flavonoids are common in the human dietary intake of vegetables, fruits, spices and beverages.[20] Catechins, which are major cocoa polyphenols, comprise a subgroup of flavanols including catechin, epicatechin, gallocatechin, epigallocatechin and the respective gallic acid esters at the 3-OH position (catechin gallate, epicatechin gallate, gallocatechin gallate and epigallocatechin gallate) as major representatives.[21] Procyanidins (synonymic proanthocyanidins) are oligomers of parent catechins, among which are the bioactive dimeric procyanidins B1, B2, B3 and B4. Higher polymers are also found in natural products with increasingly complex structures. Catechins and procyanidins are secondary plant components in the human diet when consumed in tea, apples, grapes and wine, as well as in cocoa. The pattern and levels of catechins and procyanidins vary in different food and food products, but they are generally present in the parts per million range.

Cocoa butter, the oil obtained from cacao beans, is mainly composed of triglycerides with a mixture of saturated and monounsaturated fatty acids. The majority of the saturated fatty acids are palmitic acid and stearic acid. Oleic acid is the dominant unsaturated fatty acid therein. Each fatty acid contributes about a third to this pattern. The di-unsaturated n-6 linoleic acid contributes about 3%. Further lipids in cocoa butter are phytosterols including beta-sitosterol and sigmasterol, both of which are present in minor amounts. Cocoa butter has a low melting point and is used in the production of perfumes, decorative cosmetics (*e.g.* lipsticks) and soaps.

10.5 Topical Effects of Cocoa Products

External application of chocolate is promoted for the maintenance of wellness and beauty, and it is claimed to prevent skin aging and wrinkling. Several scientific studies that have evaluated the benefit of topical cocoa on skin properties are compiled below. Dermal application of cocoa has been reported to influence skin moisture, soothe burns and disinfect wounds.[22] Cocoa butter has been suggested for use in preventing or minimizing the stretch marks that occur during pregnancy.

In a randomized, double-blind, placebo-controlled trial, 150 pregnant women received topically applied cocoa butter cream and were followed-up from week 16 of pregnancy through to delivery in order to assess the development of stretch marks (*striae gravidarum*). *Striae gravidarum* were found in only 44% of participants using cocoa butter cream compared with 55% of those 150 women who received a placebo cream. However, this difference was not statistically significant. Rather, the incidence of stretch marks was more closely associated with younger-aged mothers and higher-weight neonates.[23]

No significant effects of cocoa butter on the development of postpartum stretch marks over the abdomen, breasts and thighs were reported in a study enrolling 175 women (91 *versus* 84 controls). Topical application of a lotion containing cocoa butter does not appear to reduce the likelihood of developing *striae gravidarum*.[24]

Davis and Perez[25] reported preliminary findings of using pure or hydrolyzed cocoa butter in healing burn wounds using a porcine model. Epithelization of the wounds was promoted upon treatment with cocoa butter. In this study, the pure form of cocoa butter was found to be more efficient than the hydrolyzed product.

Several lipophilic ingredients of cosmetics have been shown to exhibit comedogenic properties. Comedones (blackheads) are dilated hair follicles filled with keratin squamae, bacteria and sebum. They are primary signs of acne and have been attributed to the prolonged use of cosmetics. Following the rabbit ear assay, comedogenicity was studied using different commercial batches of cocoa butter.[26]

Polyphenols only reveal their bioactivity in deeper skin layers when they permeate the skin barrier, in particular the *stratum corneum*. The release of polyphenols from emulsions that were loaded with mixtures of catechin, epigallocatechin gallate, resveratrol, quercetin, rutin and protocatechuic acid was investigated *ex vivo* in pig skin and in a cellulose membrane model.[27] All substances passed through the *stratum corneum* barrier and were found mostly in the epidermis and dermis.

Human skin explants maintained in survival were used as a model system to examine the influence of cocoa butter and polyphenols on various parameters related to skin restructuring.[28] After treatment with cocoa butter or cocoa butter plus polyphenols, histological samples were prepared, stained and analyzed for glycosaminoglycans and collagen types I, III and IV. After 5 days of treatment with cocoa polyphenols, a moderate increase in epidermal thickness and collagen density was determined. Application of cocoa polyphenols was associated with elevated glycosaminoglycans and collagen type I. Collagen type III and type IV were clearly increased on day 5, and the effects were more pronounced with polyphenols than with cocoa butter alone. Topical application of green tea extract containing catechins and catechin gallate derivatives has also been successfully used in cell culture, animal and human studies.[29]

10.6 Methods to Determine Skin Properties and Function

A number of objective methods that are useful in determining skin properties and function are described below. These methods have been validated in dermatological studies and have been used to evaluate the cutaneous effects of dietary supplements and functional food in humans. The suitable instruments described herein are commercially available and frequently used in dermatological practice and research.

10.6.1 Photoprotection Against UV-induced Erythema

The UV irradiation test has been shown to be a useful method for testing the photoprotective effects of nutritional supplements.[30] In this test, erythema are generated using a blue-light solar simulator (Sol 3, Hönle, Munich, Germany) in the scapular region of test persons. The redness of the erythema is measured by means of chromametry before UV irradiation and 24 h after irradiation (erythema maximum). Color measurements (reddening index *a*-value) are performed with a Minolta Chromameter CR 300 (Minolta CR 300, Ahrensburg, Germany). Before intervention, the erythema threshold is determined for each test subject using different light intensities in order to determine the individual minimal erythemal dose (MED). In most studies, erythema were induced with 1.25 MED. The intensity of the erythema is examined before supplementation and during/after the period of the supplementation of the respective test substances (antioxidants). Decreasing *a*-values (redness of the skin) indicate a photoprotective effect.

10.6.2 Cutaneous Blood Flow and Oxygen Saturation of Hemoglobin

This method involves applying the Oxygen-To-See system (O2C system; Lea Instruments, Giessen, Germany), whereby perfusion parameters including relative blood flow and blood flow velocity are determined.[31] Furthermore, the oxygen saturation of hemoglobin and the relative hemoglobin amount are analyzed photometrically. The O2C measuring device uses a laser light for determining perfusion parameters in the tissue. The movement of the erythrocytes triggers a Doppler shift in the laser light, and this value represents the parameter "flow". In addition, a white-light source is used to detect the hemoglobin parameter "oxygen saturation". Oxygen saturation is determined from blood color. The color changes correspond to the degree of oxygen saturation of the hemoglobin. The probes are built in such a way that the perfusion parameters can be measured simultaneously in two separate layers (or depths).

10.6.3 Skin Structure by Ultrasound Measurements

The use of an ultrasound device with a frequency of 20 MHz (Derma Scan C, Vers. 3) and a two-dimensional configuration (Cortex Technology, Hadsund, Denmark) allows the non-invasive differentiation of individual tissue structures.[32] A total of 256 randomly chosen colors are assigned to the different echo amplitudes. Here, the lighter colors correspond to a strong reflection and dark colors to a weak reflection. The measuring head provides an axial resolution of *ca.* 50 μm, analogous to demodulation technology. The length of the image is 12.1 mm and the penetration depth is 7 mm. High-frequency ultrasound B-scanning is applied in order to analyze tissue structures and obtain information on skin density (pixel density) and skin thickness (mm).

10.6.4 Evaluation of the Skin Surface

The measuring principle of the SELS method is based on a photograph of living skin under defined illumination conditions, as well as on electronic processing, and automatic analysis of the picture.[33] In this process, the skin surface is described in terms of four different skin parameters: roughness, scaling, smoothness and wrinkles. The SELS device consists of a measuring head (VisioScan C, Courage & Khazaka, Cologne, Germany) that contains two special metal-halogenide lamps arranged on opposite sides and uniformly illuminating the 15×17 mm measuring area of the skin. The spectrum of the lamps and their intensity, as well as their arrangement, are selected in such a way that only the skin surface, without reflections from deeper layers, can be monitored. A CCD camera, also installed in the measuring head, records a picture of the illuminated skin area, which is then transferred as a gray-value bitmap file. With the use of the additional software "SELS", the specific skin parameters are calculated.

10.6.5 Skin Hydration Measured by Corneometry

Determining skin hydration by means of the condensator method has been scientifically established for many years and accepted worldwide as an efficient technique to determine the water content of the *stratum corneum* under different experimental conditions.[34] The Corneometer® CM 825 (Courage & Khazaka, Cologne, Germany) used in the studies described below measures skin moisture by means of a capacitive method. The measuring head with a diameter of 10 mm is placed—applying constant pressure—on the skin area of interest, whereby the skin functions as a dielectric of the condensator. In this procedure, the water content of the externals applied to the skin is detailed in the measurement. All values are given as arbitrary units.

10.6.6 Skin Barrier Function Evaluated by the Measurement of TEWL

TEWL of the skin, given in $g\ h^{-1}\ m^{-2}$, is frequently used to evaluate barrier function, a measurement that is useful in cutaneous therapy procedures. In addition to this, TEWL measurements of the skin are applied in dermatology and cosmetics to assess the therapeutic effects of drugs or the efficacy of cosmetic treatments.[35] One practicable method measuring TEWL employs the use of a Tewameter® TM 300 (Courage & Khazaka, Cologne, Germany). With this device, the evaporation of water is measured directly on the skin surface with a special probe. The method is suitable to directly measure water evaporation on the skin surface and assess the water balance. With this device, TEWL can be recorded with sufficient precision and is reproducible in a range from 0 to 300 $g\ h^{-1}\ m^{-2}$.

10.7 Human Studies on Systemic Effects of Cocoa

Dietary antioxidants contribute to endogenous photoprotection and are important for the maintenance of skin health. Such effects have been also been attributed to flavonoids, which are found in cocoa products. Two different original studies were performed by the authors of this chapter identifying the effects of cocoa on skin parameters in women by using natural products with either a high or low flavanol content.

The study on long-term ingestion of cocoa flavanols was performed as a double-blind, randomized controlled study.[36] Two groups of volunteers consumed either a high-flavanol (326 mg day^{-1}) or low-flavanol (27 mg day^{-1}) cocoa powder (control), dissolved in 100 ml of water and were followed over a period of 12 weeks. Epicatechin (61 mg day^{-1}) and catechin (20 mg day^{-1}), the major flavanol monomers in the high-flavanol product (HF), were provided in doses of 61 and 20 mg day^{-1}, respectively. Approximately similar amounts are found in 100 g of dark chocolate. Only 6.6 mg of epicatechin and 1.6 mg of catechin per day were ingested with the low-flavanol product (LF). Making up the difference of total flavanols, the daily dose of procyanidins (oligomers) was 245 mg in the HF group and 18.8 mg in the LF group.

A total of 24 female volunteers between 18 and 65 years old with healthy, normal skin of type II according to Pathak[37] were included in the study. Participants ($n=12$ per group) were randomly assigned to either the HF group or the LF group and were advised not to change their dietary habits. No further supplements (vitamins, carotenoids or polyphenols) were allowed during the study. The cocoa beverage was provided as a dry powder and was dissolved in 100 ml of hot water just prior to ingestion. The drink was consumed every morning with a meal during the 12 weeks of this study. Both preparations were provided by Mars, Inc. (Hackettstown, NJ, USA).

On week 0, at the end of week 6 and at the end of week 12, the following parameters related to photoprotection and skin health were determined: sensitivity towards UV irradiation, cutaneous blood flow, skin structure and texture, skin hydration and TEWL. Details of the different measurements to evaluate photoprotection and skin physiological parameters were noted in part 10.6 of this chapter. Compliance was assessed by interview and by counting of the remaining packages. Written informed consent was obtained from each participant, and the study design was approved by the Ethical Committee of the University of Witten/Herdecke, Germany.

For all of the parameters, descriptive statistics (mean, standard deviation, minimum, lower quartile, median, upper quartile and maximum) were calculated for the three time points (week 0, week 6 and week 12). Also, pre–post differences for each combination of two time points were calculated from the data. Within the two treatment groups, each combination of two time points was compared using the Wilcoxon signed-rank test. The pre–post differences of the two treatment groups were compared using the Wilcoxon rank-sum test.

Photoprotection by dietary cocoa flavanols against UV-induced skin responses (erythema) was measured in terms of a decrease in reddening following exposure of selected skin areas to 1.25 MED of solar-simulated radiation. Reddening after UV exposure was determined by chromametry. Chromametry a-values 24 h after irradiation and the difference between chromametry a-values after and before irradiation (Δ-a values) were taken as measures for the UV response of the skin. In the HF group, the Δ-a values were significantly lower after week 6 and week 12 than at the beginning of the study ($P = 0.001$ and 0.012, respectively). The a-values determined 24 h after irradiation were approximately 15 and 25% lower at week 6 and week 12, respectively, compared to the beginning of the study. No significant changes in the 24 h a-values were observed in the LF group during the 12 weeks of treatment. Thus, it was concluded that the consumption of a flavanol-rich cocoa beverage provides photoprotection, whereas a similar cocoa beverage low in flavanols does not.

Following supplementation with the HF cocoa beverage, an increase in cutaneous blood flow was observed in cutaneous (1 mm depth) and subcutaneous (7–8 mm depth) tissue. In comparison to the starting value, peripheral blood flow was significantly increased at 1 mm depth and 7–8 mm depth after 12 weeks of treatment. No change in blood flow was found in the LF group. The values for cutaneous blood flow differed significantly between the HF and LF groups on week 6 and week 12.

Upon supplementation with the HF cocoa beverage, skin structure and texture were modulated. A moderate but statistically significant increase in density and thickness of the skin was observed. For both parameters, no change was found in the LF group. Typical examples of ultrasound B-scans before and 12 weeks after supplementation are shown in Figure 10.1. Using the SELS method, a statistically significant decrease in skin roughness and scaling was measured in the HF group, whereas no change was found in the LF group. Skin hydration was significantly increased after 12 weeks of supplementation with the HF cocoa beverage, whereas it was not affected in the LF group. In the HF group, TEWL was significantly decreased on week 12 as compared to baseline; no difference between baseline and week 12 was found in the LF group.

These data show that ingesting dietary flavanols from cocoa contributes to endogenous photoprotection and improves dermal blood circulation. Furthermore, cocoa flavanols affect cosmetically relevant parameters of skin surface and hydration.

Long-term ingestion of cocoa rich in flavanols increases cutaneous blood flow and improves skin condition in humans. Further study was undertaken to investigate the acute effects of a single dose of cocoa rich in flavanols on dermal microcirculation.

In a crossover design study, ten healthy women ingested a cocoa drink with a high (326 mg) or low (27 mg) content of flavanols.[38] Dermal blood flow and oxygen saturation were examined by laser Doppler flowmetry and spectroscopically at a 1 mm skin depth at time points of 0, 1, 2, 4 and 6 hours

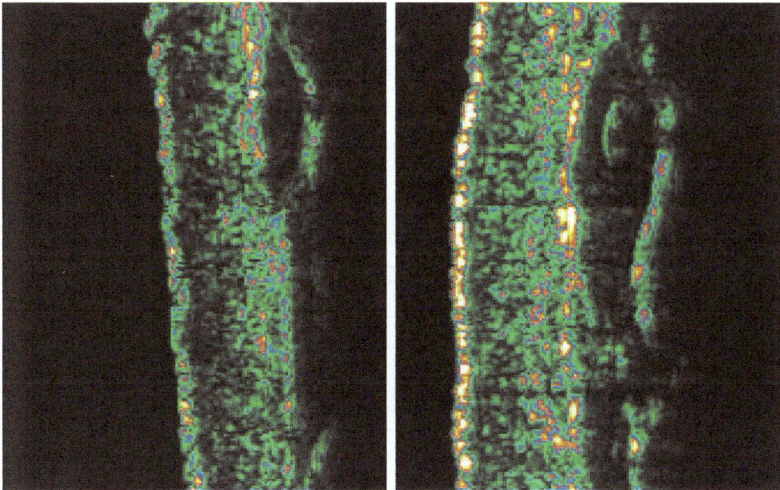

Figure 10.1 Changes in skin density, measured by ultrasound before (Left) and after 12 weeks (Right) of supplementation with high-flavanol cocoa.

after ingestion. At the same time points, plasma levels of total epicatechin (free compound plus conjugates) were measured by means of HPLC. Peripheral blood flow in the skin and oxygen saturation of hemoglobin were determined with the O2C system at a 1 mm depth. After supplementation with the HF cocoa drink, an increase in blood flow was observed in cutaneous tissues (Figure 10.2). Compared to the baseline value, peripheral blood flow was significantly elevated by about 1.7-fold at 2 hours. Concomitantly, oxygen saturation increased from 25% at baseline to 45% at 2 hours. No significant change of these parameters was determined in the LF group. At baseline, the levels of total epicatechin were similar in the HF (11.6 ± 7.4 nmol l^{-1}) and the LF groups (9.5 ± 1.7 nmol l^{-1}). Plasma levels significantly rose after ingestion of a single dose of the HF cocoa drink, reaching a maximum of 62.9 ± 35.8 nmol l^{-1} at 1 hour (Figure 10.2B). The response differed between individuals, with the maximum varying between 37.0 and 144.9 nmol l^{-1}. Total epicatechin levels decreased continuously thereafter and returned almost to baseline values at 6 hours. No significant changes in plasma levels of total epicatechin were determined in the LF group. This study provides further evidence that flavanol-rich cocoa improves dermal blood flow and oxygen saturation. In addition to the long-term effect, a short-term change of cutaneous blood flow can also be achieved with cocoa polyphenols.

Photoprotective effects of cocoa were also observed in other studies. Williams *et al.* conducted a study that substantiated the photoprotective effect of eating chocolate.[39] In this double-blind *in vivo* study, 30 healthy subjects were included. Fifteen subjects were randomly assigned to either a HF or LF chocolate group and consumed a 20 g portion of their allocated chocolate daily. The HF chocolate contained a minimum of 3% flavanols

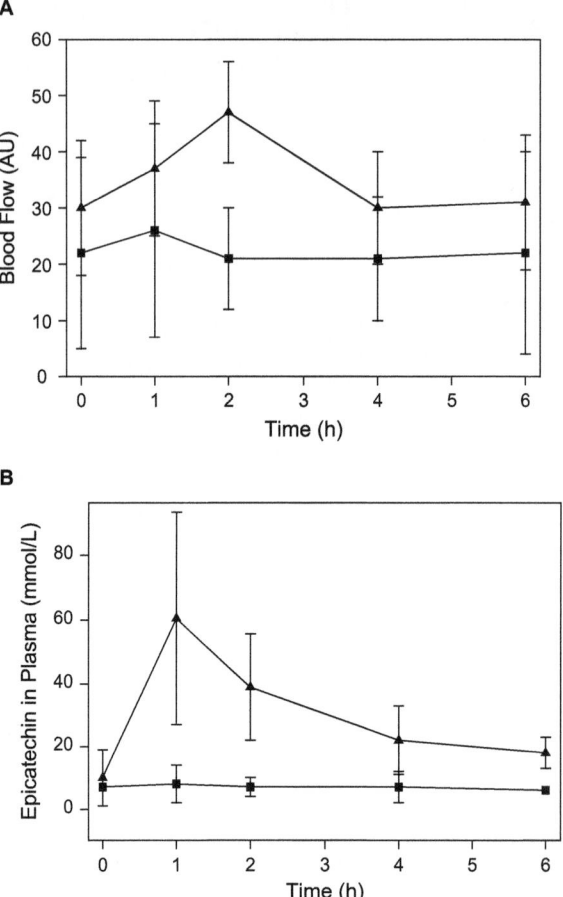

Figure 10.2 Effect of high- and low-flavanol cocoa on cutaneous blood flow and plasma levels of total epicatechin. (A) Peripheral blood flow in skin (1 mm depth) after ingestion of a single dose of high-flavanol (▲) or low-flavanol (■) cocoa drink ($n = 10$). (B) Plasma levels of total epicatechin (free epicatechin plus glucuronate and sulfate conjugates) after ingestion of a single dose of high-flavanol (▲) or low-flavanol (■) cocoa drink ($n = 10$).

(>600 mg flavanols), while the LF chocolate contained <30 mg of flavanols per 20 g portion. MED was assessed at baseline and after 12 weeks. In the HF group, the mean MED was more than doubled after 12 weeks, whereas in the LF group, the MED remained unchanged. The authors concluded that regular consumption of chocolate rich in flavanols confers significant photoprotection and can thus be effective at protecting human skin from harmful UV effects, whereas conventional chocolate has no effect.

Antioxidant properties are likely involved in the mechanism of photoprotection provided by antioxidants such as polyphenols. Katz *et al.* reported that cocoa consumption may stimulate changes in the redox-sensitive signaling pathways involved in the regulation of gene expression and the immune response.[40] They concluded that cocoa can protect nerves from injury and inflammation, protect skin from oxidative damage from UV radiation in topical preparations and provide beneficial effects on satiety, cognitive function and mood.

10.8 Compounds and Biochemical Mechanisms

As reported in the literature and derived from the human intervention studies discussed above, skin properties can be modulated by the dietary polyphenols provided in cocoa products. Cosmetically relevant parameters are related to the structure of skin surface, water homeostasis and barrier function. Thus, appearance is determined by characteristic parameters such as hydration, water loss, thickness, density, texture, color, blood flow and other physiological properties. Several endogenous and environmental factors, including cellular signaling and intrinsic aging, UV exposure or external chemical and mechanical damage influence skin properties. Protective effects against UV-induced damage and inflammation contribute to skin health over the long term. Polyphenolic compounds reveal a number of different biological properties that are likely to be responsible for the effects observed in skin. However, it is unlikely that a single mechanism can be correlated with all of the observed effects. Major possible mechanisms and theories of action are discussed below, including antioxidant activity, absorption of UV light, impact on inflammatory signaling and nitric oxide (NO)-mediated vasodilation.

10.8.1 Antioxidant Activity

In plants, micronutrients play an important role in the protection against excess light. Based on their structural features, flavonoids, carotenoids and vitamins E and C are compounds that are suitable for photoprotection in humans. Upon UV irradiation, a sequence of reactions is induced that leads to the formation of singlet oxygen, superoxide, hydrogen peroxide and reactive intermediates of lipid oxidation. Reactive oxygen species pose a photo-oxidative stress that may damage or chemically modify biologically important molecules including DNA and proteins. A network of antioxidant enzymes and molecules are involved in defense. The antioxidant properties of polyphenols have been suggested to play a role in their bioactivity. Polyphenols' antioxidant activity is determined by the redox-sensitive hydroxyl groups attached to the same aromatic ring.[41] The aromatic hydroxyl groups are oxidized in a two-electron oxidation process to the corresponding quinone. Thus, the antioxidant activity of flavonols and, in particular, catechins is related to the number and localization of the OH groups.[42] Most efficient is a substitution pattern with two hydroxyl groups in the B ring orientated in the ortho position. *In vitro*, flavonoids are scavengers of reactive intermediates, including the superoxide anion, peroxyl radicals and hydroxyl radicals. Antioxidant activities have also been shown for peroxynitrite and NO. Flavonoids quench singlet molecular oxygen with quenching rate constants between 10^5 and 10^7 mol^{-1} s^{-1},[43] but it is not proven yet that this process is important *in vivo*. However, their antioxidant properties *in vivo* have been challenged since flavonoids are efficiently glucuronidated, sulfated or methylated during gut and liver passage, and the amount of the parent compound circulating in the blood is quite low. Free hydroxyl groups are required for radical scavenging, and it has been discussed whether the direct antioxidant activity of flavonoids *in vivo* plays a major role in their mode of action. Indirect antioxidant effects could be mediated *via* the inhibition of pro-oxidant enzymes[44] or stimulation of enzymes involved in antioxidant defense.

10.8.2 UV Absorption: Inflammation

Due to the aromatic ring systems, flavonoids, including cocoa polyphenols, absorb light in the UVA and UVB range. Their absorption spectra and absorption coefficients at selected wavelengths depend on the arrangement of the ring system, the presence or absence of non-aromatic double bonds and the ring substituent. Photoprotection is mediated by the absorption of light, which interferes with the light absorption of sensitive molecules like DNA or proteins. The mechanism is related to that of topical sunscreen, but the effect is less pronounced. Sunburn is an inflammatory event associated with complex biochemical processes involved in such a tissue response. Diminished erythema (reddening) may be related to interference with the signaling pathways of inflammation. A number of dietary antioxidants, including flavonoids, modulate intra- and intercellular signaling. Certain flavonoids inhibit the endogenous synthesis of prostaglandins while they inhibit key enzymes involved in prostaglandin biosynthesis. Modulation of the UV-dependent induction of erythema by cocoa polyphenols measured in terms of a decrease of redness after irradiation may be based their antioxidant activity, UV absorption properties or interference with inflammatory responses. It cannot be excluded that more than one mechanism is operative *in vivo*.

10.8.3 NO: Vasodilation

Increased dermal blood flow has been measured after acute and long-term application of cocoa polyphenols.[45] It has been speculated that under these conditions, the nutrient supply to the dermal layer is improved, with the dermis gaining this effect directly by the blood vessels and the epidermis *via* increased diffusion of nutritionally important macro- and micronutrients. Improving dietary conditions of the skin likely affect its function as well as its structure and appearance. Based on an increased barrier function and a decrease in TEWL, the look and shine of the skin may change. A number of natural products and plant extracts have been demonstrated to influence endothelial NO production.[46] NO is a major mediator of the vasodilatory and cardioprotective effects of cocoa. Improvements of endothelial function and the lowering of blood pressure are attributed to flavanol-mediated elevation of the NO level.[47] Vasodilatory activity was observed after consumption of pure (−)-epicatechin, similar to the response observed after the intake of a HF cocoa product. (−)-Epicatechin elevates NO in endothelial cells *via* inhibition of NADPH oxidase.[48] In an *in vitro* study with human endothelial cells, it was shown that epicatechin activates endothelial NO synthase (eNOS) *via* phosphorylation and dephosphorylation at selected hydroxyl groups.[49] It has been suggested that phosphatidylinositol 3-kinase-dependent pathways are involved in post-translational eNOS activation.

10.9 Conclusion

Cocoa polyphenols and cocoa products rich in polyphenols protect human skin against UV-induced damage at the molecular and cellular level. It has further been demonstrated that dietary intake of cocoa improves overall skin quality and appearance. Regular intake or topical application contribute significantly to photoprotection and help maintain skin health by improving skin structure and function.

It should be noted that the photoprotective effects are moderate. However, in the long term, they may well contribute to permanent, overall protection. The photoprotective effects and, to some extent, the modulation of other skin parameters are similar to those that have been reported for other dietary constituents like carotenoids or dietary lipids.

Cosmeceuticals based on cocoa and cocoa constituents continue to be developed. Their application in the field of aging and especially skin aging awaits further studies using independent scientific methods to foster additional insight into their mechanisms of action.

References

1. J. Krutmann and P. Humbert, *Nutrition for Healthy Skin*, Springer-Verlag, Berlin (Germany), 2011.
2. F. Afaq, *Arch. Biochem. Biophys.*, 2011, **508**, 144.
3. P. Elsner and H. I. Maibach, *Cosmeceuticals*, Marcel Dekker, New York, 2000.
4. M. Amer and M. Maged, *Clin. Dermatol.*, 2009, **27**, 428.
5. R. Baran and H. I. Maibach, *Textbook of Cosmetic Dermatology*, Blackwell Science Ltd., Malden, USA, 1998; P. Agache and P. Humbert, *Measuring the Skin*, Springer Verlag, Berlin, Heidelberg, 2004.
6. H. Sies and W. Stahl, *Annu. Rev. Nutr.*, 2004, **24**, 173–200.
7. U. Heinrich, H. Tronnier, W. Stahl, M. Bejot and J. M. Maurette, *Skin Pharmacol. Physiol.*, 2006, **19**, 224.
8. J. Thiele and P. Elsner, *Oxidants and Antioxidants in Cutaneous Biology*, Karger, Basel (Switzerland), 2001.

9. J. J. Thiele, C. Schroeter, S. N. Hsieh, M. Podda and L. Packer, *Curr. Probl. Dermatol.*, 2001, **29**, 26.
10. J. Fuchs and H. Kern, *Free Radical Biol. Med.*, 1998, **25**, 1006.
11. J. Y. Lin, M. A. Selim, C. R. Shea, J. M. Grichnik, M. M. Omar, N. A. Monteiro-Riviere and S. R. Pinnell, *J. Am. Acad. Dermatol.*, 2003, **48**, 866; S. F'guyer, F. Afaq and H. Mukhtar, *Photodermatol., Photoimmunol. Photomed.*, 2003, **19**, 56.
12. A. R. Brecher and S. J. Orlow, *J. Am. Acad. Dermatol.*, 2003, **49**, 171.
13. G. A. Kune, S. Bannerman, B. Field, L. F. Watson, H. Cleland, D. Merenstein and L. Vitetta, *Nutr. Cancer*, 1992, **18**, 237.
14. L. E. Rhodes, S. O'Farrell, M. J. Jackson and P. S. Friedmann, *J. Invest. Dermatol.*, 1994, **103**, 151.
15. H. S. Black and L. E. Rhodes, *Cancer Detect. Prev.*, 2006, **30**, 224.
16. E. Proksch, W. M. Holleran, G. K. Menon, P. M. Elias and K. R. Feingold, *Br. J. Dermatol.*, 1993, **128**, 473.
17. T. Brosche and D. Platt, *Arch. Gerontol. Geriatr.*, 2000, **30**, 139.
18. S. De Spirt, W. Stahl, H. Tronnier, H. Sies, M. Bejot, J. M. Maurette and U. Heinrich, *Br. J. Nutr.*, 2009, **101**, 440.
19. G. R. Beecher, *J. Nutr.*, 2003, **133**, 3248S.
20. J. M. Harnly, R. F. Doherty, G. R. Beecher, J. M. Holden, D. B. Haytowitz, S. Bhagwat and S. Gebhardt, *J. Agric. Food Chem.*, 2006, **54**, 9966; C. Manach, A. Scalbert, C. Morand, C. Remesy and L. Jimenez, *Am. J. Clin. Nutr.*, 2004, **79**, 727.
21. S. Quideau, D. Deffieux, C. Douat-Casassus and L. Pouysegu, *Angew. Chem., Int. Ed. Engl.*, 2011, **50**, 586.
22. T. L. Dillinger, P. Barriga, S. Escarcega, M. Jimenez, L. D. Salazar and L. E. Grivetti, *J. Nutr.*, 2000, **130**, 2057S.
23. K. Buchanan, H. M. Fletcher and M. Reid, *Int. J. Gynaecol. Obstet.*, 2010, **108**, 65.
24. H. Osman, I. M. Usta, N. Rubeiz, R. Abu-Rustum, I. Charara and A. H. Nassar, *BJOG*, 2008, **115**, 1138.
25. S. C. Davis and R. Perez, *Clin. Dermatol.*, 2009, **27**, 502.
26. S. H. Nguyen, T. P. Dang and H. I. Maibach, *Cutaneous Ocul. Toxicol.*, 2007, **26**, 287.
27. O. V. Zillich, U. Schweiggert-Weisz, K. Hasenkopf, P. Eisner and M. Kerscher, *Int. J. Cosmet. Sci.*, 2013, **35**, 491–501.
28. P. Gasser, E. Lati, L. Peno-Mazzarino, D. Bouzoud, L. Allegaert and H. Bernaert, *Int. J. Cosmet. Sci.*, 2008, **30**, 339.
29. S. Hsu, *J. Am. Acad. Dermatol.*, 2005, **52**, 1049; N. Pazyar, A. Feily and A. Kazerouni, *Skinmed*, 2012, **10**, 352.
30. U. Heinrich, C. Gartner, M. Wiebusch, O. Eichler, H. Sies, H. Tronnier and W. Stahl, *J. Nutr.*, 2003, **133**, 98.
31. M. P. Buise, J. van Bommel and C. Ince, in *Yearbook of Intensive Care and Emergency Medicine*, ed. J. L. Vincent, Springer Verlag, Heidelberg, Berlin, 2003.

32. U. Heinrich, H. Tronnier, W. Stahl, M. Bejot and J. M. Maurette, *Skin Pharmacol. Physiol.*, 2006, **19**, 224; P. Altmeyer, S. Gammal and K. Hoffmann, *Ultrasound in Dermatology*, Springer Verlag, Berlin, Heidelberg, 1992.
33. H. Tronnier, M. Wiebusch, U. Heinrich and R. Stute, *Adv. Exp. Med. Biol.*, 1999, **455**, 507–516; J. Kottner, M. Schario and N. A. Bartels, *Skin Res. Technol.*, 2013, **19**, 84.
34. U. Heinrich, U. Koop, M. C. Leneveu-Duchemin, K. Osterrieder, S. Bielfeldt, C. Chkarnat, J. Degwert, D. Hantschel, S. Jaspers, H. P. Nissen, M. Rohr, G. Schneider and H. Tronnier, *Int. J. Cosmet. Sci.*, 2003, **25**, 45; J. Plessis, A. Stefaniak, F. Eloff, S. John, T. Agner, T. C. Chou, R. Nixon, M. Steiner, A. Franken, I. Kudla and L. Holness, *Skin Res. Technol.*, 2013, **19**, 265.
35. L. M. Rodrigues, P. C. Pinto, J. M. Magro, M. Fernandes and J. Alves, *Skin Res. Technol.*, 2004, **10**, 257.
36. U. Heinrich, K. Neukam, H. Tronnier, H. Sies and W. Stahl, *J. Nutr.*, 2006, **136**, 1565.
37. M. A. Pathak, *J. Am. Acad. Dermatol.*, 1982, **7**, 285.
38. K. Neukam, W. Stahl, H. Tronnier, H. Sies and U. Heinrich, *Eur. J. Nutr.*, 2007, **46**, 53.
39. S. Williams, S. Tamburic and C. Lally, *J. Cosmet. Dermatol.*, 2009, **8**, 16.
40. D. L. Katz, K. Doughty and A. Ali, *Antioxid. Redox Signaling*, 2011, **15**, 2779.
41. W. Bors and C. Michel, *Ann. N. Y. Acad. Sci.*, 2002, **957**, 57–69; J. Terao, *Forum Nutr.*, 2009, **61**, 87–94.
42. K. L. Wolfe and R. H. Liu, *J. Agric. Food Chem.*, 2008, **56**, 8404.
43. C. Tournaire, S. Croux, M. T. Maurette, I. Beck, M. Hocquaux, A. M. Braun and E. Oliveros, *J. Photochem. Photobiol., B*, 1993, **19**, 205.
44. D. E. Stevenson and R. D. Hurst, *Cell. Mol. Life Sci.*, 2007, **64**, 2900.
45. U. Heinrich, K. Neukam, H. Tronnier, H. Sies and W. Stahl, *J. Nutr.*, 2006, **136**, 1565; K. Neukam, W. Stahl, H. Tronnier, H. Sies and U. Heinrich, *Eur. J. Nutr.*, 2007, **46**, 53.
46. R. Corti, A. J. Flammer, N. K. Hollenberg and T. F. Luscher, *Circulation*, 2009, **119**, 1433.
47. U. Forstermann and T. Munzel, *Circulation*, 2006, **113**, 1708; C. A. Schmitt and V. M. Dirsch, *Nitric Oxide*, 2009, **21**, 77.
48. Y. Steffen, T. Schewe and H. Sies, *Biochem. Biophys. Res. Commun.*, 2007, **359**, 828.
49. S. C. Tai, G. B. Robb and P. A. Marsden, *Arterioscler., Thromb., Vasc. Biol.*, 2004, **24**, 405.

CHAPTER 11
Chocolate and Dental Health[†]

ARMAN SADEGHPOUR

President & CEO, Theodent, The BioInnovation Center, 1441 Canal Street #411, New Orleans, Louisiana 70112, USA
Email: arman@theodent.com

11.1 Introduction

The medicinal properties of chocolate and cacao are well documented in this volume, but a paradoxical and exciting health benefit from this mysterious superfood remains to be discussed. We begin this journey with an image of a Mayan skull that was discovered in the Middle American Research Institute at Tulane University in New Orleans, Louisiana.[1] The skull (Figure 11.1) is of a Mayan elite, based upon the evidence that he was buried with a piece of ceremonial jade in the back of his mouth. Dental manipulations including three round circular jade inlays on three of his teeth also attest to his elite status. The top and bottom incisors were also filed down on four sides so that when the teeth align they form a T-shape, which is the Mayan glyph for "wind" or, in this case, quite literally, "breath".

A more obvious discovery arises when considering that this skull from the Mayan Classical Period (250–900 AD) is at least 1100 years old. My colleague, Joseph Fuselier, and I were surprised to find that this particular Mayan elite still had all of his teeth intact. The Mayans, of course, did not have a fluoridated water supply or access to fluoridated toothpastes, but something

---

[†]This chapter is dedicated to my intellectual and business mentors: Dr Tetsuo Nakamoto, Mr Joseph Fuselier, Dr Skip Simmons, Mr Alexander U. Falster and Dr Parviz Rastgoufard. It is also lovingly dedicated to my parents, Dr Bahram Sadeghpour and Dr Malektaj Yazdani, whose early work with Dr Nakamoto set the stage for my doctoral thesis research.

Chocolate and Health: Chemistry, Nutrition and Therapy
Edited by Philip K. Wilson and W. Jeffrey Hurst
© The Royal Society of Chemistry 2015
Published by the Royal Society of Chemistry, www.rsc.org

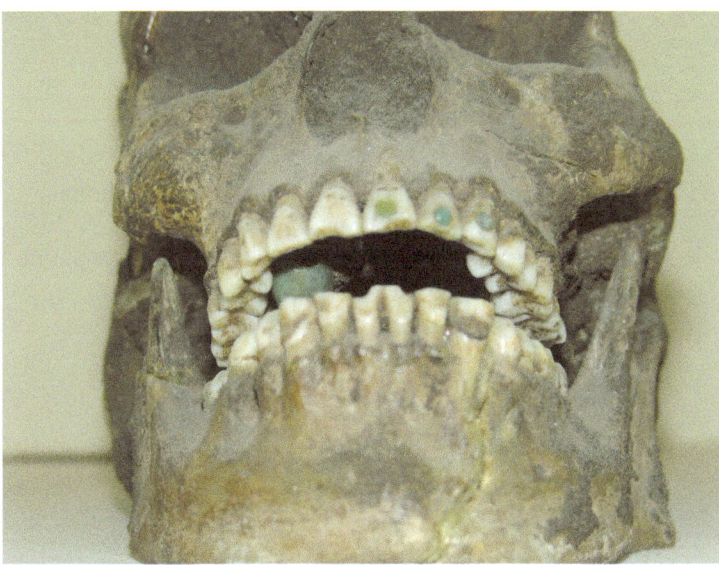

Figure 11.1 Skull of Mayan elite with notable dental manipulations.

in their diet clearly pointed to the promotion of good oral health. If you look very closely at the incisor where the circular jade inlay has fallen out, you can also see that this elite had incredibly thick enamel. This thick and robust enamel structure was primarily due to their access to cacao. Mayan historians refer to cacao as "Mayan gold", and if one was wealthy enough, as the elites were, one could afford to literally consume their wealth. Mayan elites would often pulverize and grind their cacao pods with spices, corn, hot water and chili peppers to form hot frothy beverages. The hot water served as a method by which theobromine was physically extracted from the cacao pods. Could the consumption of copious amounts of theobromine have been the hallmark of Mayan dental health? Was theobromine, in effect, the fluoride of its time? This chapter on dental health explores the health benefits of this often-overlooked methylxanthine.[2]

11.2 Theobromine Chemical Structure, Properties and Toxicity

Theobromine (Figure 11.2) is a white crystalline powder with a sublimation point of 290–295 °C and a melting point of 357 °C. It is soluble in water at a concentration of 1.0 g/2 l. In boiling water, one can dissolve 1.0 g/0.15 l and in 95% ethanol one can dissolve 1.0 g/2.2 l.[3] It is marketed under the trade names of theobrominum, theominal, théoxalvose, theoguardenal, santheose, seominal, riddospas and riddovydrin.

Theobromine is found naturally at a concentration of 1.5–3% in the cacao bean (*Theobroma cacao*) and is typically reaped from the husk of the bean,

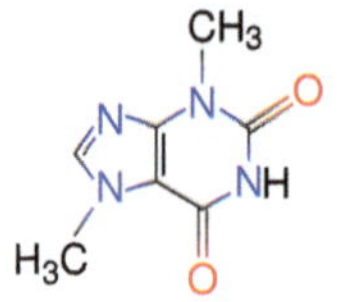

Figure 11.2 Chemical structure of theobromine (3,7-dimethylxanthine).

Theobromine Chemical and Physical Data
Chemical abstract name: 3,7-dihydro-3,7-dimethyl-1H-purine-2,6-dione
Synonym: 3,7-dimethylxanthine
Molecular formula: Figure 11.2
Molecular weight: 180.17

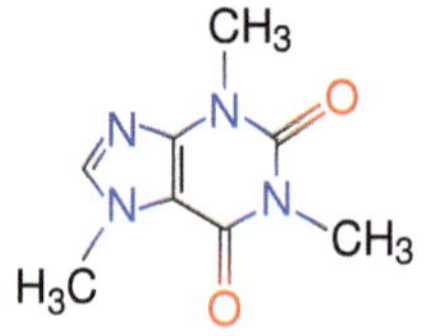

Figure 11.3 Chemical structure of caffeine (1,3,7-dimethylxanthine).

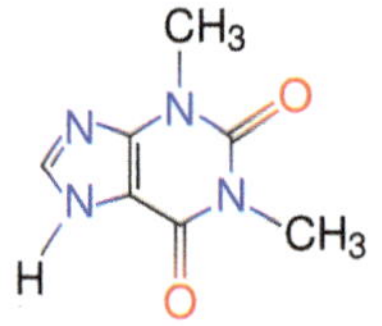

Figure 11.4 Chemical structure of theophylline (1,3-dimethylxanthine).

which contains anywhere from 0.7 to 1.2% theobromine or 15–30 g kg^{-1}.[3] It is the principal alkaloid of the cacao bean and is responsible for its bitter taste. Theobromine is found in chocolate, tea and cocoa products worldwide and was appropriately dubbed the "food of the gods" by Carl Linnaeus, in 1720, with the two Greek roots, *"Theo"* meaning God and *"broma"* meaning food. Theobromine (3,7-dimethylxanthine) bears close resemblance in its chemical composition to caffeine (1,3,7-trimethylxanthine; Figure 11.2) and theophylline (1,3-dimethylxanthine; Figure 11.3).

Theobromine is biochemically classified as a purine (3,7-dimethyl-1H-purine-2,6-dione), one of the heterocyclic aromatic organic molecules that contain a pyrimidine ring and an imidazole ring. In nature, purines and the pyrimidines comprise the base pairs that form the DNA double helix. The aromatic rings are responsible for the purines' rigid planar structure. The flat shape of these purines is of particular importance in terms of the structural building blocks of DNA. In DNA synthesis, the purines adenine and guanine form hydrogen bonds with their complementary pyrimidines, thymine and cytosine respectively.

The classification of theobromine as a purine with a rigid structure and its structural similarity to two of the four base pairs that comprise human DNA warrant it as a molecule worth closer examination. Theobromine is notable for its structural similarity to caffeine and theophylline with only small differences in methylation (see Figures 11.3 and 11.4). These differences, however, are key to the attenuation of this molecule's dental health benefits. Although the Mayans discovered chocolate's promotion of dental health long ago, the similarity of theobromine to our DNA's basic building blocks may well suggest a new frontier of nutriceutical care, in dentistry and beyond.

With only one methyl group difference in structural comparison to caffeine, it should come as no surprise that theobromine is primarily used, commercially, in the production of caffeine.[4] Though part of the same family, the stimulant effects of theobromine are distinguishable from those of caffeine. Caffeine acts relatively quickly and its main effect on human beings is increased awareness. Theobromine's effect is more subtle and produces a milder but more prolonged mood elevation than caffeine. Theobromine's plasma half-life in the bloodstream is 50% at 6 hours after consumption, whereas caffeine's is 50% at only 2 hours after consumption. Another difference is that theobromine is not physiologically addictive, showing no withdrawal symptoms. Conversely, caffeine has been proven to be physiologically addictive, with many cases of proven withdrawal. Caffeine has also been shown to have a deleterious effect on the hydroxylapatite mineral that comprises tooth enamel. Caffeine actually stunts the growth of the calcium and phosphate unit crystals in enamel.[5]

Two independent quantitative studies in the 1980s concluded that theobromine was present at an average level of 1.89% when comparing eight varieties of commercial cocoa powder.[6] It is particularly interesting that various concentration levels of theobromine are found in commercially ready foodstuffs (see Table 11.1).[7]

Table 11.1 Mean theobromine levels of commercially available foods.

Food	Theobromine content
Hot chocolate beverages	65 mg per 5 oz serving
Chocolate milk (from instant or sweetened cocoa powder)	58 mg per serving
Hot cocoa (average of nine commercial mixes)	62 mg per serving
Cocoa cereals[a]	0.695 mg g^{-1}
Chocolate bakery products[a]	1.47 mg g^{-1}
Chocolate toppings[a]	1.95 mg g^{-1}
Cocoa beverages[a]	2.66 mg g^{-1}
Chocolate ice creams[a]	0.621 mg g^{-1}
Chocolate milk[a]	0.226 mg g^{-1}
Chocolate pudding[a]	74.8 mg per serving
Carob products[a]	0–0.504 mg g^{-1}

[a]Theobromine content was determined by HPLC/reverse-phase column.[8]

Dark chocolate contains the highest levels of theobromine per serving of any type of chocolate, though the concentrations in chocolate vary from anywhere between 0.36 and 0.63%. To put this into perspective, a 1 oz bar of dark chocolate contains 130 mg of theobromine, while a 1 oz bar of milk chocolate only contains 44 mg of theobromine. Thus, the concentration of theobromine in a 1 oz bar of dark chocolate is approximately two-times that of a 5 oz hot chocolate beverage.[9] In order for a 143 lb human to produce a toxic level of theobromine in the bloodstream, they would have to ingest approximately 86 1 oz milk chocolate bars in one sitting. Clearly, theobromine is a very non-toxic compound to humans with no physiological addiction or withdrawal symptoms like its kin, caffeine. Theobromine, however, does make chocolate toxic for dogs.

Mayan elites potentially pushed the boundaries of theobromine consumption with their hot and frothy cacao beverages, but it was this excessive exposure to theobromine that may have had a direct correlation to enamel growth and thickness. Human enamel is the most highly mineralized and hardest substance in the body. It is 96% mineral—hydroxylapatite—and 4% water and organic materials. Hydroxylapatite is simply crystalline calcium phosphate in a ratio of 1.67 : 1.0 with the chemical formula: $Ca_{10}(PO_4)_6(OH)_2$.

The crystal symmetry of hydroxylapatite (Figure 11.5) is hexagonal in shape and dipyramidal. Human enamel is composed of hydroxylapatite unit crystals that grow on top of each other to form the characteristic enamel rods that are responsible for a tooth's incredible strength and rigidity.

11.3 Background Literature

In 1966, Strålfors reported a reduction of dental caries in hamsters that were fed diets rich in chocolate. The Strålfors study examined the effect on hamster caries by comparing cocoa powder, defatted cocoa powder and cocoa fat. Pure cocoa powder inhibited dental caries by 84, 75, 60 and 42% when the percentage of the hamster diet was 20, 10, 5 and 2%, respectively. Defatted cocoa showed a significantly higher anti-caries effect than fat-containing cocoa powder, but cocoa butter alone at 15% in the hamster's diet increased dental caries significantly.[10]

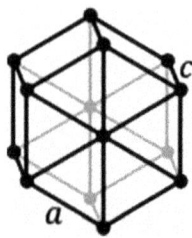

Figure 11.5 Crystal symmetry of hydroxylapatite.

In a follow-up study that same year, Strålfors attempted to study the non-fat portions of the cocoa powder, since the previous investigation showed that the caries-inhibitory effect seemed to correspond to the fat-free part of the cocoa mass extract. This study demonstrated that after being washed with water, cocoa powder had a considerably diminished caries-inhibitory effect than the unwashed cocoa powder. In the washed cocoa powder group, a considerable anti-caries phenomenon was found to persist, "indicating an existence of a non-water soluble cariostatic factor". It is interesting to note that in his discussion, Strålfors alludes to "two caries-inhibitory substances in cocoa: one water-soluble and another, which is sparingly soluble in water".[11]

Theobromine appears to be the sole factor responsible for the anti-caries activity of cocoa powder.[12] The fact that theobromine is only slightly soluble in colder-temperature water is the reason that Strålfors' study concluded the existence of "two caries-inhibitory substances in cocoa". Washing the cocoa powder with water simply dissolved away some of the theobromine, thereby diminishing the caries-inhibitory effect of the washed cocoa powder.

In another study in 1966, Strålfors studied the effects of dialyzed, detanned and carbon-treated water extracts of cocoa on golden hamster caries. The "diffusate" from the dialyzed water was reported to have a significant anti-caries effect. When the dialyzed water was dried on the potato starch diet of the hamsters, a 56% reduction in hamster caries was reported. The same dialyzed water extract was subsequently detanned (*i.e.*, the tannins were removed), and it still produced a 27% reduction of caries. In the final experiment, the dialyzed water extract was treated with activated carbon, and the result of this experiment showed no anti-caries results in the golden hamster.[13]

In a study from the following year, Strålfors found that theobromine and caffeine at a 0.2% concentration inhibited hamster caries, but also concurrently impaired the growth of the body. In the discussion of that same publication, Strålfors reported that when decreasing the caffeine concentration from 0.2 to 0.02%, normal growth was restored; however, the anti-caries phenomenon was lost in the lower-dosage caffeine group. Xanthine, however, was shown to inhibit caries without affecting growth. In his experiment "A1", Strålfors demonstrated that theobromine at the 0.2% concentration reduced caries by 37%.[14] This 37% decrease in hamster caries correlates with the 27% reduction as seen in the experiment performed in 1966, in which Strålfors detanned the dialyzed water extract. The 10% discrepancy between the two figures is most likely attributable to theobromine's partial solubility in water and the fact that some of it was more than likely removed when the dialyzed water extract was detanned.

A study by Ooshima *et al.* in 2000 reported that the sucrose-dependent cell adherence of *Streptococcus mutans* was significantly depressed by cocoa mass extract. It was also shown that cocoa mass extract in a 40% sucrose diet and drinking water reduced plaque and caries development in *Streptococcus sobrinus* 6715-infected rats, though not significantly. The conclusion of this experiment was that cacao mass extract contained some sort of anti-caries potential, though it was not significant enough to reduce the cariogenic strength of sucrose.[15]

It is interesting to note from my own studies that when looking at the scanning electron micrographs of control groups *versus* the theobromine groups, the theobromine-treated teeth have less bacterial attachment and growth, while the control samples show significant signs of bacterial attachment and growth. These results support the findings of Ooshima *et al.* from 2000 (see Figures 11.6 and 11.7).[16]

Figure 11.6 Control (no theobromine exposure) showing considerable bacterial attachment and growth.[17]

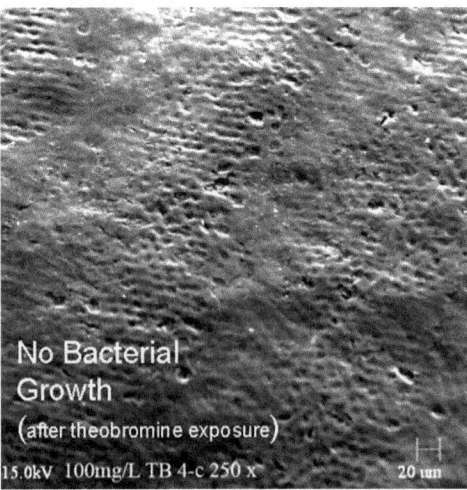

Figure 11.7 Enamel treated with theobromine showing no bacterial attachment on the enamel surface.[18]

11.4 An Alternative to Fluoride?

Fluoride is the standard additive in most commercially available toothpastes. Fluoride works by replacing the hydroxyl group in hydroxylapatite, making the tooth less susceptible to decay. Normal enamel dissolves at an acidic pH of 5.5, but when fluoride is incorporated into the tooth structure, it requires a more acidic pH of 4.5 to dissolve the hydroxylapatite in the enamel.

In normal enamel, the unit crystals that comprise the tooth are about 0.5 microns in size. Tetsuo Nakamoto discovered that in the presence of theobromine, these unit crystals in the enamel could grow to up to 2.0 microns in size. This increased size of the calcium and phosphate unit crystals in the teeth also makes the enamel less susceptible to the decay and acid erosion caused by the bacterial fermentation of sugars to acid. Though studied previously, no one had determined this effect on hydroxylapatite size until Nakamoto worked with mineralogists William B. Simmons and Alexander U. Falster.[23,24]

In one of my most compelling studies, enamel samples were exposed daily for 30 minutes to varying experimental levels of sodium fluoride (NaF) and theobromine, which were dissolved in artificial saliva formulation as the base. Dosage concentrations examined in the study included a wide range for each substrate. NaF was examined at 0.15, 0.25, 0.5 and 1.1%. This range of NaF represents the lower end of NaF concentrations that are readily available in commercial dentifrices all the way up to 1.1%, which is a prescription-strength level of NaF. The objective of the study was to find an optimal effective concentration of theobromine for a toothpaste formulation. Since the effective dosage concentration of theobromine for this toothpaste formulation was unknown, a wider range of theobromine was examined in our experiment. Theobromine was examined at the following concentrations: 1, 5, 10, 25, 50, 100, 200 and 500 mg l^{-1}.

Lower-concentration dosages of theobromine were of particular interest in these experiments because the Food and Drug Administration (FDA) and American Dental Association (ADA) are more likely to approve theobromine dosages that are well below what they deem to be potentially "harmful". After exposure to NaF or theobromine, samples were then rinsed with distilled water and placed in a solution of only artificial saliva for the remainder of the 24 hour period (rich in calcium, phosphate and magnesium). Control samples were concurrently exposed to artificial saliva for each day of the experiment, rinsed with distilled water and also held in the artificial saliva for the remainder of the 24 hour period. A positive control scenario was expected, as the artificial saliva also acts as a re-mineralization agent. Standard Knoop microhardness tests were performed each day for a period of 8 days on each sample. Day 1 measurements represent the baseline hardness for each tooth specimen. It is important to note that since each tooth was sectioned into four specimens, comparisons were readily made between the reactivity of different substrates on the same, or relatively similar, enamel surface. For groups in which all four sectioned specimens from one tooth had flat surfaces that were suitable for Knoop microhardness testing, additional theobromine dosages were examined. The results of theobromine exposure on enamel hardness were quite compelling. As presented in the data below, theobromine at much lower concentrations has a more significant effect than fluoride does on enamel hardness (Figure 11.8). The implications of this work point to the use of theobromine as a safe alternative to fluoride in commercially available toothpastes.

Another original study sought to examine the effects of theobromine on acid de-mineralization. Since larger hydroxylapatite crystals mean a harder enamel surface, it was logical to hypothesize that this stronger enamel would also be less susceptible to bacterial acid de-mineralization. Samples were divided into the same two experimental groups, 15% NaF

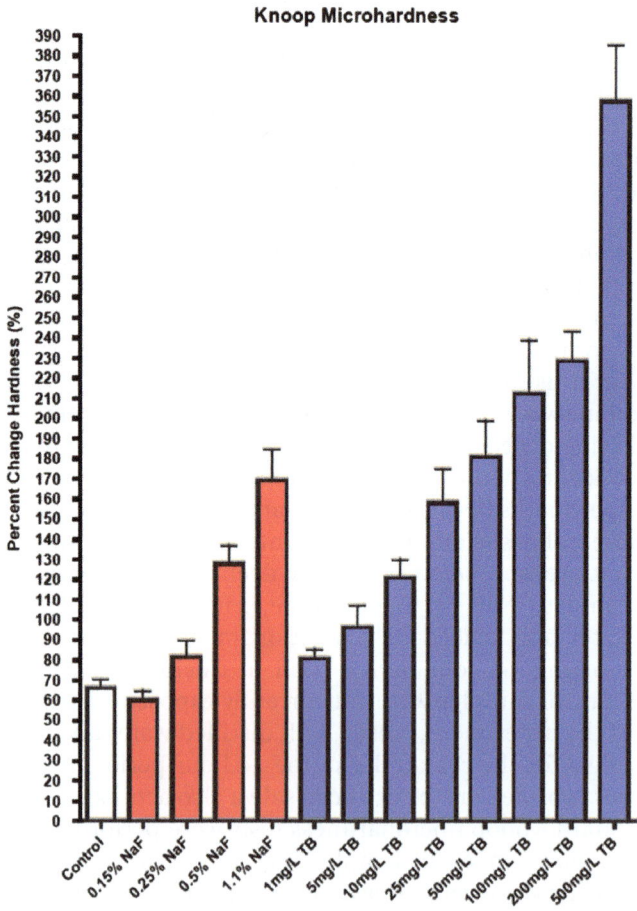

Figure 11.8 Percentage change in enamel microhardness upon exposure to varying concentrations of sodium fluoride or theobromine.

and 100 mg l^{-1} of theobromine. When available, multiple teeth extracted from the same patients were used laterally for each experimental group. Samples from both groups were soaked in their respective treatments overnight. Samples were rinsed with distilled water and were then individually exposed to 20 ml of 0.001 N hydrochloric acid (HCl) for a period of 10 minutes. The 20 ml of resulting HCl was then bottled, capped and marked appropriately for di-calcium phosphate (DCP) analysis of calcium leakage from the exposed tooth surface area.

DCP analysis was performed on a Beckman Spectraspan-5 with a calcium wavelength of 393.366 nm, a plasma position of 0, 0.1–10 ppm calibration range and a 30 seconds count time. The average calcium leakage from the 0.15% NaF group was 0.9269 ppm. The average calcium leakage was 0.8572 ppm for the 100 mg theobromine group. One ppm is equivalent to 1 mg of solute per liter of solution, or in percentage by mass, it is 0.0000926% for NaF and 0.0000857% for theobromine. This difference significantly indicates that after a 10 minutes exposure to 0.001 N HCl, the 100 mg l^{-1} theobromine group actually released 8.17% less calcium than the 0.15% NaF group. Thus, the enamel samples that were exposed to 100 mg l^{-1} of theobromine for a 24 hours period were less susceptible to calcium loss by acid dissolution.[19]

11.5 The Human Mouth

The human mouth is filled with a range of bacterial types; four in particular are correlated with the incidence of dental caries: *Streptococcus mutans*, *Lactobacillus* spp., *Nocardia* spp. and *Actinomyces viscosus*. Oral bacteria convert sugars such as glucose, fructose and sucrose to lactic acid through fermentation. Bacteria like *Lactobacillus* spp.—named as such because it grows in lactic acid-rich conditions—thrive in a sugar- and acid-rich environment. Bacteria begin to collect around the gums and teeth in creamy masses known as dental plaque.

Acid demineralization occurs when bacterial plaques are left unattended for periods of time. The process of bacterial acid demineralization is cyclical in nature. As bacteria ferment sugars and convert them to lactic acid, the bacteria thrive in the acidic environment. As the bacteria proliferate, they are able to convert even more sugar into harmful enamel-eroding acid and thus the cycle persists. The pH level of the mouth consequently drops in these areas. When it reaches a critical acid level (pH 5.5), demineralization and weakening of the tooth structure in the areas of bacterial plaque attachment are noted. Acid demineralization is the loss of the actual mineral content of the tooth, mainly the calcium, phosphate and magnesium contained in hydroxylapatite. Remineralization can occur if the acid concentration of the oral cavity is neutralized or buffered by a basic compound such as calcium carbonate, which is found in the enamel surface of teeth and in virtually all commercially available dentifrices. Minerals found in saliva and dentifrices also aid in the remineralization of the acid and bacterial plaque-compromised parts of the tooth.

Given the acidic environment that persists in the oral cavity as a result of bacterial flora and fauna, acid demineralization is the main concern related to the problem of dental caries. The results of the acid dissolution study lead us to speculate that theobromine is a particularly effective agent in helping the enamel surface of human teeth resist the effects of bacterial acid demineralization.

The research discussed herein clearly indicates that theobromine is ultimately a safer and more active ingredient for commercial dentifrices than fluoride. A growing body of literature also supports this claim, and recent concerns have been raised by the National Research Council about the Environmental Protection Agency (EPA)'s standards for public water fluoridation in the publication entitled *Fluoride in Drinking Water, A Scientific Review of the EPA's Standards*.[20] It is interesting to note that the USA is the only country that still widely fluoridates its public water supply despite the controversy that persists regarding mandatory water fluoridation. It is also interesting to note on 7 January 2011, the Department of Health and Human Services actually lowered its recommended dose for public water fluoridation by 30%, decreasing the amount from 1.0 to 0.7 ppm, citing dental fluorosis as the reason for the decrease. Dental fluorosis is an unsightly spotting and weakening of the enamel caused by excess exposure to fluoride during the development of teeth. Fluoride exposure is of particular concern for children under the age of 8 years, and all fluoride toothpastes are required to carry federally mandated "do not swallow" warning labels on the packaging.

11.6 Room for Innovation

Since the 1950s, there has been little to no innovation in commercial toothpaste in terms of reconsidering which anti-caries preventative to use. The standard ingredient is either 0.24% NaF (0.15% fluoride ion) or 0.76% sodium monofluorophosphate. Most "innovations" in toothpaste are based on false assertions and gimmick advertising that help brand products appear "unique" in a marketplace of seemingly endless options. A 1998 *Consumer Reports* rating of 35 brands of toothpaste gave an "excellent" rating to 30 of the brands examined.[21] The lack of alternatives to fluoride in the marketplace led my colleagues and myself towards the development of Theodent™ in May 2007. Based on current remineralization data and an additional 2 years of research and development, we designed a toothpaste formulation that would remineralize the enamel structure on contact. The company Theodent™ was founded by Tetsuo Nakamoto, Skip Simmons, Joseph Fuselier and myself. The fruits of our labor resulted in a proprietary formulation known as Rennou™, which

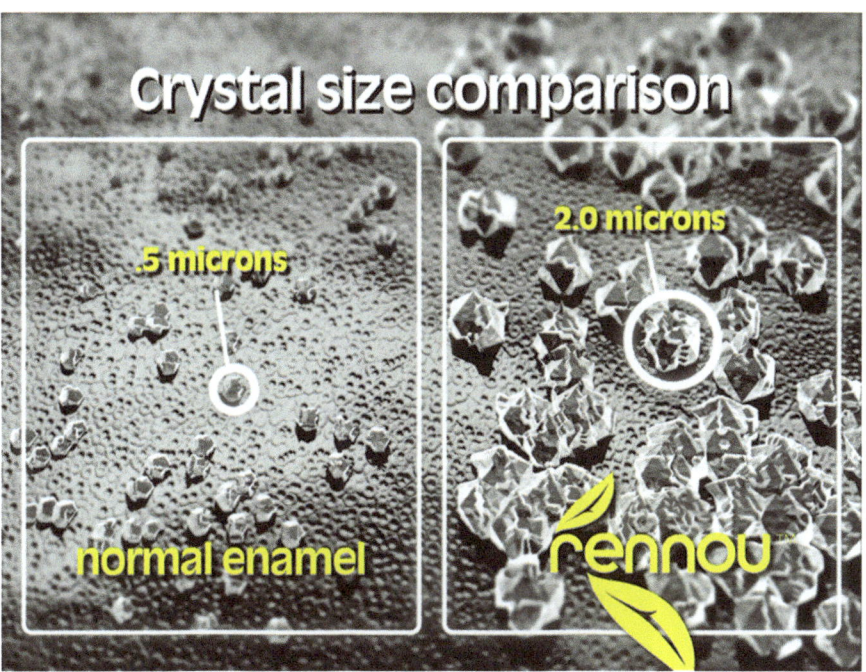

Figure 11.9 The effect of Rennou™ on enamel crystal remineralization.

contained theobromine, a source of calcium and a source of phosphate. The effect of Rennou™ on crystal size is illustrated in Figure 11.9. Aptly named, Rennou™ works by renewing the enamel surface on contact and by building larger hydroxylapatite crystals on the surface of human teeth.

In January 2012, Theodent™ toothpaste containing Rennou™ was launched in 171 whole foods stores nationwide and the company began working with researchers around the world to independently test and verify our revolutionary new oral care technology. Bennett Amaechi, Director of Cariology at the University of Texas Health Science Center in San Antonio, has produced extensive independent research regarding theobromine in oral care today. A 2013 study published in *Caries Research* revealed that theobromine at 71-times less than the concentration of fluoride has a remineralization effect on enamel lesions that is comparable to that of fluoride.[22]

Theodent™ toothpaste (see Figure 11.10) won the most prestigious international design award—the Red Dot–for its innovative new look and packaging. Its bold brown coloring is an homage to chocolate, which lies at the origins of its science. The Mayan secrets of better dental health will remain a mystery, but are now available in a product that definitely holds itself to a new and higher standard.

Figure 11.10 Theodent™ toothpaste.

References

1. The author discovered this skull in 2008 upon completion of my doctoral research work at Tulane.
2. D. D. Drouillard, E. S. Vessel and B. H. Dvorchik, *Clin. Pharm. Ther.*, 1978, **23**(3), 296–302; J. H. Fries, *Ann. Allergy*, 1978, **41**(4), 195–207; C. Berlin, Jr. and C. Daniel, *Pediatr. Res.*, 1981, **15**, 313; J. S. Bonvehi and F. V. Coll, *J. Agric. Food Chem.*, 1998, **46**, 620–624; A. T. Borchers, C. L. Keen, S. M. Hannum and E. M. Gershwin, *J. Med. Food*, 2000, **3**(2), 77–105.
3. M. Windholz, *The Merck Index*, Merck Publishing Group, Rahway, New Jersey, 10th edn, 1983, p. 1327.
4. G. F. McCutheon, "Caffeine", in *Encyclopedia of Industrial Chemical Analysis*, ed. F. D. Snell, C. L. Hilton and L. S. Ettre, Interscience Publishers, New York, London, Sydney, 1969, 8, pp. 55–71.
5. T. Nakamoto and R. Shaye, *J. Nutr.*, 1986, **116**, 633–640; T. Nakamoto, S. L. Cheuk, S. Yoshino, A. U. Falster and W. B. Simmons, *Arch. Oral Biol.*, 1993, **38**(10), 919–922.

6. C. A. Shively and S. M. Tarka Jr., "Methylxanthine Composition and Consumption Patterns of Cocoa and Chocolate Products", *The Methylxanthine Beverages and Foods: Chemistry, Consumption, and Health Effects*, Alan R. Liss, New York, 1984, pp. 149–178; B. L. Zoumas and W. R. Kreiser, *J. Food Sci.*, 1980, **45**, 314–316.
7. B. L. Zoumas and W. R. Kreiser, *J. Food Sci.*, 1980, **45**, 314–316; J. L. Blauch and S. M. Tarka Jr, *J. Food Sci.*, 1983, **48**, 745–750; C. A. Shively and S. M. Tarka Jr, "Methylxanthine Composition and Consumption Patterns of Cocoa and Chocolate Products", *The Methylxanthine Beverages and Foods:* Chemistry, Consumption, and Health Effects, Alan R. Liss, New York, 1984, pp. 149–178.
8. W. J. Craig and T. T. Nguyen, *J. Food Sci.*, 1984, **49**, 302–304.
9. C. A. Shively and S. M. Tarka Jr, "Methylxanthine Composition and Consumption Patterns of Cocoa and Chocolate Products", *The Methylxanthine Beverages and Foods: Chemistry, Consumption, and Health Effects*, Alan R. Liss, New York, 1984, pp. 149–178.
10. A. Strålfors, *Arch. Oral Biol.*, 1966, **11**, 149–161.
11. A. Strålfors, *Arch. Oral Biol.*, 1966, **11**, 323–328.
12. A. Sadeghpour, *A Neural Network Analysis of Theobromine vs. Fluoride on the Enamel Surface of Human Teeth: An Experimental Case Study with Strong Implications for the Production of a New Line of Revolutionary and Natural Non-Fluoride Based Dentifrices*. 2007. 150 p., *Dissertation Abstracts International*, 68(07), Section B, p. 4642.
13. A. Strålfors, *Arch. Oral Biol.*, 1966, **11**, 609–615.
14. A. Strålfors, *Arch. Oral Biol.*, 1967, **12**, 321–332.
15. T. Ooshima, Y. Osaka, H. Sasaki, K. Osawa, H. Yasuda and M. Matsumura, *Arch. Oral Biol.*, 2000, **45**, 639–645; T. Ooshima, Y. Osaka, H. Sasaki, K. Osawa, H. Yasuda and M. Matsumoto, *Arch. Oral Biol.*, 2000, **45**, 805–808.
16. T. Ooshima, Y. Osaka, H. Sasaki, K. Osawa, H. Yasuda and M. Matsumura, *Arch. Oral Biol.*, 2000, **45**, 639–645; T. Ooshima, Y. Osaka, H. Sasaki, K. Osawa, H. Yasuda and M. Matsumoto, *Arch. Oral Biol.*, 2000, **45**, 805–808.
17. A. Sadeghpour, *A Neural Network Analysis of Theobromine vs. Fluoride on the Enamel Surface of Human Teeth: An Experimental Case Study with Strong Implications for the Production of a New Line of Revolutionary and Natural Non-Fluoride Based Dentifrices*. 2007. 150 p., *Dissertation Abstracts International*, 68(07), Section B, p. 4642.
18. A. Sadeghpour, *A Neural Network Analysis of Theobromine vs. Fluoride on the Enamel Surface of Human Teeth: An Experimental Case Study with Strong Implications for the Production of a New Line of Revolutionary and Natural Non-Fluoride Based Dentifrices*. 2007. 150 p., *Dissertation Abstracts International*, 68(07), Section B, p. 4642.
19. A. Sadeghpour, *A Neural Network Analysis of Theobromine vs. Fluoride on the Enamel Surface of Human Teeth: An Experimental Case Study with*

Strong Implications for the Production of a New Line of Revolutionary and Natural Non-Fluoride Based Dentifrices. 2007. 150 p., *Dissertation Abstracts International*, 68(07), Section B, p. 4642.
20. National Research Council, *Fluoride in Drinking Water: A Scientific Review of the EPA's Standards*, The National Academies Press, Washington, D.C., 2006.
21. S. Stevenson, "Paste Test", *Consumer Reports*, 7 Oct. 1998.
22. B. T. Amaechi, N. Porteous, K. Ramalingam, P. K. Mensinkai, R. A. Ccahuana Vasquez, A. Sadeghpour and T. Nakamoto, *Caries Res.*, 2013, **47**, 399–405.
23. T. Nakamoto, W. B. Simmons and A. U. Falster, Apatite-forming systems: Methods and Products, *US Pat.*, 5919426, 1997.
24. T. Nakamoto, W. B. Simmons and A. U. Falster, Apatite-forming systems: Methods and Products, *US Pat.*, 6183711 B1, 2001.

Epilogue

Throughout the late 20th century, inquiries flooded into psychologists regarding what might best be termed "chocolate psychology". As pop culture scholar and Worcester State College media professor emerita Linda Fuller reminds us, chocolate has gained at least a folk reputation across a wide range of psychological conditions. It has been viewed as: a cure for agoraphobia, a security blanket, an appeal to the senses, a strategy of control, a means to excite passions and curiosities, an excuse gratification, an antidote to guilt, a step toward intimacy, a cause of dependence, a stress reliever, a restorer of strength, a reward, or maybe just a simple consolation in a world of contracting opportunities.[1]

Questions continue to rise over chocolate's so-called addictive nature, either in terms of emotional or physical addiction. Unlike other reputed addictions, chocoholism is not typically touted as something to be avoided. Integrative medicine guru, Dr Andrew Weil, and Winifred Rosen addressed chocolate "addiction" in their highly-acclaimed, *From Chocolate to Morphine: Everything You Need to Know about Mind-Altering Drugs* (1983). They claim that "cases of chocolate dependence are easy to find". Most chocoholics are women, they argue, and "many of them crave chocolate most intensely just before their menstrual periods. Women who develop an addictive relationship with chocolate usually eat it in cyclic binges rather than continually, and often say that it acts on them like an instant anti-depressant".[2]

Despite anecdotal evidence that may say "Yes, Yes, Yes", little modern biomedical evidence supports the claim that one can become addicted to chocolate. Since depriving one of chocolate fails to produce scientifically significant signs of withdrawal, it is thereby not technically classed as a physically addictive agent. Further, scientists have not shown a state of dependence regarding chocolate's use.[3] Recent trends tend to avoid chocolate—in favor of investigations into the "psychological drivers of chocolate consumption"[5], or eating chocolate as an encouraging "habit" that may be "good" and may have excellent outcomes, including the ability to improve one's positive mental attitude.[6] Such habits may well be driven by needs to re-establish some inner balance or homeostasis. These needs, which certainly seem to have a psychological basis, may also be biologically, evolutionarily and culturally driven as well.[7]

Some psychologists have argued that chocolate may pharmacologically stimulate behaviors of compulsive eating, while others claim that this finding may just as well be the result of a more generalized aesthetic craving for the sweetness, oily richness and complete orosensory experience that chocolate provides. Indeed, chocolate's rich natural complexity—a complexity that rivals any other food—makes the actual source of perceived cravings or chocoholism exceedingly difficult to ascertain.[8] In other words, complexities in distinguishing causal factors regarding qualities that include preferring, liking and craving chocolate may well be inexorably tied to the complexities of the substance itself. Despite the lingering concern over categorizing chocolate's cravings in the bioscience literature, many will likely follow the folk wisdom that Khodorowsky and Robert shared in *The Little Book of Chocolate*: whether addictive, habitual or merely craved, chocolate "will always be assured a place of honor in the pharmacopoeia of pleasure".[9]

Another recent and exciting potential health benefit of chocolate has come from its use in the care of dementia patients. As part of the comfort-centered care offered in facilities such as the Beatitudes Nursing Home in Phoenix, Arizona, chocolate is regularly administered to patients, and its effects are noted in the nursing charts. If, as behavioral research suggests, creating facilities for people with significant dementia that are filled with positive emotional environments can diminish patient's inexplicable yet measurable signs of distress (and other behavioral problems such as late-afternoon "sundowning"), then perhaps more positive outcomes are achievable simply through chocolate.[10]

Many people become less stressed when they experience something that once produced great joy in their lives. Chocolate has been shown to be strongly "linked to memories of childhood, the maternal instinct, and affection". Overall, it mentally triggers feelings of—or at least associations with—"warmth and protection, reminding us of situations that are pleasant and familiar".[11] Thus, if chocolate had once been a favored comfort food, why not offer it as needed to people with dementia who, as a particular group of humans, need added human comfort? Additionally, as Beck and Damkjaer have suggested, chocolate may help curb the weight loss and decrease in the body mass index commonly seen among those living in extended care facilities, including nursing homes.[12] In terms of nutritional intervention, chocolate may be a quality of life extender as it boosts the nutritional, physical and mental health of the growing number of elderly individuals who are admitted to long-term care facilities. More recent studies by Jun Wang, W. Jeffrey Hurst, Giulio Maria Pasinetti and others in their research team have shown that cocoa extracts interfere with amyloid peptide-β and tau protein oligomerization into aggregates as seen in Alzheimer's dementia, thereby suggesting chocolate's promising role as a dietary intervention for these patients.[13]

More broadly, investigators are increasingly delving into what has hitherto been one of the least explored aspects of chocolate and health: its potential benefit regarding human cognition. In part, investigators have been limited by tools for reliably measuring cognition. This complication has been further compounded given that those in the field have difficulty in uniformly describing "cognition". In science, cognition refers to the mental processing that includes the attention of working memory, comprehending and producing language, calculating, reasoning and problem solving. Various disciplines, such as psychology, philosophy and linguistics, all study cognition. However, the term's usage varies across disciplines; for example, in psychology and in cognitive science, "cognition" usually refers to an information processing view of an individual's psychological functions.

One major advance in our understanding of cognition has come through the expanding utility of magnetic resonance imaging (MRI), functional MRI (fMRI) and computerized axial tomography (CAT) scanning. MRI, a variant of nuclear magnetic resonance, is a technique that has been around for decades. Basically, MRI is a tool that uses a magnetic field and pulses of radio wave energy to create images of organs and structures *in situ*. fMRI provides the ability to evaluate certain sections of the brain depending on function whereby the resulting output in color is a measurement of respective brain activity. fMRI works by detecting the changes in blood oxygenation and flow that occur in response to localized neural activity. When a particular brain area is more active, it consumes more oxygen. To meet this increased demand, blood flow increases to the active area. fMRI can be used to produce activation maps showing which parts of the brain are involved in a particular mental process. For example, if someone smells a particular scent, the section of the brain focusing on olfaction becomes highlighted. CAT scanning is essentially a three-dimensional X-ray of a selected bodily region. It is a form of tomography in which a computer controls the motion of an X-ray source and detectors, processes the data and produces the image.

One of the first investigators to examine cacao and cognition was Hendrik J. Smit from the University of Bristol's Department of Experimental Psychology. Smit *et al.* published a series of papers beginning in 2004, including one in *Psychopharmacology,* in which he investigated a group of 20 self-proclaimed "chocoholics".[14] In one study, a battery of measurements was taken before and after a series of treatments that included 11.6 g of cocoa and a caffeine and theobromine combination of 19 and 250 mg, respectively. In another study, Schmidt investigated the effects of the xanthine levels following consumption of either milk or dark chocolate.[14] The results indicated improvements in cognitive function and energy arousal from both cocoa and the methylxanthine mixture and improvements in cognitive function from dark and milk chocolate. Studies utilizing white chocolate produced a response that was only comparable to that of water.

A different approach was taken by Parker et al. in an article entitled "Mood State Effects of Chocolate".[15] Here, chocolate's wide acclaim as a stimulant, relaxant, aphrodisiac, tonic, euphoriant and antidepressant was acknowledged. Parker et al. outlined the composition and manufacture of chocolate, chocolate craving and various psychoactive properties, noting chocolate's stimulation of the release of the neurotransmitter anadamide, a chemical analogue to cannabinoid. In describing various neurotransmitter systems, including those identified by dopamine, serotonin and opioids, special attention was drawn to chocolate cravings. Conclusions from this review posited that though chocolate does not act precisely as an antidepressant, it seems to play at least a transient comforting role.

Elsewhere, Bisson et al. described the outcome of a study on rats that were fed a cocoa polyphenolic extract based on a high-polyphenol cocoa powder called Acticoa developed by Barry Callebeaut.[16] In this study, 24 mg kg^{-1} of this powder was fed per day to rats for 15–27 months. The outcomes that were monitored included dopamine levels in the urine and lifespan in addition to cognition. The results notably indicated that the Acticoa powder-fed animals showed significantly improved cognition and an increased lifespan. In another study, the cognitive performance of 2031 humans was assessed based upon the consumption of flavanoids from chocolate, wine or tea.[17] These studies indicated that individuals who consumed these flavonoid-rich sources demonstrated improved cognitive performance compared to those who had not, and the level of improved cognitive performance was directly dose dependent.

Overall, the connections between chocolate and health seem promising. An increasing number of studies strongly suggests that further explorations into chocolate's potential ability to improve cognitive function is warranted.[18] Although cognition may not be the final frontier of chocolate's possible nutritive and health benefits, it is currently the most rapidly expanding area of biopsychomedical investigation into chocolate. Indeed, by any measure, this is one of the newest and most exciting and promising areas of experimental exploration in this field!

The two most common classes of biochemical compounds being investigated regarding chocolate and cognition are the methylxanthines and the polyphenols. Most investigators are utilizing high-polyphenol cocoa or cocoa extracts in their studies. Measurements of cognition are increasingly being determined based upon the experimental tools mentioned above, though much clarification of the growing pools of data remains to be done. Whereas the majority of research has produced positive outcomes, some investigators argue that chocolate shows stronger connections with affecting mood rather than cognition *per se*. Just where the future cartography of chocolate's benefits to nutrition and health will lead us remains unknown, but increasingly the focus is being drawn to mapping the possible ways in which chocolate affects brain performance and function. Chocolate has been promoted as "brain food" off and on for over a century. Now, scientifically sound evidence for such claims seems to be mounting.

Philip K. Wilson
W. Jeffrey Hurst

References

1. L. K. Fuller, *Chocolate Fads, Folklore & Fantasies: 1,000 + Chunks of Chocolate Information*. Haworth Press, New York, 1994, p. 17.
2. A. Weil and W. Rosen, *From Chocolate to Morphine: Everything You Need to Know About Mind-Altering Drugs*, Revised Edition, Houghton Mifflin, Boston, 2004, p. 48; Although a true gendered history of chocolate awaits an author, pharmacologist and ecofeminist scholar Cat Cox's *Chocolate Unwrapped: The Politics of Pleasure* (Women's Environmental Network, London, 1993) provides a remarkable starting point. Cultural anthropologist Emma Robertson adds another poignant yet startling view of women and chocolate in her *Chocolate, Women and Empire: A Social and Cultural History*, Manchester University Press, Manchester, UK, 2009; Marcy Norton's *Sacred Gifts, Profane Pleasures: Tobacco and Chocolate in the Atlantic World* (Cornell University Press, Ithaca, New York, 2008) illustrates the importance of chocolate in the lives of women in Mesoamerica and throughout the Atlantic World, especially regarding their marriageability, menstruation and social networking. In cultures where women secured their livelihood by becoming skilled in a desired craft, the ability to make good chocolate could raise a woman from the situation of a slave slated for death to a noblewoman. In this sense, chocolate literally became central to life or death.
3. K. Bruinsma and D. L. Taren, *J. Am. Dietetic Assoc.*, 1999, **99**, 1249–1256.
4. C. S. Elliott, Curing Irrationality with Chocolate Addiction, in *Chocolate: Food of the Gods*, ed. A. Szogyi, Greenwood Press for Hofstra University, Westport, CT, 1997, pp. 29–30.
5. E. Molinari and E. Callus, Psychological Drivers of Chocolate Consumption, in *Chocolate and Health*, ed. R. Paoletti, A. Poli and A. Conti, Springer Verlag Italia, Milan, 2012, pp. 137–146.
6. L. D. Reid, Delicious or Addictive?, in *Chocolate, Fast Foods and Sweeteners: Consumption and Health*, ed. M. R. Bishop, Nova Science Publishers, New York, 2010, pp. 313–317.
7. Pennsylvania's Susquehanna University psychologist Debra A. Zellner chides the view of chocolate cravings merely in biophysiological terms, favoring instead conceptualizing cravings as culturally defined; S. L. Vélez, *The Virtues and Delights of Chocolate*, Compañia Nacional de Chocolates, Medellín, Colombia, 2003, pp. 108–109.

8. Jeff Morgan (1994) focuses upon chocolate's "flavor and texture unlike any other" (*Am. J. Clin. Nutr.*, 1994, **60**, 1065S–1067S). Many have claimed, with considerable evidence, that chocolate is the most commonly craved food; M. M. Hetherington and J. I. Macdiarmid, *Appetite*, 1993, **21**, 233; D. Benton, The Biology and Psychology of Chocolate Craving, in *Coffee, Tea, Chocolate, and the Brain*, ed. A. Nehlig, CRC Press, Boca Raton, 2004, p. 205.
9. K. Khodorowsky and H. Robert, *The Little Book of Chocolate*, Flammarion, Luzon, France, 2001, p. 26.
10. P. Belluck, *Giving Alzheimer's Patients Their Way, Even Chocolate*, The New York Times, December 31, 2010, http://www.nytimes.com/2011/01/01/health/01care.html?_r=1&pagewanted=print, accessed August 31, 2014; W. A. McIntosh, K. S. Kubena and W. A. Landmann, Chocolate and Loneliness Among the Elderly, in *Chocolate: Food of the Gods*, ed. A. Szogyi, Greenwood Press for Hofstra University, Westport, CT, 1997, pp. 3–10, found that chocolate benefited the general aging population in terms of stress relief.
11. E. Molinari and E. Callus, Psychological Drivers of Chocolate Consumption, in *Chocolate and Health*, ed. R. Paoletti, A. Poli and A. Conti, Springer Verlag Italia, Milan, 2012, p. 137.
12. A. M. Beck and K. Damkjoer, Chocolate: A Significant Part of Nutrition Intervention among Elderly Nursing Home Residents, in *Chocolate, Fast Foods and Sweeteners: Consumption and Health*, ed. M. R. Bishop, Nova Science Publishers, New York, 2010, pp. 252–253.
13. J. Wang, *et al.*, *J. Alzheimer's Dis.*, 2014, **41**, 643–650.
14. H. J. Smit, E. A. Gaffan and P. J. Rogers, *Psychopharmacology*, 2004, **176**, 412–419.
15. G. Parker, I. Parker and H. Brotchie, *J. Affect. Disord.*, 2006, **92**, 149–159.
16. J.-F. Bisson, *et al.*, *Br. J. Nutr.*, 2008, **100**, 94–101.
17. A. N. Sokolova, M. A. Pavlovab, S. Klosterhalfena and P. Enck, *Neurosci. Biobehav. Rev.*, 2013, **37**, 2445–2453.
18. Among the key investigations in this field are the following, which are listed chronologically: N. D. L. Fisher, F. A. Sorond and N. K. Hollenberg, *J. Cardiovasc. Pharm.*, 2006, **47**, S210–S214; P. Rozan, *et al.*, *J. Food Sci.*, 2007, **72**, S203–S206; W. D. Crews, Jr, D. W. Harrison and J. W. Wright, *Am. J. Clin. Nutr.*, 2008, **87**, 872–880; J.-F. Bisson, *et al.*, *Br. J. Nutr.*, 2008, **100**, 94–101; J. P. E. Spencer, *Genes Nutr.*, 2009, **4**, 243–250; E. Nurk, *et al.*, *J. Nutr.*, 2009, **139**, 120–127; G. Dilip, *Agro Food Ind. Hi-Tech*, 2009, 27–28; C. Alex, *Agro Food Ind. Hi-Tech*, 2009, **20**, 10–13; A. B. Scholey, *et al.*, *J. Psychopharmacol.*, 2010, **24**, 1505–1514; D. T. Field, C. M. Williams and L. T. Butler, *Physiol. Behav.*, 2011, **103**, 255–260; E. S. Mitchell, *et al.*, *Physiol. Behav.*, 2011, **104**, 816–822; D. L. Katz, K. Doughty and A. Ali, *Antioxid. Redox Signaling*, 2011, **15**, 2779–2811; R. J. Williams and J. P. E. Spencer, *Free Radical Biol. Med.*, 2012, **52**, 35–45; L. Fernández-Fernández, *et al.*, *Behav. Brain. Res.*, 2012, **228**, 261–271; L. Fruson, S. Dalesman and K. Lukowiak, *J. Exp. Biol.*, 2012, **215**,

3566–3576; S. T. Shulman, *Pediatr. Ann.*, 2012, **41**, 486–487; K. Hristova, *et al.*, *Open Nutraceuticals J.*, 2012, **5**, 207–212; C. F. Haskell, F. L. Dodd, E. L. Wightman and D. O. Kennedy, *Nutr. Res. Rev.*, 2012, **26**, 49–70; A. Nehlig, *Br. J. Clin. Pharmacol.*, 2013, **75**, 716–727; M. P. Pase, *et al.*, *J. Psychopharmacol.*, 2013, **27**, 451–458; M. J. Baggott, *et al.*, *J. Psychopharmacol.*, 2013, **28**, 109–118; D. A. Judelson, *et al.*, *J. Clin. Psychopharmacol.*, 2013, **33**, 499–506; F. Dal Moro, *Food Chem. Toxicol.*, 2013, **59**, 808; F. A. Sorond, *Neurology*, 2013, **81**, 904–909; L. Calderón-Garcidueñas, *et al.*, *Front. Pharmacol.*, 2013, **4**, 104; R. Latif, *Curr. Opin. Clin. Nutr. Metab. Care*, 2013, **16**, 669–674; U. Uysal, *et al.*, *Curr. Pharm. Des.*, 2013, **19**, 6094–6111; A. N. Sokolov, M. A. Pavlova, S. Klosterhalfen and P. Enck, *Neurosci. Biobehav. Rev.*, 2013, **37**, 2445–2453; C. J. Lowe, P. A. Hall, C. M. Vincent and K. Luu, *Front. Hum. Neurosci.*, 2014, **8**, 267.

APPENDIX 1

Brief Historical Timeline of the Early Mentions of Chocolate in terms of Science, Nutrition and Medicine

Ca. 1100 BCE Analysis of pottery from northern Honduras reveals presence of theobromine.
1526   Gonzalo Fernández de Oviedo y Valdéz publishes *La Historia general y natural de las Indias*; identifies chocolate as healthy.
1528   Cortés returns to Spain from Mexico with cocoa beans and the utensils necessary for their preparation.
1575   In Girolamo Benzoni's *History of the New World*, Mexican chocolate is called "more a drink for pigs than a drink for humanity". During a wine shortage, however, Benzoni found that bitter chocolate "satisfies and refreshes the body".
1585   The first shipment of cacao beans from plantations in the Americas is delivered to Spain.
1590   *Florentine Codex* identifies medicinal uses of chocolate.
1590   José de Acosta identifies medicinal uses of chocolate.
1591   Juan de Cárdenas identifies medicinal uses of chocolate.
1592   Augustin Farfan identifies medicinal uses of chocolate.
1609   *Libro en el cual se trata del chocolate* is published in Mexico, probably the first book devoted entirely to the subject of chocolate.
1631   Antonio Colmenero de Ledesma publishes what is commonly noted as the first book dedicated entirely to chocolate: *Curioso Tradado de la Naturaleza y Calidad del Chocolate, Dividido en Quatro Puntos*.

1636 Antonio de Leon Peinelo publishes *Question Moral se el Chocolate Quebranta el Ayuno Eclesiastico*, discussing moral issues regarding whether or not chocolate is a food, whether it can be consumed during Lent and so on.

1648 Thomas Gage, an English Dominican friar who visited Cortés in the New World, reports on the uses of cacao and chocolate in his book *A New Survey of the West Indies*.

1652 J. Wadsworth translates Colmenero de Ledesma's text into English under his own name as: *Chocolate. Or an Indian Drinke*.

1661 The French medical establishment endorses the use of chocolate.

1662 Henry Stubbe identifies medicinal uses of chocolate in *The Indian Nectar or a Discourse Concerning Chocolata*.

1671 Philippe Sylvestre Dufour reviews the publication by René Moreau and publishes *De l'Usage du Caphé, du Thé, et du Chocolat*.

1672 William Hughes identifies medicinal uses of chocolate, with specific comments on asthma: *The American Physitian [sic]...Whereunto is added a Discourse on the Cacao-Nut Tree, and the Use of Its Fruit, with All the Ways of Making Chocolate*.

1680 Sir Hans Sloane blends chocolate with milk for use as a medicinal beverage.

1684 François Foucault in his dissertation *An Chocolatae usus salubris?* argues for the medicinal benefits of chocolate.

1685 Henry Mundy publishes *Opera Omnia Medico-Physica de Aere Vitali, Esculentis et Potelentis cum Appendice de Parergis in Victu et Chocolatu, Thea, Caffea, Tobacco*.

1685 J. Chamberlaine publishes a translation of Philippe Sylvestre Dufour's book, entitling it *The Manner of Making Coffee, Tea and Chocolate*.

1687 Nicolas de Blégny identifies medicinal uses of chocolate in *Le bon Usage de Thé, du Caffé, et du Chocolat pour la Preservation et pour la Guerison des Malades*.

1693 White's Chocolate House established in London.

1695 Marcus Mappus publishes *Dissertationes Medicae Tres de Receptis Hodie Etiam in Europa, Potus Calidi Generibus Thée, Cafe, Chocolata*.

1706 [Dr] Duncan publishes *Wholesome Advice Against the Abuse of Hot Liquors, Particularly of Coffee, Tea, Chocolate, etc.*

1724 R. Brooks translates and publishes the work of De Chélus (1719), entitling it *Natural History of Chocolate*.

1732 Frenchman Monsieur Dubuisson invents a table for grinding cacao that is heated underneath to hasten the process.

1738 F. E. Bruckman publishes *Relatio de Cacao*.

1739 H. T. Baron publishes *An Senibus Chocolatae Putus?*

1741 Carl von Linné (Linneaus) identifies various medicinal uses of chocolate.

1747 R. Campbell publishes *The London Tradesman*, which contains an account of chocolate processing.

1750	Henri Lekain blends chocolate with coffee, a drink that Voltaire takes daily.
1753	Linnaeus assigns the classification *Theobroma cacao* to the chocolate tree.
1754	Etienne Bachot publishes a dissertation entitled *An chocolatae usus salubris?*
1760	Richard Saunders in *Poor Richard's Almanack* recommends chocolate to smallpox patients.
1765	Walter Baker and Company open the first chocolate mill in colonial North America.
1780	Completion of Diderot's *Encyclopedia*, which contains images of chocolate production.
1790	From *Concise Observations on the Nature of Our Common Food*, "CHOCOLATE, Which is the cocoa-nut mixed with flour or sugar, is very wholesome".
1795	J. S. Fry and Sons in England grind cacao beans using steam power.
1796	Antonio Lavadan identifies medicinal uses of chocolate.
1800	Sulpice Debauve and Antoine Gallais, pharmacists to King Louis XVI, produce medicinal chocolates.
1803	M. Parmentier publishes *On the Composition and Use of Chocolate*.
1804	The descriptive qualities of pure, unadulterated chocolate are defined in the *American Register*.
1825	Jean Anthelme Brillat-Savarin identifies medicinal uses of chocolate.
1827	Antoine Gallais publishes *Monographie du Cacao, ou Manuel de l'Amateur de Chocolat*.
1828	Dutchman Coenraad Van Houten develops a cocoa press and the "Dutching" process, in which alkali is added to cocoa powder to mellow the flavor and enable it to be easily mixed with liquids.
1828	Patent entitled "Method for pressing the fat from cocoa beans" filed by Caspar Van Houten, April 4.
1828	Thomas Graham identifies medicinal uses of chocolate.
1841	Alexander Woskresensky identifies theobromine in cacao beans.
1846	Auguste Saint-Arroman identifies medicinal uses of chocolate.
1847	J. S. Fry and Sons make a "chocolate bar" for eating.
1849	George and Richard Cadbury introduce a line of milk chocolate prepared after the "Sloane Recipe" (Sir Hans Sloane's recipe of 1687).
1849	*Scientific American* reports that individuals habitually drinking chocolate do not experience attacks of cholera or "dysenteric affections", but that other family members who drink coffee, tea or cold water experience these diseases.
1851	William Alcott, in *The Young Mother, or, Management of Children with Regard to Health*, notes that "Chocolate can quench thirst, but not as well as water due to its mucilage and nutriment properties".
1859	Alfred Mitscherlich publishes *Der Kakao und die Schokolade*.

Brief Historical Timeline

1861   Army of Virginia and Florence Nightingale publish *Directions for Cooking by Troops in Camp and Hospital*: "Cocoa is often recommended to the sick in lieu of tea or coffee…".

1873   Catherine Beecher, in *Miss Beecher's Housekeeper and Healthkeeper, Containing Five Hundred Recipes for Economical and Healthful Cooking*, identifies the usefulness of cocoa and chocolate for sick and young children.

1875   Swiss inventor Daniel Peter creates milk chocolate using Henri Nestlé's condensed milk product.

1876   Daniel Peter introduces a line of milk chocolate designed for eating.

1879   Rodolphe Lindt of Switzerland invents conching, a process that creates a smoother-eating chocolate and revolutionizes chocolate making.

1880   J. M. Fothergill, in *Food for the Invalid*, provides recipes using cocoa nibs and for the preparation of chocolate cream.

1880   James William Holland, in *Diet for the Sick. Notes: Medical and Culinary*, notes that chocolate is an irritant to dyspeptics.

1884   A. W. Nicholson, in *Food and Drink for Invalids*, claims that chocolate is hard to digest.

1891   Walter Baker and Company publish *The Chocolate Plant and its Products*.

1892   Mary L. Clarke, in *Cooking for the Sick and Convalescent*, provides recipes for cocoa and chocolate.

1892   Historicus (Richard Cadbury) publishes *Cocoa: All About It*.

1893   Mary Boland, in *A Handbook of Invalid Cooking for the Use of Nurses in Training Schools, Nurses in Private Practice, and Others Who Care for the Sick*, identifies cocoa as an easily digested food.

1899   Edward Munson prepares *Emergency Diet for the Sick in the Military Service*, in which he identifies the importance of medical stores, including chocolate.

1901   Helena Viola Sachse, in *How to Cook for the Sick and Convalescent*, includes many chocolate recipes.

1902   Ellen Duff, in *A Course in Household Arts*, includes extensive information on cocoa and chocolate.

1907   Amended sections of the United States Food Law defining adulteration of candies and chocolates.

1909   Julius Friedenwald, in *Diet in Health and Disease*, identifies "good uses of chocolate".

1910   Alida Frances Pattee, in *Practical Dietetics with Reference to Diet in Disease*, features medicinal uses of cocoa and chocolate.

1925   The New York Cocoa Exchange, modeled after the Chicago Board of Trade, opens to sell cocoa futures, and in 1979, this exchange joins the Coffee and Sugar Exchange.

1928   School Hygiene in Athens, Greece, recommends milk with coffee or chocolate for children.

APPENDIX 2

Theobroma cacao's Reputed Medicinal Properties[†]

The numbers in parentheses indicate how many separate chemicals *Theobroma cacao* has for that activity. For example, "Analgesic (13)" indicates that this plant species has 13 separate chemicals that are known to have analgesic activity. Question marks following activities imply likely but not yet proven effects.

(+)-Inotropic (2)
11B-HSD inhibitor (2)
5-Alpha-reductase inhibitor (4)
5-HT inhibitor (1)
5-Lipoxygenase inhibitor (4)
ACE inhibitor (4)
ATPase inhibitor (1)
Abortifacient (1)
Absorbent (1)
Acaricide (3)
Acetylcholinergic (1)
Acidifier (1)
Acidulant (3)
Additive (1)
Adenosine antagonist (1)
Adrenergic (2)
Aggregant (1)

Aldehyde-oxidase inhibitor (2)
Aldose-reductase inhibitor (12)
Allelochemic (7)
Allelopathic (4)
Allergenic (11)
Alpha-amylase inhibitor (1)
Alpha-glucosidase inhibitor (1)
Alpha-reductase inhibitor (1)
Analeptic (1)
Analgesic (13)
Analgesic synergist (1)
Androgenic (1)
Androgenic? (1)
Anemiagenic (1)
Anesthetic (2)
Angiogenic (1)
Angiotensin receptor blocker (2)

[†]Drawn from the Phytochemical and Ethnobotanical Database of the Agricultural Research Service/United States Department of Agriculture (USDA)'s Dr Jim Duke (http://www.ars-grin.gov/duke/).

Chocolate and Health: Chemistry, Nutrition and Therapy
Edited by Philip K. Wilson and W. Jeffrey Hurst
© The Royal Society of Chemistry 2015
Published by the Royal Society of Chemistry, www.rsc.org

Anorexic (2)
Anthelmintic (2)
Anti-ADD (1)
Anti-AGE (1)
Anti-CFS (1)
Anti-CTS (1)
Anti-CVI (2)
Anti-Crohn's (3)
Anti-EBV (2)
Anti-GTF (1)
Anti-HIV (7)
Anti-*Legionella* (3)
Anti-Lyme (4)
Anti-MS (1)
Anti-MSG sensitivity (1)
Anti-Meniere's (1)
Anti-morning sickness (1)
Anti-PMS (4)
Anti-Raynaud's (2)
Anti-TMJ (1)
Anti-Tourette's (1)
Anti-UTI (1)
Antiacetylcholinesterase (1)
Antiacid (1)
Antiacne (3)
Antiacrodynic (1)
Antiadenomic (1)
Antiadenoviral (1)
Antiadhesive (1)
Antiaflatoxin (2)
Antiaggregant (13)
Antiaging (4)
Antiakathisic (1)
Antialcoholic (3)
Antialdosteronic (1)
Antialkali? (1)
Antialkalotic (1)
Antiallergic (11)
Antialopecic (5)
Antialzheimeran (6)
Antialzheimeran? (1)
Antiamblyopic (1)
Antiamebic (1)
Antianaphylactic (5)
Antiandrogenic (5)
Antianemic (3)
Antianginal (1)
Antiangiogenic (4)
Antianorectic (1)
Antianxiety (1)
Antiaphthic (1)
Antiapneic (2)
Antiapoplectic (1)
Antiapoptotic (2)
Antiarabiflavinotic (1)
Antiarrhythmic (6)
Antiarteriosclerotic (2)
Antiarthritic (8)
Antiasthmatic (9)
Antiasthmatic? (1)
Antiataxic (1)
Antiatherogenic (4)
Antiatheromic (1)
Antiatherosclerotic (8)
Antiautistic (1)
Antibackache (1)
Antibacterial (25)
Antiberiberi (1)
Antibiotic (1)
Antibradiquinic (1)
Antibradyarrhythmic (1)
Antibronchitic (1)
Antibrucellosic (1)
Anticalculic (1)
Anticancer (6)
Anticancer (breast) (2)
Anticancer (cervix) (2)
Anticancer (colon) (2)
Anticancer (forestomach) (2)
Anticancer (kidney) (1)
Anticancer (liver) (3)
Anticancer (lung) (1)
Anticancer (prostate) (1)
Anticancer (skin) (2)
Anticanker (2)
Anticapillary fragility (3)
Anticarcinogenic (6)
Anticarcinomic (2)
Anticarcinomic (breast) (1)
Anticardiospasmic (1)
Anticariogenic (4)
Anticarpal tunnel (2)

Anticataract (8)
Anticellulitic (2)
Anticephalagic (2)
Anticervicaldysplasic (3)
Anticheilitic (3)
Antichilblain (2)
Anticholinesterase (1)
Anticirrhotic (1)
Anticlastogen (3)
Anticlaudificant? (1)
Anticlimacteric (2)
Anticoagulant (1)
Anticold (1)
Anticolic (1)
Anticolitic (2)
Anticollegenase (1)
Anticomplementary (5)
Anticonvulsant (4)
Anticoronary (2)
Anticystitic (2)
Antidecubitic (3)
Antideliriant (1)
Antidementia (5)
Antidepressant (8)
Antidepressant? (1)
Antidermatic (1)
Antidermatitic (8)
Antidiabetic (14)
Antidiabetic? (1)
Antidiarrheic (2)
Antidiuretic (1)
Antidote (1)
Antidote (aluminum) (2)
Antidote (cadmium) (1)
Antidote (hydrazine) (1)
Antidote (hypoglycin-A) (1)
Antidote (iodine) (1)
Antidote (lead) (2)
Antidote (paraquat) (1)
Antidote (pesticides) (1)
Antidysenteric (2)
Antidyskinetic (2)
Antidysmenorrheic (1)
Antidyspeptic (1)
Antidysphagic (1)
Antieczemic (3)

Antiedemic (10)
Antielastase (3)
Antiemetic (2)
Antiemphysemic (1)
Antiencephalitic (2)
Antiencephalopathic (6)
Antiencephalopathic? (1)
Antiendometriotic (1)
Antiendotoxic (1)
Antienteritic (1)
Antiepileptic (3)
Antierythemic (2)
Antiescherichic (3)
Antiesherichic (1)
Antiestrogenic (4)
Antiexudative (1)
Antifatigue (5)
Antifeedant (8)
Antifertility (3)
Antifibrinolytic (2)
Antifibrosarcomic (1)
Antifibrotic (1)
Antiflu (5)
Antigallstone (2)
Antigastric (2)
Antigastrisecretogogic (1)
Antigastritic (5)
Antigenotoxic (1)
Antigingivitic (2)
Antiglaucomic (2)
Antiglossitic (2)
Antiglutamaergic (1)
Antigonadotrophic (1)
Antigonadotropic (4)
Antigranular (1)
Antihangover (4)
Antiheartburn (1)
Antihematuric (1)
Antihemolytic (2)
Antihemorrhagic (3)
Antihemorrhoidal (1)
Antihepatitic (3)
Antihepatoadenomic (1)
Antihepatotoxic (13)
Antiherpetic (13)
Antihiccup (1)

Antihistaminic (13)
Antihomocystinuric (1)
Antihydrophobic (1)
Antihyperactivity (1)
Antihyperammonemic (1)
Antihypercholesterolemic (3)
Antihyperglycemic (3)
Antihyperinsulinemic (1)
Antihyperkeratotic (1)
Antihyperkinetic (2)
Antihyperleptinemic (1)
Antihyperlipoproteinaemic (1)
Antihypertensive (7)
Antihyperthyroid (3)
Antihyperventilation (1)
Antihypotensive (1)
Antihysteric (1)
Anti-ichythyotic (1)
Anti-ielus? (1)
Anti-implantation (1)
Anti-impotence (1)
Anti-infertility (3)
Anti-inflammatory (28)
Anti-insomniac (2)
Anti-insomniac? (1)
Anti-insomnic (1)
Anti-ischemic (2)
Antikeratitic (1)
Antiketotic (2)
Antilactagogue (1)
Antileishmanic (2)
Antilepric (1)
Antileukemic (11)
Antileukoplakic (1)
Antileukorrheic (1)
Antileukotriene (4)
Antileukotriene-D4 (2)
Antilipolytic (1)
Antilipoperoxidant (3)
Antilithic (5)
Antilupus (1)
Antilymphedemic (1)
Antilymphocytic (1)
Antilymphomic (3)
Antimaculitic (2)
Antimalarial (2)

Antimange (1)
Antimanic (2)
Antimastitic (1)
Antimeasles (1)
Antimelanogenic (3)
Antimelanomic (4)
Antimenopausal (2)
Antimenorrhagic (2)
Antimetastatic (4)
Antimigraine (4)
Antimite (1)
Antimitotic (2)
Antimononucleotic (1)
Antimorphinistic (1)
Antimutagenic (23)
Antimycoplasmotic (1)
Antimyocarditic (2)
Antimyocontractant (1)
Antimyopic (1)
Antinarcotic (1)
Antinauseant (1)
Antineoplastic (1)
Antinephritic (3)
Antinesidioblastosic (1)
Antineuralgic (4)
Antineuramidase (1)
Antineurasthenic (1)
Antineuritic (2)
Antineurogenic (1)
Antineuropathic (2)
Antinitrosaminic (5)
Antinitrosic (1)
Antinociceptive (6)
Antinyctalopic (1)
Antiobesity (7)
Antiophidic (5)
Antiophthalmic (1)
Antiorchitic (1)
Antiosteoarthritic (1)
Antiosteoporotic (4)
Antiotitic (1)
Antioxaluric (1)
Antioxidant (38)
Antioxidant synergist (4)
Antioxidant? (2)
Antiozenic (1)

Antipancreatitic (1)
Antipapillomic (1)
Antiparkinsonian (6)
Antiparotitic (1)
Antipellagric (3)
Antiperiodontal (2)
Antiperiodontic (1)
Antiperiodontitic (2)
Antiperistaltic (1)
Antipermeability (1)
Antiperoxidant (9)
Antiperoxynitrite (1)
Antipharyngitic (1)
Antiphenylketonuric (1)
Antiphotophobic (2)
Antipityriasic (1)
Antiplaque (3)
Antiplasmodial (1)
Antiplatelet (1)
Antipneumonic (1)
Antipodriac (2)
Antipolio (3)
Antipoliomyelitic (2)
Antiporphyric (1)
Antipredatory (1)
Antiprogestational (1)
Antiprolactin (2)
Antiproliferant (7)
Antiprostaglandin (2)
Antiprostanoid (1)
Antiprostatadenomic (1)
Antiprostatitic (5)
Antiprotozoal (1)
Antipruritic (1)
Antipsittacotic (1)
Antipsoriac (4)
Antipurpuric (2)
Antipyretic (4)
Antirachitic (1)
Antiradiation (1)
Antiradicular (15)
Antirenitic (1)
Antireserpine (1)
Antiretardation (1)
Antiretinopathic (1)
Antiretinotic (1)

Antirheumatic (5)
Antirheumatitic? (1)
Antirhinitic (3)
Antisalmonella (1)
Antischizophrenic (2)
Antisclerodermic (1)
Antiscoliotic (1)
Antiscorbutic (1)
Antiscotomic (1)
Antiseborrheic (4)
Antiseptic (16)
Antiserotonergic (1)
Antiserotonin (2)
Antiserotoninic (1)
Antishingles (1)
Antishock (2)
Antisickling (5)
Antispasmodic (16)
Antisprue (1)
Antistaphylococcic (4)
Antistomatitic (2)
Antistreptococcic (1)
Antistress (3)
Antistroke (1)
Antisunburn (3)
Antisyncopic (1)
Antisyndrome-X (1)
Antithiamin (3)
Antithrombic (2)
Antithrombogenic (2)
Antithyreotropic (2)
Antithyroid (4)
Antithyrotoxic (1)
Antitic (1)
Antitoxoplasmotic (1)
Antitrypanosomic (2)
Antitubercular (1)
Antitumor (20)
Antitumor (bladder) (1)
Antitumor (brain) (1)
Antitumor (breast) (4)
Antitumor (central nervous system) (1)
Antitumor (cervix) (2)
Antitumor (colon) (7)
Antitumor (forestomach) (2)

Antitumor (gastrointestinal) (1)
Antitumor (gastric) (1)
Antitumor (kidney) (2)
Antitumor (liver) (3)
Antitumor (lung) (7)
Antitumor (mouth) (1)
Antitumor (ovary) (2)
Antitumor (pancreas) (1)
Antitumor (prostate) (3)
Antitumor (skin) (6)
Antitumor (stomach) (2)
Antitumor (thyroid) (1)
Antitumor promoter (7)
Antitussive (3)
Antiulcer (15)
Antiulcerogenic (1)
Antiuremic (1)
Antiuricosuric (1)
Antivaccinia (2)
Antivaginitic (2)
Antivaricose (3)
Antivertigo (2)
Antiviral (23)
Antivitiligic (1)
Antixerophthalmic (1)
Antixerotic (1)
Anxiolytic (4)
Aphidifuge (2)
Aphrodisiac (1)
Apoptotic (8)
Aromatase inhibitor (2)
Arrhythmigenic (1)
Artemicide (2)
Arteriodilator (3)
Ascaricide (1)
Asthma preventive (1)
Astringent (1)
Ataxigenic (1)
Atherogenic (1)
Autotoxic (1)
Bacteristat (2)
Barbituate synergist (1)
Beta-adrenergic receptor blocker (3)
Beta-blocker (1)
Beta-glucuronidase inhibitor (3)
Bradycardiac (1)

Bronchoconstrictor (1)
Bronchodilator (2)
Bronchorelaxant (1)
Bruchiphobe (1)
CNS active (2)
CNS depressant (2)
CNS paralytic (1)
CNS stimulant (5)
COMP inhibitor (1)
COMT inhibitor (2)
COX-1 inhibitor (2)
COX-2 inhibitor (5)
Calcium antagonist (3)
Calcium channel blocker (1)
Calmodulin antagonist (1)
Cancer preventive (39)
Candidicide (3)
Candidistat (1)
Capillariprotective (3)
Capillaritonic (1)
Carcinogenic (7)
Cardiac (2)
Cardiodepressant (2)
Cardioprotective (7)
Cardiotonic (6)
Cardiovascular (3)
Carminative (1)
Caspase-8 inducer (1)
Catabolic (3)
Chelator (1)
Chemopreventive (10)
Cholagogue (3)
Choleretic (14)
Cholesterolytic (1)
Choline sparing (1)
Cholinergic (1)
Circulotonic (1)
Clastogenic (2)
Co-carcinogenic (2)
Coagulant (1)
Cold preventive (1)
Collagen sparing (2)
Collagenase inhibitor (1)
Collagenic (2)
Collyrium (1)
Colorant (1)

Comedolytic (1)
Contraceptive (1)
Convulsant (1)
Copper chelator (2)
Coronary dilator (1)
Corrosive (1)
Cosmetic (3)
Counterirritant (1)
Culicide (1)
Cyclooxygenase activator (1)
Cyclooxygenase inhibitor (3)
Cytochrome-P450-1A2 inhibitor (1)
Cytoprotective (2)
Cytotoxic (8)
DME inhibitor (1)
DNA active (1)
DNA protective (1)
Deiodinase inhibitor (2)
Demulcent (2)
Deodorant (1)
Depressant (1)
Dermatitigenic (3)
Detoxicant (2)
Diagnostic (1)
Diaphoretic? (3)
Differentiator (2)
Disinfectant (1)
Diuretic (14)
Dopamine blocker (1)
Dopaminergic (1)
Dye (2)
Elastase inhibitor (1)
Emetic (2)
Emollient (1)
Energizer (1)
Epidermal stimulant (1)
Ergotamine enhancer (1)
Essential (6)
Estrogenic (6)
Estrogenic? (1)
Euphoriant (1)
Expectorant (2)
Flavor (22)
Fatal (1)
Febrifuge (1)
Fetotoxic (2)
Fibrinolytic (1)

Fistula preventive (1)
Flatugenic (3)
Flatulent (1)
Fungicide (17)
Fungistat (2)
Fungitoxic (1)
GABA-nergic (1)
Gastroprotective (1)
Glucosyl-transferase inhibitor (2)
Glutathione depleting (1)
Goitrogenic (2)
Gonadotrophic (1)
Gram(+)icide (1)
Gram(−)icide (1)
HIV-RT inhibitor (2)
Hemolytic (1)
Hemorrhagic (1)
Hemostat (7)
Hemostatic (1)
Hepatocarcinogenic (1)
Hepatomagenic (2)
Hepatoprotective (14)
Hepatotonic (2)
Hepatotoxic (4)
Hepatotropic (2)
Herbicide (5)
Histamine inhibitor (2)
Hyaluronidase inhibitor (3)
Hydrocholerectic (1)
Hypercholesterolemic (2)
Hyperglycemic (2)
Hypertensive (7)
Hyperuricemic (1)
Hypnotic (2)
Hypoammonemic (1)
Hypoarginanemic (1)
Hypocholesterolemic (19)
Hypoglycemic (15)
Hypolipidemic (3)
Hypotensive (9)
Hypothermic (1)
Hypouricemic (1)
ICAM-1 inhibitor (2)
IKK inhibitor (1)
Immunomodulator (5)
Immunostimulant (9)
Immunosuppressant (3)

Inotropic (3)
Insecticide (4)
Insectifuge (9)
Insectiphile (2)
Insulinase inhibitor (1)
Insulinogenic (2)
Insulinotonic (1)
Interferon synergist (1)
Interferonogenic (2)
Iodothyronine-deiodinase inhibitor (2)
Irritant (14)
JNK inhibitor (1)
Juvabional (4)
Keratitigenic (1)
Keratolytic (1)
Larvicide (1)
Larvistat (4)
Laxative (4)
Laxative? (1)
Leukotriene inhibitor (2)
Lipolytic (2)
Lipotropic (1)
Lipoxygenase-1 inhibitor (1)
Lipoxygenase inhibitor (11)
Lithogenic (1)
Litholytic (1)
Lubricant (2)
Lyase inhibitor (1)
Lymphocytogenic (1)
Lymphokinetic (1)
Lypolytic (1)
MAO-A inhibitor (2)
MAO inhibitor (3)
MAPK inhibitor (1)
MMP-9 inhibitor (2)
Mast cell stabilizer (1)
Memory enhancer (1)
Metal chelator (3)
Metal chelator (copper) (2)
Metalloproteinase inhibitor (2)
Metastatic (1)
Monoamine precursor (2)
Mosquitofuge (1)
Motor depressant (1)
Mucogenic (1)
Mucolytic (2)
Musculotropic (1)
Mutagenic (4)
Mycobactericide (1)
Myocardiotonic (2)
Myocontractant (1)
Myoprotective (1)
Myorelaxant (8)
Myostimulant (2)
NADH-oxidase inhibitor (1)
NEP inhibitor (2)
NF-κB inhibitor (2)
NO-genic (2)
NO inhibitor (3)
NO scavenger (1)
NO synthase inhibitor (1)
Narcotic (2)
Natriuretic (2)
Nematicide (3)
Neoplastic (1)
Nephrotoxic (1)
Neuroinhibitor (1)
Neuroprotective (3)
Neurotoxic (4)
Neurotransmitter (2)
Odontolytic (1)
Ornithine-decarboxylase inhibitor (6)
Osmoregulator (1)
Osteogenic (1)
Osteolytic (1)
Ovicide (1)
Oviposition stimulant (2)
Ovulant (1)
Oxidant (2)
Oxytocic (1)
P450 inducer (1)
P450 inhibitor (1)
PAF inhibitor (2)
PGE2 inhibitor (1)
PTK inhibitor (2)
Pancreatogenic (1)
Paralytic (1)
Percutaneostimulant (1)
Perfumery (10)
Peristaltic (1)
Pesticide (41)
Phagocytotic (4)
Phosphodiesterase inhibitor (1)

Phospholipase inhibitor (1)
Phytoalexin (1)
Pigment (1)
Piscicide (1)
Pituitary stimulant (1)
Plasmodicide (1)
Poultice (1)
Preservative (3)
Priapistic (1)
Pro-oxidant (5)
Progesteronigenic (1)
Proliferant (2)
Propecic (6)
Prostaglandigenic (5)
Prostaglandin secretor (1)
Prostaglandin synthesis
  inhibitor (3)
Prostaglandin synthetase
  inhibitor (1)
Protein kinase C inhibitor (2)
Protisticide (3)
Psychotropic (1)
Pyrogenic (1)
Quinone-reductase inducer (3)
Radioprotective (1)
Refrigerant (1)
Renotoxic (1)
Respirastimulant (1)
Respirodepressant (1)
Reverse transcriptase inhibitor (1)
Rodenticide (1)
Rubefacient (1)
Schizophrenigenic (1)
Secretogogue (2)
Sedative (8)
Sequestrant (1)
Serotoninergic (1)
Soap (1)
Spasmogenic (1)
Spermicide (2)
Spermigenic (1)
Stimulant (3)
Succinic-dehydrogenase inhibitor (1)
Succinoxidase inhibitor (1)
Sunscreen (6)
Suppository (1)

Sweetener (3)
Sympathomimetic (2)
TNF-alpha inhibitor (3)
Tachycardic (2)
Teratogenic (5)
Teratologic (2)
Termitifuge (1)
Thymoprotective (1)
Thyroid-peroxidase inhibitor (1)
Topoisomerase-I inhibitor (4)
Topoisomerase-II inhibitor (4)
Toxic (2)
Tranquilizer (2)
Tremorigenic (1)
Trichomonicide (1)
Triglycerigenic (1)
Tumor promoter (2)
Tumorigenic (2)
Tyrosinase inhibitor (4)
Tyrosine kinase inhibitor (2)
UV screen (1)
Ubiquiot (6)
Ulcerogenic (3)
Uricogenic (1)
Uricosuric (3)
Urinary acidulant (1)
Uterosedative (1)
Uterotrophic (1)
VEGF inhibitor (2)
Varroacide (1)
Vasoactive (2)
Vasoconstrictor (3)
Vasodilator (16)
Vasopressor (4)
Vasoprotective (1)
Vasorelaxant (1)
Verrucolytic (1)
Vulnerary (4)
Xanthine-oxidase inhibitor (8)
cAMP inhibitor (2)
cAMP-phosphodiesterase
  inhibitor (7)
cGMP inhibitor (1)
cGMP-phosphodiesterase
  inhibitor (3)
iNOS inhibitor (3)

Subject Index

References to figures are given in *italic* type. References to tables are given in **bold** type.

abscisic acid, 75
absorption, 106
Acacia catechu, 58
acetic anhydride, 58
acetylcholine, 91
acid demineralization, 205
acne, 161, 182
ACRI *see* American Cocoa Research Institute
Actinomyces viscosus, 205
ADA *see* American Dental Association
addiction, 211
adenine, 198
adenosine monophosphate-activated protein kinase (AMPK), 141
adenosine receptors, 90–1
adipokines, 118
adiponectin, 119
Africa, 30, **31**
African Cocoa Initiative, 44
aging, 13, 181, 183, 191, 193
aglycones, 106
agricultural
    extension services, 47
    farmer education, 47
    non-governmental organizations (NGO), 35, 38, 52
AHA *see* American Heart Association
albuminoids, 56
aldehydes, 94

alkali, 58, 83, 94
alkalinizing, 3
alkaloids, 90–1, 95
almendras, 29
Alzheimer's Disease, 213
Amazon, 29, 33, 72
amelonado, 69, 71–2
American Cocoa Research Institute (ACRI), 34–5
American Dental Association (ADA), 203
American Heart Association (AHA), 89
Americas, 31, **31**, 33
amines, 94
amino acids, 59, 61, 87, 94, 148–9
    n-phenylpropenoyl, 64
amphetamine, 163
AMPK *see* adenosine monophosphate-activated protein kinase
anandamide, 17, 61, 92, 163
Andes, 29
anhedonia, 171
anthocyanidins, 76
anthocyanins, 70, 94, 103, 183
anthrocyanidins, 59
antioxidants, 17, 86, 92–3, 95, 135, 149
    activity, 103–5, 123, 191–2
    skin health, and, 187, 192
    women's health, and, 162, 167, 171

antioxidation
    chocolate and postexercise,
        155–6
    cocoa, and, 124–8
antiplatelet, 17
apigenin, 64
appetite, 138
apples, 183
Arabidopsis, 68, 70, 75
Arabidopsis thaliana, 75
arachidonic acid, 164
area under the concentration (AUC),
    127
arginine, 87
arteriosclerosis, 17
asthma, 91
atherosclerosis, 17, 105, 122–3,
    125–6, 135, 161
    lipids, and, 139–40, 142
atrophy, 148
AUC *see* area under the
    concentration
Aztec, 3, 12, 32, 83, 132, 160, 167

bacteria
    intestinal, 121
    oral, 205
Bacteroides, 117, 122
Badnavirus, 5
Beatitudes Nursing Home, 212
beetles, 5
benzene, 114
    ring, 59, 183
benzoic acid, 122
berries, 127
bifidobacteria, 121
Bifidobacterium, 117, 122
Bill and Melinda Gates
    Foundation, 43
bioactive compounds, 94–5, 171
bioavailability, 87, 95, 162
    trials of health benefits from
        cocoa, 105–6, **107–13**,
        114–18
biogenic amines, 61, 92, 162–3
blackberries, 58

blackheads, 184
black pod rot, 5, 76
blood pressure (BP), 86, 90, 105, 118,
    120, 126–7, 136, 155, 171
blueberry powder, 93
Bolivia, 72
BP *see* blood pressure
brain lipid, 61
bran, 135
Brazil, 6, 29–30, 34, 72
breeding, 73–4
Brillat-Savarin, Jean Anthelme,
    167
British Chocolate Cookie and Cake
    Association, 35
bronchodilation, 91
Burma, 58

Ca *see* calcium
cacahuatl, 132, 160
cacao (*see also* cocoa)
    beans, 2, 6, 9, 56, 83, 135, 160,
        197
    bioavailability, vii (*see also*
        bioavailability)
    biochemistry of, vii, 67
    chemistry of, vii, 56–64, **57, 65**
        anandamide, 61, *61*
        biogenic amines, 61
        flavanols, 56–60
        methylxanthines, 60–1
        non-flavanol
            polyphenolics, 63–4, **64**
    cultivation of, vii
    diet, incorporating into, 88–9
    genome analysis, recent
        advances in, 71–7
        breeding, 73–4, 77
        evolution, domestication
            and germplasm
            collections, 71–3
        functional genomics,
            74–5
        genomics of cacao health
            and nutrition, 75–6
        pathogens, 74, 76

Subject Index 233

genomics applications, vii, 67–77
    Cacao Genome Sequencing Project, 69–71
    genomics, future of, 77
health benefits of, vii, 82, 133
historical nutritional and medicinal benefits of, 3, 161
history of cacao consumption, 83–4, 133–4
metabolism, vii
nutrient composition of, vii, 84, 86–8
pests and diseases, 4–5, 28, 34–7, 39, 51, 76–7
physiological functions of cacao's bioactive compounds, 89–93
    alkaloids, 90–1
    biogenic amines, 92
    endocannabinoids, 92
    opioids, 91–2
    oxalates, 93
    polyphenolics: flavanols, 92–3
processing, 94
    changes in nutrient and bioactive compounds, 94–5
seeds, 132, 134
sustainability of (*see under* cocoa)
Cacao Genome Sequencing Project, 69–71
Cacao Nacional Boliviano (CNB), 72
Cadbury, 48
    Dairy Milk bar, 48
C.A.F.E. (Coffee and Farmer Equity) Practices, 48
caffedyme, 64
caffeic acid, 103, 115, 162
caffeine (1,3,7-trimethylxanthine), 60, 90–1, 134, 162, 168, 198–9, 201, 213
calcium (Ca), 86, 199, 203
    phosphate, 200

Cameroon, 39–44
cancer, 105, 116, 118, 135–6, 161
    anti-carcinogenic properties, 62
cannabinoids, 61, 163
capsaicin, 91
capsids, 5
carbohydrates (CHO), 12, 16, 56, 87–8, 147, 154, 157
    beverage, 150–2
    snacks, 139
    women's health, and, 164
β-carbolides, 92
cardiovascular
    disease (CVD), 86–8, 105, 118, 122–5, 127, 135–7
        women's health, and, 165, 170
    health and chocolate, 60, 132–42
    system, 60, 161, 163
carob, 28
carotenoids, 179, 181–2, 191
Casanova, Giacomo Girolamo, 167
casein, 148
cashews, 38
catabolism, 148
catechin gallate, 183
catechins, 57–9, 63, 84, 93–5, 135
    digestive tract, diabetes, and, 104–5, 114, 121–2, 124
    skin health, and, 183–4, 187
    tea, 123
    women's health, and, 162
catecholaminergic transmission, 163
catecholamines, 121, 163
CATIE *see* Centro Agronómico Tropical de Investigación y Enseñanza
CAT scanning *see* computerized axial tomography scanning
Caulimoviridae, 5
Cecidomyiid, 4
Central America, 6, 30
central nervous system (CNS), 60, 90

Centre de coopération Internationale en Recherche Agronomique pour le Développement (CIRAD), 69
Centro Agronómico Tropical de Investigación y Enseñanza (CATIE), 34, 51, 73
Ceratocystis wilt, 73
Ceratopogonid, 4
cerebral vasomotor reactivity (CVR), 141
CGS *see* Cocoa Genome Consortium
chemokines, 118
cherries, 58
child labor, 39–42, 47, 49–50
    monitoring system, 42, 46
    surveys, 40–2, **41**, 47
child trafficking, 44–5
chlorogenic acid, 162
CHO *see* carbohydrates
chocoholism, 17, 164, 211–12
chocolate, 1–2, 28, 104–5, 133
    addiction, 164
    adulteration, 16
    aphrodisiac effects, of, 17
    bars, 83, 146, 154–5, 157
    cake, 104
    cardiovascular health, and, 60, 132–42
    chemistry of, 59
    cookies, 104
    craving, 164, 166, 214
    dark chocolate (*see* dark chocolate)
    definition of, 2–3, 160
    dental health, and (*see* dental health)
    exercise recovery, and, 146–57
        chocolate and postexercise antioxidation, 155–6
        chocolate and postexercise mood state, 156–7
        chocolate bars and cocoa, 154–5
        chocolate milk and postexercise recovery, 149–54
        physiology and role of nutrients, 147–9
    flavanol-rich, 125, 127
    food, as, 7–8, 157
    frosting, 104
    -health-beer, 18
    health benefits of, 82, 84, 96, *165,* 214
        bioavailability trials, 105–6, **107–13**, 114–18
    historical timeline of chocolate in science, nutrition and medicine, 218–21
    houses, 83
    industry, 28
    liquor, 28, 32, 132
    milk, 9, 146, 157
        postexercise recovery, and, 149–54
    milk chocolate (*see* milk chocolate)
    nutrient composition from cacao to chocolate, 83–4, **85**, 86–8
    nutrition, as, 6–12, 95
    nutritional values, of, *11,* 11–12
    nutrition and biochemistry as related to health and medicine, 134–5
    preventative medicine, as, 2, 12
    products, 10, 94–5, 155
        health benefits of, 82
    psychology, 211
    science, in, 2–6
    semisweet baking chips, 104
    skin health, and (*see* skin health)
    syrup, 104
        nutrient composition of, **85**
    therapy, as, 12–18
    tree, 1–2, *5,* 68
    white chocolate (*see* white chocolate)

Subject Index

wine, 14
women's health, and
  (*see* women's health)
Chocolate Manufacturers of
  America, 35
chocol ha, 132
cholesterol, 17, 84, 86, 88, 93, 134, 139
  serum, 123
  women's health, and, 162, 165, 171
cholesterolemia, 134
CIRAD *see* Centre de coopération Internationale en Recherche Agronomique pour le Développement
citrus fruits, 64, 118
CLA *see* conjugated linoleic acid
Clostridia, 121
Clostridium histolyticum, 117, 122
Clostridium perfringens, 117
clovamide, 64, 92
CMA *see* U.S. Chocolate Manufacturers Association
CNB *see* Cacao Nacional Boliviano
CNS *see* central nervous system
cocoa, 3, 16, 104–5, 133 (*see also* cacao)
  antioxidation, inflammation, and, 124–8
  beans, 28, 32, 60, 103–4
    composition after fermentation and drying, **57**
  blood lipids, diabetes, and, 122–4
  certification, 47–9
    government certification, 47
    self-certification, 47
    third-party certification, 47
  current view of cocoa sustainability, *49,* 49–50
    develop and apply technology to improve cocoa yields, 51–2

progress and future directions, 50
relief from onerous cocoa export taxes, 50
drink, 9, 104, 114, 187–8
exercise recovery, and, 154–5
flavor chemistry, 57, 64, **65**
genome, 51, 67 (*see also* cacao: genomics applications)
health benefits of, 103, *165*
  bioavailability trials, 105–6, **107–13**, 114–18
history of, 160–1
ingredients in chocolate, 160
liquor, 104–5, 133
mirid bugs, 76
-nut cream, 116
pod borer, 35
prebiotic environment, and, 121–2
press, 83
production, early problems with, 32–3
  cocoa demand in *Nuevo Espana,* 32–3
  plant disease, 33
production, recent concerns with, 33–49
  child labor incident broadcast on TV in UK, 39
  child labor issues undertaken by U.S. Congress: Harken-Engel Protocol, 40–1
  child labor surveys in West Africa, 40–2, **41**
  child trafficking in West Africa, Payson analysis of, 44–5
  collapse of Malaysian plantation cocoa, 33–4
  pest spread from Malaysia to Indonesia, 35–6
  "slave ship" reported in Gulf of Guinea, 40

cocoa (*continued*)
    witches' broom
        devastation of Brazil
        crop, 34–5
    sustainable production, of, vii, 28
        1998 view of cocoa sustainability, *37*, 37–8
        certification of the cocoa supply chain, 47–9
        child/youth-directed sustainability projects, 42–3
        formation of World Cocoa Foundation, 39
        Panama Conference, 36–7
        Sustainable Tree Crops Meeting in Washington, D.C., 38–9
    today's production, 30–2, **31**
    tree, 28–9
    where originated, 29–30
    women's health, and (*see* women's health)
    yield in world's regions, **33**, 50
cocoa butter, 9, 15, 17, 28, 32, 104–5, 115, 132–3
    biosynthesis, 70
    chemistry of, **62**, 62–3, 75
    health benefits of, 62–3, 123
    ingredients in chocolate, 160
    skin care, and, 171, 183–4
Cocoa Communities Project, 45–7
Cocoa Genome Consortium (CGS), 69
Cocoa Livelihoods Program, 43–4, 51
cocoa powder, 28, 32, 60, 62, 132–3, 187, 199–201
    digestive tract, diabetes, and, 104–5, 115, 118
    nutrient composition of, 83, **85**, 90, 93
cocoa powder extract (CPE), 120
cocoa products (CP), 121, 125, 133, 193, 198
    cardiometabolic effects of, 135–42

cocoa swollen shoot virus (CSSV), 5, 74, 76
coffee, 38, 138
Coffee and Farmer Equity Practices *see* C.A.F.E. Practices
cognition, 141–2, 162, 213–14
colic, 134
collagen, 180
colon, 106, 115–16, 125
Columbia, 29, 72, 160, 167
Columbus, Christopher, 32
comedones, 184
comfort foods, 212
compulsive eating, 17, 212
computerized axial tomography (CAT) scanning, 213
conching, 3
confections, 114
conjugated linoleic acid (CLA), 62, *63*
Conopomorpha cramerella, 5
copper (Cu), 84
    toxicity, 74
corneometry, 186
corporate social responsibility (CSR), 47–8
Cortés, Hernán, 3, 32, 83
cortisol, 121
cosmetics, 171, 179–80, 193
CP *see* cocoa products
CPE *see* cocoa powder extract
cranberries, 58
    powder, 93
C-reactive protein (CRP), 117
creatine kinase, 150
criollo (native born), 2, 33, 83
    beans, 64
    genetic group, 68–9, 71–2
CRP *see* C-reactive protein
CSR *see* corporate social responsibility
CSSV *see* cocoa swollen shoot virus
Cu *see* copper
CVD *see* cardiovascular disease
CVR *see* cerebral vasomotor reactivity
cysteinyl leukotrienes, 105

cytokines, 118–19
cytosine, 198

daily value (DV), 86
d'Annunzio, Gabriele, 167
dark chocolate, 17, 104–5, 121, 141, 162, 187, 213
    diet, as part of, 88
    flavanol-rich, 114, 120, 126
    ingredients of, 133
    nutrient composition of, 84, **85**, 86–8
    nutrition and biochemistry of, 134–5
    theobromine, in, **199**, 200
DBP *see* diastolic blood pressure
dementia, 92, 212
dental caries, 161, 200 (*see also* tooth decay)
dental health
    chocolate, and, 196–207
        background literature on reduction in dental caries, 200–2, *202*
        human mouth, the, 205–6
        theobromine as alternative to fluoride, 203–5
        theobromine chemical structure, properties and toxicity, 197–200, *198*, **199**
deoxyclovamide, 64
depression, 166
dermatitis, 182
dermis, 180–1
diabetes, 105, 118–20, 135–6, 138
    anti-diabetic properties, 62
    cocoa and, 122–7, 161
diadzein, 123
diarrhea, 134
diastolic blood pressure (DBP), 90, 118, 136, 169
diet, 8–9
    healthy, 82, 165–6, 171
    high-fat, 119
    incorporating cacao into, 88–9

Dietary Guidelines Advisory Committee, 88
Dietary Guidelines for Americans, 88
dietary supplements, 180–1
digestive tract *see* gastrointestinal tract
3,4-dihydroxyphenylacetic acid, 106, 115
diketopiperazines, 118
1,3-dimethylxanthine, 198
3,7-dimethylxanthine, 60, *60*
3,7-dimetilxantina, 163
1,3-diphenylpropane, 59
diseases and pest *see* cacao: pests and diseases
Distentiella theobroma, 5
diuresis, 134
DNA, 67–8, 93, 191–2, 198
    damage, 182
    sequencing, 68–9, 71
    synthesis, 105
    testing, 73
Dogfish Head's Theobroma Ale, 18
Dominican Republic, 72
dopamine, 91–2, 163
Dutching, 3, 16, 83, 103
Dutch Process, 3, 83, 94–5
DV *see* daily value
dysentery, 134
dysglycemia, 138
dyslipidemia, 118–19, 135, 138, 161
dyspepsia, 134

ECHOES *see* Empowering Cocoa Households with Opportunities and Educational Solutions
Ecuador, 30, 33, 72
eicosanoids, 119
Ek Chuah, 134
elastin, 180
electrolytes, 149
emotional eaters, 166
Empowering Cocoa Households with Opportunities and Educational Solutions (ECHOES), 43
enamel, 197, 199, 203–6

endocannabinoids, 92, 163
endocrine system, 60
endorphins, 92, 164
endothelial, 86–7, 105, 120, 125–6, 136, 162–3, 171
    dysfunction, 169–70
    nitric oxide synthase (eNOS), 137
endothelium, 87, 125
energy, 10, 95, 147, 179, 213
Engel, Eliot, 40
eNOS *see* endothelial nitric oxide synthase
EPI *see* epicatechins
epicatechin gallate, 183
epicatechins (EPI), 9, 17, 57–8, 63, 84, 86, 94–5, 156
    cardiovascular health, and, 135, 142
    digestive tract, diabetes, and, 104–6, 114, 120–2, 124–5
    skin health, and, 183, 187, 193
    women's health, and, 162
epidermis, 180
epigallocatechin, 103, 183
    gallate, 183–4
epimerization, 94–5
ergogenicity, 156
erythema, 181–2, 185, 188, 192
Escherichia coli, 117, 122
estrogen, 170
ethanol, 94
Eubacterium rectale-Clostridium coccoides, 117, 122
euphoria, 164

Fair Trade, 45, 48
fat, 9, 16, 104, 118, 122
    saturated, 123
fatty acids, 76, 88, 119, 122–3, 138
    monounsaturated, 134, 164
    palmitic, 134, 164
    polyunsaturated, 182
    saturated, 134, 164
    stearic, 134, 164
FDA *see* United States Food and Drug Administration

Fe *see* iron
fermentation, 56, 90, 94–5, 103, 205
fertility, 166
ferulic acid, 106, 115
fiber, 56, 86, 95
flavan-3-ols, 57–9, *59*, 104, 116–17, 122, 124–5, 127
flavanols, 56–60, 104–6, 114–16, 120–2, 124, 127, 135, 137
    cocoa, 155–6, 170
    nutrient composition of chocolate, 83, 86, 92–3, 95
    skin health, and, 183, 187–8, 190
flavanones, 183
flavones, 59, 64, 114, 183
flavonoids, 17, 59, 64, 84, 86, 103, 116–17, 123–4
    biosynthesis, 70
    cardiovascular health, and, 135, 138
    cocoa, 75, 87, 162–3
    skin health, and, 179, 181, 183, 191–2
    women's health, and, 162–3, 171
flavonols, 64, 183
flavonones, 59, 64
flavor chemistry, 57, 64, **65**
flow mediated dilation (FMD), 88, 125–6
fluoride, 196–7, 203–6
FMD *see* flow mediated dilation
fMRI *see* functional MRI
food
    allergens, 122
    energy, 11–12
forastero (foreign born), 2, 33, 83
    genetic group, 71–2
frosty pod rot, 5, 36, 51, 76
fructose, 205
Fry, Joseph, 83
functional MRI (fMRI), 213
fungi, 5, 33
Fuselier, Joseph, 206

galactose, 59
gallic acid esters, 106, 183
gallocatechin, 183
    gallate, 183
γ-linolenic acid (GLA), 182
gastrointestinal (GI) tract, 60, 106, 114, *115,* 116, 121
genetic diversity, 72, 76
genistein, 123
Ghana (Gold Coast), 6, 30–2, 34, 39–47, 50
GI tract *see* gastrointestinal tract
GLA *see* γ-linolenic acid
glaucoma, 105
glucanase, 75
glucose, 118–20, 124, 138, 205
    blood, 149
    regulation, 127–8
    tolerance test, 126
glucuronides, 105–6, 114
glutamine, 87
glutathione peroxidase, 93
glycemia, 136, 138–9
glycerides, 9, 62
glycogen, 147–8, 151–2
glycosides, 106
glycosylation, 114
Gold Coast, 6
grapes, 58, 183
guanine, 198
Guyana, 29

Harkin, Tom, 40
Harkin–Engel Protocol, 40–1, 49
Harpoon's Chocolate Stout, 18
HDL cholesterol *see* high-density lipoprotein cholesterol
health benefits
    cacao, cocoa or chocolate, of (*see under* cacao; chocolate; cocoa)
heart failure (HF), 136, 140–1
hemoglobin, 185
hemostasis, 127
Hershey Company, vii, 9–10, 61–2
hesperetin, 123

HF *see* heart failure
high-density lipoprotein (HDL) cholesterol, 63, 84, 118, 122–3, 139, 162
hippuric acid, 115
HOMA-IR *see* Homeostasis Model Assessment of Insulin Resistance
homeostasis, 87, 119, 121, 124, 212
Homeostasis Model Assessment of Insulin Resistance (HOMA-IR), 88–9
Honduras, 71
hops, 58
hormones, 166, 170
human genome sequencing project, 69
hydration, 186, 191
hydrogen peroxide, 191
hydrolysate, 148
hydrophobicity, 114
4-hydroxybenzoic acid, 115
3-hydroxybenzopyran, 58
4-hydroxyhippuric acid, 115
hydroxylapatite, 200, *200,* 203
hydroxyl groups, 59
3-hydroxyphenylpropionic acid, 121
hydroxyphenylvaleric acids, 118
hydroxyphenylvalerolactones, 118
hypercholesterolemia, 139
hyperglycemia, 124–6, 138
hyperproliferation, 137
hypertension, 118–20, 123, 127, 134–8, 161
    maternal, 168–9
hypodermis, 180–1

ICGS consortium *see* International Cacao Genome Sequencing consortium
ichthyosis, 182
ICI *see* International Cocoa Initiative of Geneva
IFESH *see* International Foundation for Education and Self Help
IGT *see* impaired glucose tolerance

ILO *see* United Nations International Labor Organization
imidazole rine, 198
immune, 74–6, 105, 122, 191
impaired glucose tolerance (IGT), 120
Inca, 83
India, 58
Indonesia, 30–2, 35–7, 50
Indonesian Pod Borer Project, 39
inflammation, 14, 91, 105, 118–20, 122, 136–7
    anti-inflammation, 163
    cocoa, and, 124–8, 191
    UV absorption, 191–2
insulin, 124–5, 162
    resistance, 118–19, 127–8, 136, 138–9, 142
    sensitivity, 105, 120, 124, 136
International Cacao Genome Sequencing (ICGS) consortium, 68–70
International Cocoa Initiative of Geneva (ICI), 45
International Cocoa Organization, 17, 84
International Cocoa Research and Education Foundation, 17
International Cocoa Research Conference, 69, 73
International Coffee and Cocoa Organization, 30
International Diabetes Foundation, 118
International Foundation for Education and Self Help (IFESH), 43
International Group for the Genetic Improvement of Cacao, 69
International Programme to Eliminate Child-labor (IPEC), 42
intestinal bacteria, 121
intestine, 114, 121
IPEC *see* International Programme to Eliminate Child-labor
iron (Fe), 84, 86
    deficiency, 86

isoflavones, 183
isoflavonoids, 59
isoprostanes, 155
Ivory Coast, 6, 30–1, 39–47, 49–50

K *see* potassium
kagaw, 160
kakaw, 133, 160
keratinocytes, 180
keratodermatosis, 182
kidney(s), 93
    stones, 13–14, 93
Kuna Indians, 123–4, 135–6, 161

LA *see* linoleic acid
lactic acid, 94, 205
lactobacilli, 121
Lactobacillus, 117, 122, 205
lactose, 87
    sugar, 149
LDL-C *see* low-density lipoprotein cholesterol
LDL oxidization *see* low-density lipoprotein oxidization
leptin, 119
leucine, 87, 148
leukotrienes, 163
Liberia, 43
Linneaus, Carl, 1, 160, 198
linoleic acid (LA), 62, *63*, 88, 164, 182
lipid(s), 75, 119, 134, 164
    atherosclerosis, and, 139–40
    brain, 61
    cocoa, 122–4, 182
    epidermal, 182
    oxidation, 182, 191
    profile, 136, 142
lipophilic compounds, 59, 62
lipoproteins, 17
liver, 123
low-density lipoprotein cholesterol (LDL-C), 88, 122–4, 139, 162
low-density lipoprotein (LDL) oxidization, 105
luteolin, 64
lycopene, 181

Subject Index

macronutrients, 95, 192
magnesium (Mg), 84, 163, 203
magnetic resonance imaging (MRI), 213
maize, 75
Malaysia, 6, 31, 33–6, 50
manos, 3
MAO *see* monoamine oxidase
Maya, 3, 72, 83, 132–4, 196, 199, 207
mealybugs, 5
melanin, 181
melanocytes, 181
menopause, 170
Mesoamerica, 29, 32, 132–4, 160
metabolic syndrome, 118–20, 123–4
metabolism, 121
metabolites, 105–6, 114, 122
metate, 3
methoxy groups, 59
methylation, 58
methyl ether, 58
methylxanthines, 60–1, 90, 93, 134, 162–3, 214
Mexican pepperleaf, 134
Mexico, Southern, 29
Mg *see* magnesium
m-hydroxybenzoic acid, 106
m-hydroxyphenylacetic acid, 106
m-hydroxyphenylpropionic acid, 106
microbiome, 121
microbiota, 116–17, 121–2, 128
microflora, 106, 121
micronutrients, 95, 149, 180, 191–2
midges, 4
migraine headaches, 61
milk, 9, 84, 114–15, 133, 152
    bovine, 149
    protein, 115
    skimmed, 118
    sugar, 89
milk chocolate, 15, 104, 114, 139, 141, 162, 213
    bar, 9
    ingredients of, 133

nutrient composition of, 84, **85**, 86–9
theobromine, in, **199**, 200
milling, 3, 94
minerals, 56, 162, 205
mitochondriogenesis, 137
Mixtec, 3
Moctezuma, 32
Moniliophthora (Crinipellis) perniciosa, 5
Moniliophthora roreri, 5, 51, 76
monoamine oxidase (MAO), 90
Montezuma II, 167
mood, 156–7, 163, 166, 171, 214
moths, 5
MRI *see* magnetic resonance imaging
muscle, 140, 147–52, 157
    protein, 151–2, 157
myoglobin, 150

Na *see* sodium
N-acylethanolamines, 92, 163
NaF *see* sodium fluoride
Nakamoto, Tetsuo, 206
napthalene, 58
N-arachidonylethanolamine, 61
narigenin, 64
naringenin, 123
Nestlé, Henri, 83
neurotransmission, 60
neurotransmitters, 91–2, 95
Neuvo Espana, 32–3
NGO *see* agricultural non-governmental organizations
Nicaragua, 71–2
Nigeria, 39–44, 49
nitric oxide (NO), 17, 86, 105, 125, 137, 170, 192
    enzyme synthase, 163
    -mediated vasodilation, 191–3
nitroxidation, 137
N-linoleoylethanolamine, 92
NO *see* nitric oxide
Nocardia, 205
noradrenaline, 91–2

norepinephrine, 163
nuclear magnetic resonance, 213
nutricosmeticals, 181
nutrients, 8, 12, 82, 94–5, 146–7
nutrition, 181–2

obesity, 118–20, 127, 137–8, 142, 161, 170
oleic acid, 88, 123, 134, 164–5
oligomers, 104, 106, 114
Olmec, 3, 133
3-*O*-methylglucuronide, 114
opiods, 91–2
ORAC *see* oxygen radical absorbance capacity
oro negro (black gold), 3
ovaries, 170
oxalates, 93
oxalic acid, 162
oxidation, 94, 105, 138, 156, 162
oxidative stress, 75, 103, 124–8, 136, 155, 167, 170
oxygen, 185, 191
oxygen radical absorbance capacity (ORAC), 104

P *see* phosphorus
palmitic acid, 88, 134, 164
Panama Cocoa Sustainability Principles, 36–7
Panama Conference, 36–7
pancreatic lipase, 139
pathogens, 74, 76, 122
Payson Center for International Development and Technology Transfer, 44–5
PEA *see* phenylethylamine
pentacetate, 58
pepe de oro (seeds of gold), 3
peroxynitrite, 192
Peru, 29, 72
pesticides, 47
pests and diseases *see* cacao: pests and diseases
Peter, Daniel, 84
phenolcarboxylic acids, 117

phenolic(s), 16, 103–5, 114, 116, 120
   acids, 115–16, 121, 128
   compounds, 92, 95, 103, 105–6
phenols, 162
phenylacetic acid, 117, 122
phenylalanine, 59
phenylethylamine (PEA), 17, 92, 162–3
2-phenylethylamine, 61, *61*
phenyl group, 59
phenylpropenoic acid amide, 64
phenylpropionic acid, 122
phloroglucinol dimethyl ether, 58
phosphate, 9, 199, 203
phosphodiesterase, 91
phosphorus (P), 84, 95
phosphorylation, 151–2, 156
photo-oxidation, 182
photo-oxidative stress, 191
photoprotection, 181, 185, 188–93
phytic acid, 162
Phytophthora, 73
Phytophthora megakarya, 5, 76
piceid, 64
Piper auritum, 134
placental disease, 169
Planococcoides njalensis, 5
platelet, 105, 162
polymerization, 114
polyphenolic(s), 92–3, 104, 106, 114, 116–17, 124, 127
   compounds, 103, 128
   non-flavanol, 63–4
polyphenols, 57–9, 94, 135, 162, 214
   biosynthesis, 70
   digestive tract, diabetes, and, 104–6, 114, 116–18, 120, 122–8
   red wine, 123
   skin health, and, 179, 181–4, 191–3
pomegranate powder, 93
potassium (K), 84, 86
pre-eclampsia, 168–9
pregnancy, 168–71
preservatives, 16

proanthocyanidins, 76, 94, 183
procyanidins, 58-9, 60, 63, 92, 94
    digestive tract, diabetes, and,
        103-6, 114, 116-18, 124
    skin health, and, 183
    women's health, and, 162-3
progesterone, 170
prostacyclins, 105, 163
proteins, 74-6, 93, 105, 114, 148-9,
    157, 191-2
    muscle, 151-2, 157
protocatechuic acid, 115, 184
prunin, 64
pseudochromosomes, 69
psoriasis, 182
purine, 198
pyrimidines, 198

quercetin, 58, 64, 184

Rainforest Alliance, 45, 48
randomized control trials (RCT), 88
raspberries, 58, 86
RCT see randomized control trials
reactive oxygen species (ROS), 93, 105
renal system, 60
Rennou™, 206-7, 207
reproduction, 166-8
respiratory system, 60
resveratrol, 64, 184
retinoic acid, 182
retinoids, 182
retinol, 182
rhamnose, 59
RNA, 93
    sequencing, 69
roasting, 3, 64, 94-5
Rogue's Ale's Chocolate Stout, 18
ROS see reactive oxygen species
Roundtable for Sustainable Palm
    Oil, 48
rutin, 184

SAH see systemic arterial
    hypertension
Sahlbergella singularis, 5

salsolinol, 92
Samuel Adams' Chocolate
    Bock, 18
satiation, 138
SBP see systolic blood pressure
Se see selenium
selenium (Se), 182
Selenothrips rubrocinctus, 5
self-incompatibility (SI), 73
SELS see surface evaluation of living
    skin
serotonin, 17, 91-2, 162-3
SI see self-incompatibility
Simmons, Skip, 206
simple sequence repeats (SSR), 72
sitosterol, 134
skin
    aging, 181, 183
    cancer, 182
    care, 170-1
    health and chocolate, 179-93
        cocoa constituents with
            dermal activity, 182-3
        compounds and
            biochemical
            mechanisms, 191-3
                antioxidant activity,
                    191-2
                NO: vasodilation,
                    192-3
                UV absorption:
                    inflammation,
                    192
        human studies on
            systemic effects of
            cocoa, 187-91
        methods to determine
            skin properties and
            function, 184-6
                cutaneous blood
                    flow and oxygen
                    saturation of
                    hemoglobin, 185
                evaluation of the
                    skin surface,
                    186

skin (continued)
    photoprotection against UV-induced erythema, 185
    skin barrier function evaluated by measurement of TEWL, 186
    skin hydration measured by corneometry, 186
    skin structure by ultrasound measurements, 185
  skin and nutrition, 181–2
  skin structure and function, 180–1
  topical effects of cocoa products, 183–4
  wrinkling, 182–3
slave labor, 49
social responsibility report (SRP), 47
sodium (Na), 134
sodium fluoride (NaF), 203
sodium monofluorophosphate, 206
Sonoran's White Chocolate Ale, 18
sorghum, 58
South America, 6, 29–30, 38, 83
Southeast Asia, 30, **31**, 38
Southern Tier's Chokolat, 18
soy, 148–9
Spain, 32
sports
  beverage, 146
  performance bars, 146
SRP *see* social responsibility report
SSR *see* simple sequence repeats
Staphylococcus, 117, 122
STCP *see* Sustainable Tree Crops Programme
stearic acid, 17, 62, *62*, 88, 134, 164
stilbene, 64
stimulation, 91, 163, 167

stratum basale, 180–1
stratum corneum, 180–2, 184
strawberries, 58, 86
Streptococcus mutans, 201, 205
Streptococcus sobrinus, 201
streptozotocin (STZ), 127
stress, 165–6
stroke, 135–6, 141–2
STZ *see* streptozotocin
sucrose, 87, 205
sugars, 59, 87, 89, 94, 115, 133, 149, 160, 205
sulfates, 105–6
sulfoglucuronides, 106
sunburn, 182, 192
superoxide, 191
surface evaluation of living skin (SELS), 182
Sustainable Tree Crops Programme (STCP), 38–9, 51
systemic arterial hypertension (SAH), 136–7, 142
systolic blood pressure (SBP), 90, 118, 136, 162, 169

tannins, 9, 17
TC *see* total cholesterol
tea, 84, 138, 183, 198
  green, 93, 118, 127, 138
terpenes, 75
terpenoids
  biosynthesis, 70
tetrahydro-β-carbolines (THβC), 90
tetramethyl ether, 58
TEWL *see* transepidermal water loss
THβC *see* tetrahydro-β-carbolines
Theobroma, 83
Theobroma cacao (food of the gods), v, 1–3, 29, 68, 83, 132, 160, 197
  reputed medicinal properties, 222–30
theobromine (3,7-dimethylxanthine), 9, 17, 60, *60*, 84, 90–1, 134, 213
  anti-caries activity with no bacterial attachment on enamel, 200–2, *202*

Subject Index

chemical structure, properties and toxicity, 197–200, *198,* **199**
dental health, and, 197–206
women's health, and, 162–3
Theodent™, 206–7, *208*
theophylline (1,3-dimethylxanthine), 90, 198–9
theosterols, 9
thrips, 5
thrombosis, 105
thymine, 198
tlacoxochitl, 134
Toltec, 3
tooth decay, 17 (*see also* dental caries)
total cholesterol (TC), 88, 162, 165
transepidermal water loss (TEWL), 181
triglyceridemia, 124
triglycerides, 118, 139, 164
1,3,7-trimethylxanthine, 60
Trinidad, 71
trinitario (sent from heaven), 2, 33, 83
genetic group, 71–2
tropical
fruits, 38
hardwoods, 38
tryptamine, 92, 162–3
type 2 diabetes mellitus *see* diabetes
tyramine, 92, 162–3
tyrosine, 59

ultrasound, 185
ultraviolet (UV) radiation, 171, 181, 188, 190–3
UN-FAO *see* United Nations Food and Agriculture Organization
United Nations, 50
Food and Agriculture Organization (UN-FAO), 35
International Labor Organization (ILO), 41
United States Agency for International Development (US-AID), 38, 40–1, 44

United States Department of Agriculture (USDA), 11, 68
National Nutrient Database, 84
United States Department of Labor (US DOL), 41, 45, 50
United States Food and Drug Administration (FDA), 203
US-AID *see* United States Agency for International Development
U.S. Chocolate Manufacturers Association (CMA), 12
USDA *see* United States Department of Agriculture
US DOL *see* United States Department of Labor
U.S. Pure Food, Drug and Cosmetic Act, 16
U.S. Pure Food and Drug Act, 16
UTZ, 45, 48
UV radiation *see* ultraviolet radiation

Van Houten, Coenraad, 83
vanilla, 133
flowers, 134
vanillic acid, 106
vasodilation, 105, 125, 134, 141, 155
nitric oxide, 191–3
vasoprotection, 105
Venezuela, 29
veratric acid, 58
virus *see* cacao swollen shoot virus
vitamin(s), 9, 162, 180–1
vitamin A, 182
vitamin C, 86, 181, 191
vitamin D, 86
vitamin E, 181–2, 191

WACAP *see* West African Cocoa Agriculture Program
water, 56, 149–50
availability and quality, 47
Watts, James, 3
WCF *see* World Cocoa Foundation

West Africa, 30–1, 33, 38, 41–3, 48, 50–1, 76
    child labor incident, 39
    child labor surveys, 40–2, **41**
    child trafficking, 44–5
    individual company-funded efforts, 46
West African Cocoa Agriculture Program (WACAP), 42, 47
West Indies, 6
whey, 148
white chocolate, 87–9, 126, 162
    ingredients of, 133
    nutrient composition of, **85**
wine, 183
    red, 106
winnowing, 3
witches' broom disease, 5, 33–5, 75–6
women's health
    chocolate, cocoa, and, 160–72
        cocoa and beauty, 170–1
        cocoa and menopause, 170
        cocoa and pregnancy, 168–71
        cocoa and reproduction, 166–8
        female mental well-being, 165–6
        health, cocoa and chocolate: scientific evidence, 162–4
World Bank, 50
World Cocoa Foundation (WCF), 39, 43–4
wrinkles, 181

xanthine, 164, 201
xerosis, 182
xocolatl, 83, 132
xylose, 59

yeast, 94
Young's Double Chocolate Stout, 18

Zapotec, 3
zinc (Zn), 84
Zn *see* zinc